Achieving sustainable cultivation of rice

Volume 1: Breeding for higher yield and quality

It is widely recognised that agriculture is a significant contributor to global warming and climate change. Agriculture needs to reduce its environmental impact and adapt to current climate change whilst still feeding a growing population, i.e. become more 'climate-smart'. Burleigh Dodds Science Publishing is playing its part in achieving this by bringing together key research on making the production of the world's most important crops and livestock products more sustainable.

Based on extensive research, our publications specifically target the challenge of climate-smart agriculture. In this way we are using 'smart publishing' to help achieve climate-smart agriculture.

Burleigh Dodds Science Publishing is an independent and innovative publisher delivering high quality customer-focused agricultural science content in both print and online formats for the academic and research communities. Our aim is to build a foundation of knowledge on which researchers can build to meet the challenge of climate-smart agriculture.

For more information about Burleigh Dodds Science Publishing simply call us on +44 (0) 1223 839365, email info@bdspublishing.com or alternatively please visit our website at www.bdspublishing.com.

Related titles:

Achieving sustainable cultivation of rice Volume 2: Cultivation, pest and disease management
Print (ISBN 978-1-78676-028-9); Online (ISBN 978-1-78676-030-2, 978-1-78676-031-9)

Rice insect pests and their management
Print (ISBN 978-1-78676-196-5); Online (ISBN 978-1-78676-198-9, 978-1-78676-199-6)

Achieving sustainable cultivation of wheat Volume 1: Breeding, quality traits, pests and diseases
Print (ISBN 978-1-78676-016-6); Online (ISBN 978-1-78676-018-0, 978-1-78676-019-7)

Achieving sustainable cultivation of wheat Volume 2: Cultivation techniques
Print (ISBN 978-1-78676-020-3); Online (ISBN 978-1-78676-022-7, 978-1-78676-023-4)

Achieving sustainable cultivation of maize Volume 1: From improved varieties to local applications
Print (ISBN 978-1-78676-008-1); Online (ISBN 978-1-78676-010-4, 978-1-78676-011-1)

Achieving sustainable cultivation of maize Volume 2: Cultivation techniques, pest and disease control
Print (ISBN 978-1-78676-012-8); Online (ISBN 978-1-78676-014-2, 978-1-78676-015-9)

Chapters are available individually from our online bookshop: https://shop.bdspublishing.com

climate•SMART•publishing
IN AGRICULTURAL SCIENCE

BURLEIGH DODDS SERIES IN AGRICULTURAL SCIENCE

NUMBER 03

Achieving sustainable cultivation of rice

Volume 1: Breeding for higher yield and quality

Edited by Professor Takuji Sasaki,
Tokyo University of Agriculture, Japan

burleigh dodds
SCIENCE PUBLISHING

Published by Burleigh Dodds Science Publishing Limited
82 High Street, Sawston, Cambridge CB22 3HJ, UK
www.bdspublishing.com

Burleigh Dodds Science Publishing, 1518 Walnut Street, Suite 900, Philadelphia, PA 19102-3406, USA

First published 2017 by Burleigh Dodds Science Publishing Limited

Library of Congress Control Number: 2016962110

British Library Cataloguing in Publication Data
A catalogue record for this book is available from the British Library

ISBN 978-1-78676-024-1 (print)
ISBN 978-1-78676-026-5 (online)
ISBN 978-1-78676-027-2 (online)
ISSN 2059-6936 (print)
ISSN 2059-6944 (online)

Typeset by Deanta Global Publishing Services, Chennai, India
Printed by Lightning Source

Contents

Series list

Title	Series number
Achieving sustainable cultivation of maize - Vol 1 From improved varieties to local applications Edited by: Dr Dave Watson, CGIAR Maize Research Program Manager, CIMMYT, Mexico	001
Achieving sustainable cultivation of maize - Vol 2 Cultivation techniques, pest and disease control Edited by: Dr Dave Watson, CGIAR Maize Research Program Manager, CIMMYT, Mexico	002
Achieving sustainable cultivation of rice - Vol 1 Breeding for higher yield and quality Edited by: Prof. Takuji Sasaki, Tokyo University of Agriculture, Japan	003
Achieving sustainable cultivation of rice - Vol 2 Cultivation, pest and disease management Edited by: Prof. Takuji Sasaki, Tokyo University of Agriculture, Japan	004
Achieving sustainable cultivation of wheat - Vol 1 Breeding, quality traits, pests and diseases Edited by: Prof. Peter Langridge, The University of Adelaide, Australia	005
Achieving sustainable cultivation of wheat - Vol 2 Cultivation techniques Edited by: Prof. Peter Langridge, The University of Adelaide, Australia	006
Achieving sustainable cultivation of tomatoes Edited by: Dr Autar Mattoo, USDA-ARS, USA & Prof. Avtar Handa, Purdue University, USA	007
Achieving sustainable production of milk - Vol 1 Milk composition, genetics and breeding Edited by: Dr Nico van Belzen, International Dairy Federation (IDF), Belgium	008
Achieving sustainable production of milk - Vol 2 Safety, quality and sustainability Edited by: Dr Nico van Belzen, International Dairy Federation (IDF), Belgium	009
Achieving sustainable production of milk - Vol 3 Dairy herd management and welfare Edited by: Prof. John Webster, University of Bristol, UK	010
Ensuring safety and quality in the production of beef - Vol 1 Safety Edited by: Prof. Gary Acuff, Texas A&M University, USA & Prof.James Dickson, Iowa State University, USA	011
Ensuring safety and quality in the production of beef - Vol 2 Quality Edited by: Prof. Michael Dikeman, Kansas State University, USA	012
Achieving sustainable production of poultry meat - Vol 1 Safety, quality and sustainability Edited by: Prof. Steven C. Ricke, University of Arkansas, USA	013
Achieving sustainable production of poultry meat - Vol 2 Breeding and nutrition Edited by: Prof. Todd Applegate, University of Georgia, USA	014
Achieving sustainable production of poultry meat - Vol 3 Health and welfare Edited by: Prof. Todd Applegate, University of Georgia, USA	015

Introduction

Rice (*Oryza sativa* L.) is a principal food for 3 billion people, providing up to 23% of calorie requirements worldwide. Rice is cultivated in more than 100 countries across the world on around 154 million hectares annually, equal to 11% of the world's cultivable land. At present, the increase in rice production rate (at under 1% per year) is lower than the increase in population of 1.5% per year. The rice community faces the challenge of increasing rice production by 25-50% over the next thirty years, in the face of a rising population, plateauing rice yields, pressures on land and other resources available for rice cultivation and an unstable, changing climate. The two volumes of *Achieving sustainable cultivation of rice* summarise some of the key research addressing those challenges. This volume (Volume 1) reviews developments in breeding to improve yields as well as research on improving nutritional and other aspects of quality. Volume 2 focusses on developments in cultivation techniques and pest management in making production more efficient and sustainable.

Part 1 Rice breeding

While modern rice breeding has been tremendously successful in raising yields to help feed a growing population, breeding strategies of the past are no longer viable for the future. It is generally agreed that the Green Revolution resulted in a relatively small number of genetically related, high-yielding semi-dwarf rice varieties, largely replacing the mosaic of lower-yielding, but genetically diverse landraces that once characterised the rice production environments in those regions. As a result, it is estimated that the modern cultivars have retained only 10 to 20% of the genome diversity that was present in the progenitor species *O. rufipogon*. This narrower genetic base has made it more difficult to sustain increases in yields, and is also vulnerable to a range of biotic and abiotic stresses exacerbated by the effects of climate change.

To address the challenges to sustainability that come with the increasing instability of rice growing environments throughout the world, rice breeders need to expand the germplasm base with which they work and consciously bring in genetic variation from diverse sources to provide durable forms of pest and disease resistance as well as tolerance to a multitude of environmental stresses, including drought, flooding, salinity, extreme soil pH, micronutrient toxicity or deficiency. In addition, the pressures of a growing population mean adopting methods that allow for greater flexibility, reduced breeding cycle time and more efficient use of genetic variation than in the past.

Chapter 1 discusses sources of genetic diversity in rice and how breeders can utilise new tools and strategies to strategically access that diversity for variety improvement. As the chapter points out, within the *Oryza* genus, there are six species complexes, containing two domesticated species: Asian rice (*O. sativa*) and African rice (*O. glaberrima*), and over 20 extant wild species, which show wide variation in plant morphology, range of habitats and tolerance to biotic and abiotic stresses. There is further variation within each complex. As an example, the five different subpopulations of *O. sativa* demonstrate different levels of genetic variation and different alleles and allele combinations that contribute to local adaptation, stress tolerance and grain quality characteristics.

As discussed in more detail in Chapter 3, many studies have identified beneficial alleles in wild *Oryza* species. To better work with wild relatives, breeders have defined primary (*O. sativa* complex), secondary (*O. officinalis* complex) and tertiary gene pools based on species crossability. Gene banks serve as one of the most accessible sources of this genetic diversity. Up until now it has not been feasible to screen large numbers of accessions either genotypically or phenotypically to identify alleles that might contribute useful variation. However, with the rapid decline in the cost of next-generation sequencing and the establishment of high-throughput genotyping facilities capable of processing tens of thousands of breeding lines, it could soon be possible to genotype the majority of accessions used in variety improvement. Genotypic information can be used to provide a unique genetic tag for every accession in a gene bank, enabling scientists to index the bank's entire contents and develop an online database of genetic signatures. The purpose of such a database is to provide a unique identifier for each accession, to quantify the genetic relationships (kinship) among accessions, and to provide the basis for unravelling genotype–phenotype relationships which are key to the discovery of useful genetic variation for breeding.

The chapter discusses initiatives to achieve this outcome such as the *Oryza* Map Alignment (OMAP) project to sequence wild relatives of rice and other projects focussed on the sequencing of diverse varieties of *O. sativa* and, in particular, *O. glaberrima* as an underexploited source of natural variation. However, it is clear that individual projects have resulted in duplication of effort and information. As the chapter argues, a public repository of rice genomic information is urgently needed to document molecular variation, to link genotypes with specific germplasm resources, and to help annotate the functional significance of discrete molecular variants. To make genomic information useful for applications in plant improvement, a significant investment in phenotyping will also be required as well as investment in a database linking the phenotypic information to specific germplasm accessions and genotypes in a way that is easily searchable. Strategic phenotyping will provide the foundation for establishing genotype–phenotype relationships and predictive models for trait improvement. A recent initiative to take this forward is the 3000 Rice Genomes Project which re-sequenced 3000 rice accessions from 89 countries to allow large-scale investigation of novel alleles by researchers around the world.

An important strategy for broadening the genetic base of breeding populations is the establishment of large, pre-breeding programmes. Pre-breeding refers to the development of new base populations or intermediate breeding populations derived from wide crosses between wild or exotic germplasm and elite breeding materials. The purpose of pre-breeding programmes is to facilitate the screening and selection of favourable recombinants in advanced breeding backgrounds without disrupting the normal breeding programme. These programmes can be accomplished using a combination of conventional and molecular plant breeding strategies, including tissue culture. Marker-assisted selection (MAS) techniques have been invaluable in pre-breeding programmes for selecting recombinant individuals that retain only very small pieces of exotic donor chromosomes, thus vastly reducing linkage drag and the overall time required for pre-breeding. The chapter also reviews the most common types of structured population crosses used for recombining natural variation, including bi-parental quantitative trait loci (QTL) mapping populations, interspecific advance backcross-QTL populations (BC-QTL/near-isogenic lines/NIL populations), chromosome segment substitution line populations and multi-parent advanced generation inter-cross (MAGIC) populations.

In rice, most traits are governed by a combination of a few genes of large effect and many genes of small effect. Historically, rice breeders have concentrated trait-based selection on favourable alleles of large effect, either via traditional pedigree breeding, phenotypic backcrossing or MAS. However, with the advent of next-generation sequencing, high-resolution genotyping and big data analysis methods, the contribution of small effect alleles to quantitatively inherited phenotypes is gaining recognition. The potential to methodically recombine many alleles of small effect using genomic selection (GS) is rapidly gaining traction as a way of enhancing the rate of genetic gain in many crops, including rice. GS offers an alternative to traditional MAS by allowing for selection based on an infinite number of loci across the genome. It has the potential to reduce breeding time by increasing the proportion of high-performing offspring in a breeding population, thus accelerating gain from selection.

Building on Chapter 1, Chapter 2 discusses developments in MAS in more detail. As the chapter shows, the sequencing of the rice genome is allowing rice geneticists and breeders to unlock the biological functions of more than 50,000 predicted genes in order to revolutionise rice breeding in the twenty-first century. The advent of molecular markers has made it possible to dissect the genetic basis of more complex traits and use functional DNA markers to monitor the introgression of the gene of interest into target cultivars for improvement. The chapter begins by reviewing the development of different types of markers, including the use of high-density single-nucleotide polymorphism (SNP) markers to locate the position of QTLs for particular traits. As the chapter shows, thousands of QTLs have been identified for traits such as grain number and size, panicle architecture, stem/lodging tolerance, submergence and drought tolerance. Developments such as SNP chips, DarT and genotype-by-sequencing technologies facilitate precise transfer of the genes with greater efficiency in less time.

MAS involves the indirect selection of target traits using molecular markers linked to genes responsible for an agricultural trait. The applicability of DNA markers in MAS depends on the reliability of the marker, template DNA requirement, means of marker assay, level of polymorphism and costs. Given the advantages of MAS compared to phenotypic screening, it has found several applications in rice including genotype identity testing, genetic diversity analysis of breeding materials, gene surveys in parental material, marker-evaluated selection, marker-assisted backcrossing (MABC) and marker-assisted pyramiding (MAP). The chapter reviews developments in MABC and MAP in particular. To date, MAS has been successfully used in rice breeding to selectively transfer major genes/QTLs controlling abiotic stress tolerance including salt, drought and submergence tolerance, pest and disease resistance, yield and grain quality.

To illustrate its contribution, the chapter includes a case study on improving flood tolerance in rice. As early as the 1970s, landraces such as FR13A and FR43B were identified to have tolerance to transient complete submergence. Use of DNA markers and linkage mapping identified a locus for submergence tolerance, designated as *Sub1*, near the centromeric region of chromosome 9 of FR13A. High-resolution mapping identified ethylene response factor-like genes such as *Sub1A*, which could then be introduced into target varieties by MAS. Before the identification of the *Sub1* locus, conventional breeding for submergence tolerance from FR13A to target varieties have been slow (up to 10 years) and had limited success. Identification of *Sub1A* made identification of lines carrying the target trait possible even at early generations, reducing the number of plants for backcrossing and accelerated the selection process to around two years, thereby increasing genetic gain per unit time.

Echoing Chapter 1, the chapter concludes by discussing genome-wide association studies (GWAS) and GS as two emerging genomics-based breeding strategies that promise to take breeding to the next stage of development. Both methods have the potential to deliver genetic gains by speeding the whole process of varietal improvement and dissemination to target regions. GWAS has emerged as an alternative to traditional linkage analysis by using a set of diverse, unrelated germplasm to identify associations between markers and traits of interest. In rice, the availability of a high-quality reference genome, abundance of germplasm and self-fertilising nature of the crop make it an ideal candidate for GWAS applications. With GWAS application becoming commonplace in gene/QTL discovery and mapping, the challenge for breeders now is in determining how information derived from such studies can be directly utilised for MAS and/or GS in applied breeding programmes.

As indicated in Chapter 1, the application of MAS in rice breeding has been mostly effective in the incorporation of traits that are controlled by major genes or QTLs with a known association to a marker. However, not all traits of agronomic importance are controlled by major genes/QTLs but by many genes with small, additive effects. Unlike MAS which utilises a defined set of markers associated with a trait for selection, GS uses all available marker (i.e. SNP) data across the genome and feeds it into a prediction model that would then estimate the genome-estimated breeding value of an individual in a breeding population. To date, results of GS simulation and cross-validation studies have not been evaluated in actual rice breeding programmes. The shift to practical application will require significant investment in resources to set up both genotyping and phenotyping pipelines needed for genome-wide selection.

Building on both Chapters 1 and 2, Chapter 3 focusses specifically on ways of improving rice yields, providing a context for Chapters 4 and 5. Yield gain can be significantly enhanced by fine-tuning aspects of plant architecture such as height, productive tiller number and main culm/stem diameter. The chapter reviews research identifying genes such as *SPL14* for primary and secondary branching of the panicle and panicle length; *SCM2* for increasing stem diameter; and the *Gn1a*, *DEP1* and *DEP3* genes for grain number and panicle size. Echoing Chapter 1, the chapter points out the particular importance of wild rice species as a source of these beneficial traits, and research to identify yield-related traits such as culm/stem length, panicle length and number, spikelets per panicle as well as grain thickness, weight and length. The chapter then focusses on genes relating to three traits which particularly influence yield: grain size and weight, number of grains per panicle and number of panicles per plant. As an example, it explores how researchers have pyramided different QTLs (including the gene *GS5*, which affects grain size, filling and weight), with F3 and F4 plants, using a set of SSR and InDel markers and NGS data, demonstrating better yield than the elite parents.

The chapter looks at developments and challenges in hybrid breeding as one method for exploiting wild varieties. Research suggests the potential to increase heterosis if more diverse parents are crossed. The chapter reviews research addressing limitations such as partial hybrid sterility and loss of vigour in future generations, requiring fresh F_1 seeds to be produced each season to maintain a high level of heterosis and the need to maintain two or three separate populations to produce hybrid seeds. The chapter discusses developments such as the identification of wide compatibility alleles and embryo rescue techniques to improve the success of wild *Oryza* breeding programmes. Other developments include the use of SNP chips and MAGIC populations to increase genetic variation and diversity through shuffling a wide range of genomes to promote identification of beneficial alleles.

Picking up from Chapter 3, Chapter 4 reviews ways of improving photosynthesis in rice. This is an important area of research not just in improving plant efficiency and yields but also in helping rice to adapt to increases in atmospheric CO_2 associated with global warming. As the chapter shows, with recent advances in the understanding of the photosynthetic biochemistry and physics of light harvesting, new strategies for improving photosynthesis are being evaluated in numerous C_3 crops, including rice. In particular, with improved understanding of photosynthetic limitation, it is now possible to target those genes that overcome these limitations. The chapter starts by summarising research on the biochemical limitations establishing the demand for CO_2 within C_3 leaves. Demand-based biochemical limitations include capacity limitations in the enzyme Rubisco which catalyses the fixation of CO_2, regeneration of ribulose biphosphate (RuBP) and the effects of photorespiration. Supply-side limitations include impediments to CO_2 diffusion through the boundary layer to the chloroplast.

The chapter goes on to review three key strategies in development for improving photosynthesis in C_3 plants. The first is to exploit existing variation in rice using QTL analysis in order to identify genetic regions that increase photosynthetic capacity. This approach is already showing promise and may lead to discoveries of unknown regulatory and development genes that could enhance photosynthesis and whole plant performance. The chapter explores developments in identifying QTL by either direct measurement of photosynthesis or indirect determination of phenotypic values of traits related to photosynthesis. These include 'phenomic' platforms which use sensors and imaging programmes to measure attributes such as leaf fluorescence, colour and temperature, and plant size and morphology which can be related to photosynthetic performance.

Since C_3 plants underperform at elevated CO_2 levels, a second strategy is to re-engineer C_3 photosynthesis in the rice plant in order to adapt it to the predicted increase in atmospheric CO_2. Approaches here include reducing Rubisco expression and overexpressing RuBP regeneration components. The final strategy reviewed is the significant challenge of introducing C_4 photosynthesis into rice. Engineering the C_4 pathway into C_3 plants has long been a dream of photosynthetic biologists since C_4 plants have been shown to demonstrate yield enhancements and resource use efficiency improvements of 40–60% relative to C_3 plants. As such, successfully engineering C_4 photosynthesis into rice represents a giant leap that could meet the world's rice demands well into the future. However, as the chapter points out, C_4 engineering is a considerable challenge given the complexity of the genetic controls over the C_4 pathway, many of which are not fully understood. To fully implement the C_4 pathway into a C_3 species such as rice, a range of changes will be required to leaf and organelle structure, metabolism and transport capabilities. The chapter reviews achievements so far such as successfully stacking the C_4 cycle genes into rice, developing a working model of Kranz leaf development, putting in place a multi-pronged gene discovery effort using the latest sequencing technologies and developing C_4 plants that can serve as genetic models. There is good reason to believe it will be possible to engineer a C_4-Rice prototype within the next decade.

Sustainable rice production requires new types of rice cultivars in specific target population environments (TPE) that combine high yield potential with greatly improved nutrient and water use efficiency (WUE) and tolerance/resistance to multiple abiotic/biotic stresses specific to each TPE. First proposed in China, Green Super Rice (GSR) is defined as rice cultivars able to produce high and stable yield with less inputs. Early breeding initiatives focussed primarily on drought tolerance/WUE and N/P use efficiency, but the scope has been broadened to promote high overall resource use efficiency to achieve more sustainable cultivation.

To address the challenge of breeding for multiple traits, the chapter describes the development of a modified backcross (BC) breeding procedure which can be easily adopted by small breeding programmes. This approach has allowed the development within a period of 5 years of large numbers of promising GSR lines with good tolerance to multiple abiotic/biotic stresses (in both rainfed and irrigated condition) from a small number of breeding populations. The shuttle breeding and molecular recurrent selection breeding strategy involves four major activities: (1) developing segregating BC populations by crossing elite widely adaptable recipients with a diverse set of donors, (2) strong phenotypic selection and progeny testing of the BCF_2 populations under multiple stresses, (3) QTL/QTL network discovery by genetic characterisation of selected introgression lines using high-throughput genotyping and (4) development of superior GSR varieties by designed QTL pyramiding or by molecular recurrent selection. As the chapter shows, the superior performance of many GSR varieties has been well demonstrated in the field trials under both irrigated and rainfed conditions at the IRRI and in many target countries across Asia and Africa. Amongst other things, this breeding strategy suggests there is sufficient diversity in the primary gene pool to improve abiotic stress tolerance, and that selecting appropriate recurrent parent(s) with high yield potential and wide adaptability to the target environments is the key for success for improving complex traits in a BC breeding programme, since a widely adaptable variety must have key adaptive traits and pathways to most environments.

The final chapter in Part 1 looks at the key topic of drought tolerance in rice. As a water-loving crop, rice can be severely affected by drought stress. It is especially sensitive to drought stress during reproductive growth and even moderate stress can result in drastic reduction in grain yield (GY). Drought resistance is a complex trait whose effect depends on the action and interaction of different morphological, biochemical and physiological responses. Research suggests that drought resistance in rice mainly relies on WUE that allows minimum water usage for maximum production and osmotic adjustment that enables plants to maintain turgor and protect the meristem.

The chapter reviews studies identifying major QTLs related to GY under water-limited conditions in rice. An example is qDTY12.1 which has an effect on a range of traits such as flowering, plant height, biomass and panicle number which influence GY under drought conditions. Physiological parameters like maintenance of water potential using deeper rooting system, synchronised modulation of metabolic processes, and the level of osmolytes and antioxidants are also key factors in selection for drought tolerance, together with endogenous levels of phytohormones. The chapter reviews research identifying QLTs relating to deep rooting and drought-protective genes such as protein kinases, late embryogenesis abundant proteins, antioxidant enzymes and osmolytes. Research has also highlighted a central role for major phytohormones such as abscisic acid, cytokinin (CK), gibberellic acid, auxin, ethylene and jasmonic acid in regulating plant growth and development as well as drought stress response. Genetic engineering by transformation of rice plants using modified gene constructs has also provided genetic evidence of drought tolerance components that can serve as models for development of cultivars with improved yield potential. The chapter looks at a range of transcription factors related to drought tolerance, such as the AP2/ERF transcription factor, *HYR* (HIGHER YIELD RICE) associated with photosynthetic carbon metabolism. Overexpression of HYR enhanced photosynthesis, contributing to higher GY under drought stressed conditions. The chapter concludes by highlighting the need for more research on the mechanisms underlying drought response in rice.

Part 2 Rice nutritional and processing quality

Chapter 7 introduces Part 2 by providing an overview of the role of rice in nutrition, providing a context for Chapters 9 and 10, in particular. As the chapter points out, malnutrition includes both under- and over-nutrition. Under-nutrition refers to dietary deficiency of key nutrients, including calories and micronutrients. Many studies have shown that unpolished rice contains high levels of minerals, many essential vitamins and a host of other phytochemicals, but once the bran is removed and the grains are white, the mineral and vitamin content is very low. Research has therefore focussed on biofortification of polished rice which remains a staple in many countries, focussing on iron, zinc and ß-carotene. Whilst there has been some success in using conventional breeding to fortify rice with zinc, for example, much research has focussed on the use of biotechnology. Examples include manipulation of the expression of NA synthase (*NAS*) genes to increase the iron content of rice grains or to increase ß-carotene to combat the serious effects of vitamin A deficiency (VAD).

The chapter also highlights the less-studied area of rice's role in over-nutrition, a condition where adequate calories are available but their dietary quality is poor. In combination with other lifestyle factors, over-nutrition can lead to obesity, diabetes and a host of other chronic diseases. Factors such as the type of carbohydrate, glycaemic index (GI) and starch digestibility are all implicated in the prevention and management of chronic disease. Research suggests, for example, that there is diversity in the GI of different rice varieties, and that low GI rice leads to a smaller glycaemic response, but that little work has been done so far to explore varieties of rice that might assist with managing blood sugar status.

Building on Chapter 7, Chapter 8 looks in more detail at nutrients and phytochemicals present in rice. As it shows, the genetic differences between cultivated varieties of rice contribute greatly to variations in nutritional content, for example, carbohydrate content, concentrations of iron and zinc, lipid, vitamin E and phenolic acid content, with pigmented rice having higher concentrations than white rice, for example. The chapter reviews current research on the major health-promoting components in rice, including sterols, oryzanols, tocopherols, tocotrienols and phenolic compounds such as flavonoids. It also discusses newly discovered phytochemical compounds such as γ-oryzanol, isoferulic, syringic and chlorogenic acids. As research shows, the differential chemopreventive bioactivities of rice against chronic diseases such as cancer are positively correlated with the presence of phytochemicals such as phenols and flavonoids, especially in the bran and hull from pigmented rice.

The chapter also looks at ways of preserving or enhancing nutritional and nutraceutical properties of rice, including the use of both conventional and transgenic techniques for biofortification, a theme which is explored in more detail in Chapters 9 and 10. It also summarises research on the effects of growing conditions and cultivation techniques on nutritional content, for example, on the way the germinating process can increase the nutritional value of rice seeds, a theme which is picked up in more detail in Chapter 12. Finally, it assesses what we know about the effects of processing techniques such as dehulling and milling as well as cooking on the nutritional value of rice, emphasising the benefits of consuming whole rice or partially milled rice. As the chapter shows, there is still much more to be done in these areas and in better understanding of health benefits *in vivo*.

As Chapter 9 points out, VAD has been recognised as a significant public health problem for more than a quarter of a century. It has been estimated that a universally available source of vitamin A could, for example, save 23–34% of all deaths of children under 5 years globally. Whilst the varied diets in industrialised countries include sufficient sources of provitamin A carotenoids, many populations in the developing world are not able to access such balanced diets. The chapter reviews the various methods of supplementation, their strengths and weaknesses, before concluding that biofortification of a staple crop such as rice is the most sustainable option. It then reviews the alternatives for producing a provitamin A rice such as using traditional breeding or mutagenesis, none of which has proved viable, leaving the option of genetic engineering. It describes the origins and development of provitamin A Golden Rice to the stage of field trials. However, as the chapter points out, continued opposition to Golden Rice on the grounds that it involves genetic modification has proved a serious impediment to its final approval and use.

Chapter 9 is complemented by Chapter 10 which focusses on the development of rice varieties with improved grain iron content. Iron deficiency anaemia (IDA), together with vitamin A, iodine and zinc deficiencies, is the most common form of micronutrient malnutrition globally. IDA is widespread in both developing and developed countries. Several possible strategies have been suggested for combating the deficiency in affected populations. These include education about appropriate dietary habits, iron supplementation of specific groups at high risk, fortification of foods and biofortification (i.e. production of nutrient-enriched crops). The latter has been suggested as a sustainable approach to address micronutrient deficiencies worldwide. However, as indicated in Chapter 9, screening has shown that rice germplasm has a very low diversity for the key micronutrients. Given this insufficient natural genetic variation in the endosperm iron levels in rice, increasing iron levels through conventional breeding has not been possible, leading researchers to explore transgenic approaches.

The chapter starts by reviewing what we know about the molecular mechanisms controlling iron uptake and the transporters and chelators necessary for effective iron translocation within the plant. Gene technology approaches have focussed on transformation of genes involved in iron uptake, transport and storage in the endosperm. Different approaches have included increasing iron stores in the rice endosperm and improving bioavailability, modifying genes involved in iron uptake and translocation and strategies focussing on concerted expression of multiple genes. However, as the chapter points out, iron biofortification of rice endosperm has proved challenging, with the majority of approaches tested so far not generating rice lines that meet the target iron levels set by the HarvestPlus programme. Studies suggest the presence of strong physiological/molecular barriers that limit iron transport into the grain and/or endosperm. Research still needs to clarify the speciation of iron forms and chelators, the contribution of each type of transporter, and the precise iron flux for each translocation step. Biofortification strategies need to explore candidate genes that can increase the availability of free iron for transport and storage into the grains, in addition to promoting effective translocation within the plant.

Rice processing quality parameters are designed to assess the suitability of particular rice varieties in meeting processing and/or consumer quality traits. There are several common challenges to ensure rice processing quality parameters and testing are relevant and robust. These include an understanding of consumer needs and technology that satisfies needs across the industry. As Chapter 11 points out, the effective measurement of quality is critical to breeders and growers if the industry is to deliver a consistent product to the consumer. As the chapter points out, there is often a relatively short period between rice harvest and rice sowing available to conduct rice quality evaluations and make selection decisions on the next generation of cultivars.

Rice quality characteristics typically assessed for uncooked grains include grain dimension, chalkiness, colour, composition and milling performance and, for cooked grains, include elongation, texture and digestibility. Composition of the grain is often limited to amylose content and nitrogen content as a proxy to protein content. Sulphur content is also measured. Functional parameters typically measured are viscosity, fragrance, texture and gelatinisation temperature, and more recently GI and antioxidant content. Chapter 11 reviews the development of newer techniques such as the use of capilliary electrophoresis to determine amylose: amylopectin ratio in rice and the development of HPLC separation techniques able to distinguish varieties of different quality classes of rice. It also reviews ongoing challenges such as the need to improve measurement of protein and lipid content, and improve the assessment of physical grain quality. There is also a need to develop good phenotype screening across a broad range of parameters to fast track the development of new and resilient varieties.

As Chapter 11 points out, it has been long established that high growth temperatures can alter grain quality attributes including increased chalk, protein content and gelatinisation temperature, and decreased amylose content. Research also shows head rice yield is influenced by factors such as nitrogen fertilisation, harvest moisture and wet/dry cycles. Chapter 12 builds on Chapter 11 by reviewing how cultivation conditions affect rice quality. Cultivation practices include sowing, transplantation and harvesting dates, as well as irrigation, fertiliser use and tillage systems. As an example, delayed sowing and transplantation times can significantly influence leaf area index, light extinction coefficient and filled spikelets growth, and thus reduce GY. Delayed planting and transplanting also results in lower amylose and lipids content as well as pasting parameters such as decreased peak viscosity and lower grain hardness for textural quality. Chapter 11 reviews the effects of sowing, transplantation and harvesting dates; the impact of irrigation; fertiliser application; and other aspects of cultivation (such as planting density). It also assesses the potential effects of elevated temperature and other aspects of climate change (such as water availability elevated atmospheric levels of CO_2). Finally, it reviews the impact of organic cultivation on rice quality as an effective, sustainable technique for producing high-quality rice.

Summary

The chapters in this volume demonstrate the impressive range of research addressing how rice cultivation can continue to feed a growing population in a more sustainable way. They show the key importance of broadening the genetic base for rice in breeding, and the need for more coordinated documentation of molecular variation, linking of genotypes with specific germplasm resources and annotation of the functional significance of discrete molecular variants. They also show the advances that have been made in exploiting the broader genetic base, for example in the use of MAS to improve pre-breeding programmes or the use of molecular techniques in back crossing to produce improved varieties of GSR. They show the potential of newer techniques such as GS in improving traits determined by a number of genes with small additive effects. The chapters demonstrate the need for a better understanding of these effects, and for more fundamental research to understand complex traits such as photosynthesis and response to drought in rice. There is clearly an urgent need for more research on ways of preserving and enhancing the nutritional and nutraceutical properties of rice to address both the problems of under-nutrition, in such areas as iron and VAD, as well as over-nutrition, an area which also takes in the wider debate about the use of genetic engineering in food production. Finally, chapters of this volume highlight key factors involved in achieving and maintaining consistent quality, including the role of cultivation. This theme is taken up in Volume 2.

Part 1

Rice breeding

Chapter 1

Ensuring and exploiting genetic diversity in rice

Jennifer Spindel and Susan McCouch, Cornell University, USA

1 Introduction

Genetic diversity is the basis for all advances in breeding and genetic improvement and represents an insurance policy for dealing with disaster. To contribute to the goal of achieving sustainable rice cultivation for the future, applied plant breeding programmes must continuously balance efforts to develop and refine elite gene pools in which favourable alleles are painstakingly accumulated over time, with the need to access novel, potentially disruptive, variation from unadapted or exotic sources for times of adversity. To dynamically manage both streams in a productive breeding programme requires a multi-tiered approach. When imposing selection in a relatively stable environment and in keeping with well-defined consumer preferences, a breeder can meet both grower and consumer demands by working within an elite gene pool. As a long-term strategy, however, this approach can have the unintended consequence of severely narrowing the genetic base, leaving breeders without access to the range of genetic variation they need to respond to new challenges as they arise.

Currently, rice breeders are struggling to cope with climate change that is affecting rice production in numerous ways. To address the challenges to sustainability that come

http://dx.doi.org/10.19103/AS.2016.0003.02

with the increasing instability of rice-growing environments throughout the world, rice breeders will need to expand the germplasm base with which they work and consciously bring in genetic variation from diverse sources. In this chapter, we will discuss where to look for useful sources of genetic diversity in rice and how breeders can utilize new tools and strategies to strategically access that diversity for variety improvement. Managing diversity wisely is key to achieving sustainable rice cultivation.

2 Genetic diversity in the genus *Oryza*

Grasses, including rice, represent an ancient lineage of angiosperms estimated to have diverged from other flowering plants ~80 million years ago (mya). Worldwide, there are 10000 grass species that are divided into two major phylogenetic clades: the Bambusoideae, Oryzoideae, Pooideae (BOP) and the Panicoideae, Arundinoideae, Chloridoideae, Micrairoideae, Aristidoideae, Danthonioideae (PACMAD) clades, estimated to have diverged ~50 mya (Glemin and Bataillon, 2009; Soreng et al., 2015). *Oryza*/rice is placed within the BOP clade, while C4 crops, including maize and millets, are placed in the PACMAD clade. The *Oryza* genus is found in the subfamily Ehrhartoidea, paraphyletic with the 'core' bamboos (Bambusoideae) which diverged from the Pooideae (oats and wheat) ~50 mya (Glemin and Bataillon, 2009; Paterson et al., 2004; Vaughan et al., 2005). Within the *Oryza* genus, there are six species complexes, containing two domesticated species: Asian rice (*Oryza sativa*) and African rice (*Oryza glaberrima*), and 21 extant wild species, all of which serve as reservoirs of natural *Oryza* variation (Klee et al., 2000, 2004; Vaughan et al., 2003, 2008b).

Tremendous variation is found across the *Oryza* genus. Ten genome types have been described (AA, BB, CC, BBCC, CCDD, EE, FF, GG, KKLL, HHJJ) and approximately half of all wild relatives are tetraploid, while the other half are diploid. Haploid genome size is highly variable, ranging from ~260 to 960 Mb. Individual species of *Oryza* are distributed across all continents, except Antarctica, and these are adapted to a wide range of habitats, including deepwater, seasonally dry, open, upland, stream sides, semi-shade, full-shade, forest-floor, river banks and rock pools (Jacquemin et al., 2013; Vaughan et al., 2003). Physical size and morphological features are also diverse for all types of traits, including vegetative traits (e.g. height, biomass, canopy architecture, leaf and sheath morphology, leaf structure, leaf physiology and tillering), reproductive traits (e.g. seed number, seed size, panicle shape, panicle architecture, and flowering time) and tolerance to both biotic and abiotic stresses (Atwell et al., 2014; Mickelbart et al., 2015; Shakiba and Eizenga, 2014).

As mentioned previously, there are two species of domesticated rice, *O. sativa* (Asian rice) and *O. glaberrima* (African rice). Asian rice is thought to be polyphyletic, with a complex domestication history involving diverse wild progenitor populations of *O. rufipogon/O. nivara* that took place approximately 8000–10000 years ago in different parts of Asia (Civáň, 2015; Huang et al., 2012a; Kovach et al., 2007; Oka, 1988; Vaughan et al., 2008a). African rice, *O. glaberrima*, thought to be monophyletic, was domesticated from its wild relative, *O. barthii*, ~3500 years ago in West Africa (Linares, 2002; Semon et al., 2005; Sakai et al., 2011; Wang et al., 2014).

Genetic distance between the two cultivated species and wild *Oryza* relatives is variable and can be vast, so to better work with wild relatives, breeders have defined primary, secondary and tertiary gene pools on the basis of species crossability (Sano, 1993). The primary gene pool, also known as the *O. sativa* complex, consists of *O. sativa*, *O. glaberrima*, *O. rufipogon*, *O. nivara*, *O. barthii* and three other wild species with the AA genome. All of these species are diploid and can be crossed with fertile results by using conventional hybridization. The secondary gene pool, also known as the *O. officinalis* complex, consists of ten wild species with non-AA genomes that are either diploid or tetraploid and are cross-incompatible with *O. sativa*; crosses between secondary gene pool species and *O. sativa* can be achieved through embryo rescue. The tertiary gene pool consists of the species most distantly related to *O. sativa*, including the *O. meyeriana*, *O. ridleyi* and *O. schlechteri* complexes. It is extremely difficult to cross primary and tertiary gene pools and any gene transfer is rare (Jena, 2010; Khush, 1997).

Sterility barriers and differentiated gene pools also exist within the domesticated *O. sativa*, as rice is subject to deep population structure. The different subpopulations of *O. sativa* harbour different levels of genetic variation and different alleles and allele combinations that contribute to the local adaptation, stress tolerance and grain quality characteristics. Five major rice subpopulation groups are widely recognized. Early isozyme studies identified similar groups (Glaszmann, 1987; Second, 1982), but they were first clearly reported by Garris et al. (2005) on the basis of simple sequence repeat (SSR) evaluation and have subsequently been confirmed on the basis of single nucleotide polymorphism (SNP) analysis (Huang et al., 2012a; McCouch et al., 2016; Zhao et al., 2011). The *indica* and *aus* subpopulations are found within the *indica* varietal group, while the *tropical japonica*, *temperate japonica* and aromatic/Group V subpopulations cluster within the *japonica* varietal group. These subpopulations are genetically well differentiated from each other, with estimates of divergence between closely related subgroups such as *indica* and *aus* or *temperate* and *tropical japonica* showing F_{st} values of ~0.3, and distantly related subpopulations such as *indica* and *temperate japonica* showing F_{st} values of ~0.5 (Zhao et al., 2010), compared to F_{st} values of 0.08–0.09 for heterotic groups in maize (Matsuoka et al., 2002).

Subpopulation structure has also been found in *O. glaberrima*. An early study of 198 diverse *O. glaberrima* accessions by using 93 SSRs identified 5 genetically distinct subpopulation groups, the most differentiated of which corresponded to a group of accessions that were admixed with *O. sativa*. The other three subpopulation groups were *O. glaberrima*-specific with pairwise F_{st} values of ~0.09–0.18, considerably less differentiated than the subpopulations of *O. sativa* (Semon et al., 2005). More recently, five subpopulation groups were also identified in *O. barthii*, the wild progenitor of *O. glaberrima*, on the basis of the re-sequencing of 94 *O. barthii* accessions (Wang et al., 2014). Pairwise F_{st} values between *O. glaberrima* and *O. barthii* subgroups 1–4 ranged from 0.055 to 0.369, while the pairwise F_{st} between *O. glaberrima* and *O. barthii* subgroup 5 was 0.009, suggesting that *O. glaberrima* was domesticated from this *O. barthii* subpopulation group (Wang et al., 2014). It is estimated that *O. glaberrima* contains ~30–40% of the variation found in *O. barthii* (Li et al., 2011; Wang et al., 2014).

3 The importance of diversity in rice

Rice diversity can have different meanings for different people. Diversity may be characterized at the genetic level, where it can be quantified in terms of molecular polymorphisms observed among individuals; it may be evaluated at the phenotypic level on the basis of observable variation in plant morphology, physiology, biochemistry and/

or development; it can be assessed at the landscape level, measured as the degree of species diversity in an ecosystem or the number of varieties that are inter-planted and grown together in a locality; or it may be considered at the geographical or ecological level, describing where different *Oryza* species or accessions are adapted and can grow, or where they originated. Diversity may also be gauged in terms of varietal ancestry, estimated as the number of genetically diverse parents contributing to the breeding pedigree of a population or lineage, or it may be reckoned in terms of the heterogeneity of stresses encountered by a rice crop over space and time, that is, within a single season in a farmer's field, across seasons in the same location or across environments in which rice is grown.

Plant breeders must take all of these dimensions of diversity into consideration. When developing new varieties, they must deliver useful genetic variation to a given ecology or landscape, and this variation must fit into a specific cultural or production environment and must be compatible with a given market or consumer niche.

The deep subpopulation structure of *O. sativa* has partitioned genetic variation so that specific types of varieties (belonging to a particular subpopulation) are known to be adapted and widely cultivated in a specific region or environment, typically characterized by water management practices (paddy rice, rainfed rice, dry or upland rice, and deep water rice), soil type, climatic conditions related to latitude and elevation, and grain quality preferences (Asch and Wopereis, 2001; Bonman, 1992; Chandre Gowda and Jayaramaiah, 1998; IRRI, 1984; Khush, 1997; Fitzgerald et al., 2009). To address the range of biotic and abiotic stresses encountered over the course of a growing season, breeders must also introduce durable forms of pest and disease resistance from diverse sources (Wang et al., 2014; Fukuoka et al., 2014, 2015; Zhang and Xie, 2014; Sardesai et al., 2002) and tolerance to a multitude of environmental stresses, including drought, flooding, salinity, extreme soil pH, micronutrient toxicity/deficiency, heat, cold, etc. (Mickelbart et al., 2015; Ren et al., 2005; Gamuyao et al., 2012; Uga et al., 2013; Famoso et al., 2011; Xu et al., 2006).

Rice varieties must frequently withstand multiple and competing forms of stress in a single growing season. In the tropics, for example, rainfed lowland rice may be intermittently subjected to both submergence and drought, with corresponding pest and disease pressure over the course of a growing season. In temperate climates, by contrast, varieties may be exposed to cold temperatures at the beginning and end of the season and heat stress at flowering. Developing rice varieties that can thrive under multiple forms of stress in the same season requires the selection of plants with genetic plasticity, plants that can respond to changing environmental conditions by adjusting their physiological growth and developmental processes. This necessitates strategic thinking on the part of the plant breeder; breeders must keep a long-term perspective in mind when deciding which crosses to make, what selection protocols to use and how to balance quality and yield potential under diverse forms of stress.

The occurrence and timing of stress conditions is becoming less predictable as the effects of climate change accelerate. Though we often refer to climate change as 'global', the effect of climate on agriculture is predominantly local, as the challenges faced by farmers and breeders are directly related to the variability of local climates. As a result, varieties that were previously bred for adaptation to one location may offer significant value in entirely different locations as local climatological conditions shift. This has important implications for germplasm exchange and the utility of genetic variation that would have been considered irrelevant a decade ago.

As new forms of climate resilience are sought, attention frequently turns to crop wild relatives, not only as immediate sources of variation for particular traits of interest, but also as repositories of information about survival strategies during times of extreme climatic fluctuations – resiliency not previously required by domesticated populations. The wave of crop domestications that took place starting approximately 10000 years ago were prefaced by major shifts in climate, particularly those that occurred during the late Pleistocene through the Holocene epochs (Zhao and Piperno, 2000; Lu et al., 2002). During the Pleistocene, the earth's climate was characterized by a prolonged ice age. The cold was ameliorated at the beginning of the Holocene, when the environment is thought to have become warm and humid, with the onset of monsoonal conditions. This early epoch of global warming favoured the expansion of wild rice populations out of the equatorial zone, and as wild rice slowly moved northward into the lower Yangtze River region in China, Mesolithic foragers, already accustomed to the gathering of wild millets, also began to gather wild rice as a source of food. A short return to glacial-like climate conditions during the Younger Dryas period in East Asia (11550–13000 years ago) resulted in many of the wild rice populations dying off, but is also believed to have hastened the domestication process as humans sought to preserve local rice populations, creating paddies to help buffer them from the cold (Zong et al., 2007). The changing climate conditions also help explain why today's *temperate japonica* gene pool is considerably narrower than the *indica* gene pool (Kovach et al., 2007; Lu et al., 2002; Zhao and Piperno, 2000; Zong et al., 2007). By tapping into the reservoir of genetic diversity present in the wild *Oryza* relatives, it should be possible to identify sources of tolerance to more extreme and/or more variable environmental conditions than have been the norm since domestication (Atwell et al., 2014; Shakiba and Eizenga, 2014; Zhang and Xie, 2014). The reality of the current epoch of global climate change adds urgency to the demands placed on plant breeders to develop crops that contribute to the long-term sustainability of cropping systems (McCouch et al., 2013a). They are being asked to simultaneously intensify agricultural systems, increasing yield per hectare, while conserving the essential resources needed to grow crops, that is, arable land, fresh water, energy and agricultural labour (Garnett and Godfray, 2012; Struik et al., 2014).

Given these current realities, there is a strong need to re-envision the way agriculture is practised to achieve a sustainable food supply (Garnett and Godfray, 2012; Struik et al., 2014; Van Noordwijk and Brussaard, 2014; Kuyper and Struik, 2014; Matson et al., 1997). Modern rice breeding efforts have been tremendously successful in raising yields to help feed a growing human population, but breeding strategies of the past are no longer viable for the future. The Green Revolution technologies that led to such rapid and spectacular increases in rice and wheat yields in South Asia and Latin America were not well adapted to the highly heterogeneous growing environments in Africa (van der Ploeg and Ventura, 2014; Vanlauwe et al., 2014). Further, because the Green Revolution varieties were so rapidly adopted by farmers, the result was that a small number of genetically related, high-yielding semi-dwarf rice varieties were planted over millions of hectares in Asia and Latin America, largely replacing the mosaic of lower yielding, but genetically diverse, landraces that once characterized the rice production environments in those regions (McCouch et al., 2012; Tilman, 1998; Zhu et al., 2000). For this reason, it is often said that an unintended consequence of the Green Revolution was that it created a modern genetic bottleneck in farmers' fields.

The overriding emphasis on yield that characterized the Green Revolution has now given way to concerns about sustainability and species diversity in the face of the new climate reality. The decrease in genetic variation in farmers' fields and the loss of species diversity across the landscape has increased the vulnerability of agriculture to pests, diseases and unpredictable weather patterns. At the same time, this loss of diversity has impacted the nutritional profiles of human populations (Ford-Lloyd et al., 2008; Matson et al., 1997). Thus, new thinking is required at many levels. Given that climate change and natural resource limitations are already wreaking havoc on farmers' livelihoods and affecting overall agricultural production around the world, it is more important than ever to develop efficient rice breeding strategies that tap the wealth of natural variation available in the diverse species of *Oryza*. For the plant breeding community, a commitment to exploring new strategies for enhancing the genetic diversity and resilience of rice cultivars must be a key component of any plan that purports to achieve sustainable rice production.

4 Sequencing and genotyping rice varieties as the basis for identifying desirable donors and traits

Understanding natural variation at the genomics level can provide valuable new insights that inform efforts to breed new varieties of climate-ready rice. Over the last two decades, the cost of next-generation sequencing (NGS) has decreased more rapidly than the cost of phenotyping, to the point where large sequencing or genotyping projects can be used in conjunction with targeted phenotyping projects to develop models for predicting the phenotypes and breeding values of germplasm. Genotypic information can be used to provide a unique genetic tag for every accession in a gene bank enabling scientists to index the bank's entire contents and develop an online database of genetic signatures (McCouch et al., 2013a). The purpose of such a database would be to provide a unique identifier for each accession, to quantify the genetic relationships (kinship) among accessions and to provide the basis for unravelling genotype–phenotype relationships which are key to the discovery of useful genetic variation for breeding applications.

Recent efforts to sequence diverse collections of both wild and cultivated rice accessions have generated an abundance of data that can be used to quantify levels of molecular variation present in the *Oryza* genus and to mine the germplasm for alleles useful for rice improvement. Efforts to phenotypically screen gene bank accessions to identify individuals carrying traits of interest have met with some success (Plucknett et al., 1987), but it is not feasible to screen large numbers of accessions for the full range of phenotypic traits, particularly those of low heritability. Nor has it been possible to detect favourable alleles that contribute to quantitative variation where the phenotype of the wild accession does not express the trait or where the wild phenotype is not obviously superior (Tanksley and McCouch, 1997).

The *Oryza* Map Alignment (OMAP) project represents a major effort to sequence wild relatives of rice. The goal of this project was to generate high-quality *de novo* reference sequences for all eight AA *Oryza* genome species and one representative reference each for the other nine genome types, and to sequence the chromosome 3 short arm for the eight AA genome species and the BB, CC, BBCC, FF, GG and *Leersia perrieri* outgroup species (Jacquemin et al., 2013). Other efforts have focused on the sequencing of diverse varieties of *O. sativa* and *O. glaberrima*. The re-sequencing of 3000 diverse *O. sativa* genomes by the International Rice Research Institute (IRRI) and the Beijing Genomics Institute (BGI) resulted in the discovery of ~20 million SNPs (The rice genome project, 2014). The materials were sequenced at 14X coverage by using Illumina technology based on NGS, and the short reads were aligned to the reference genome, Nipponbare, a *temperate japonica* variety. Of the SNPs reported by Alexandrov et al. (2014), ~98% were bi-allelic, 1.7% were tri-allelic and the remaining 0.02% had four allele variants per locus. Prior to the 3000 rice genomes project, the re-sequencing of 40 diverse *O. sativa* and 10 diverse *O. rufipogon* accessions at 15X coverage resulted in the discovery of 6.5 million high-quality SNPs (Xu et al., 2012), and the joint analysis of these 50 genomes with re-sequencing data on 21 elite rice cultivars from South America and 33 elite cultivars from North America (Duitama et al., 2015) resulted in the discovery of 23 million polymorphic sites. The re-sequencing of an additional 125 diverse *O. sativa* and wild rice accessions similarly resulted in the discovery of ~16 million SNPs (Rice SNP Consortium; unpublished data) and skim sequencing of 1083 diverse accessions of *O. sativa* and 466 accessions of *O. rufipogon* generated a database of ~8M SNPs (Huang et al., 2009, 2012b). Based on these data, it is clear that *indica* and *aus* subpopulations are significantly more diverse than *tropical* and *temperate japonica*, while the *aromatic* subpopulation is intermediate. The japonica subpopulations experienced the strongest genetic bottleneck and contain only ~20% of the variation found in the *O. rufipogon* species complex (Xu et al., 2012; Huang et al., 2012b).

A high-quality *de novo O. glaberrima* sequence was published in 2014 along with the re-sequencing of 20 *O. glaberrima* and 94 *O. barthii* accessions. A total of 8 174 678 *O. barthii* SNPs were identified, along with 4 447 424 *O. glaberrima* SNPs (Sakai et al., 2011; Wang et al., 2014). The availability of these resources is timely, as many breeders with an interest in developing rice varieties for Africa are looking towards *O. glaberrima* as an underexploited source of natural variation. In the 1990s, high-yielding rice cultivars were obtained by introducing *O. glaberrima* introgressions into *O. sativa*, varieties known as the NERICA (NEw RIce for AfriCA) (Jones et al., 1997; Sarla and Swamy, 2005). Now, there is great interest in the reverse strategy – introducing *O. sativa* introgressions into *O. glaberrima* varieties. NGS genotype data on *O. glaberrima* and *O. barthii* accessions are anticipated to facilitate these breeding goals.

How many of the SNPs discovered to date are common among the different projects has not yet been determined, but it is clear that many of the same SNPs are being discovered over and over again as new accessions are sequenced. A public repository of rice genomic information is urgently needed to document the molecular variation that has been discovered, to link genotypes with specific germplasm resources and to help annotate the functional significance of discrete molecular variants. To make genomic information maximally useful for applications in plant improvement, a significant investment in phenotyping will also be required (Araus and Cairns, 2014; Cobb et al., 2013; Furbank and Tester, 2011; Ghanem et al., 2015), as well as investment in a database linking the phenotypic information to specific germplasm accessions and genotypes in a way that is easily searchable. Strategic phenotyping will provide the foundation for establishing genotype–phenotype relationships and predictive models that are of immediate interest for trait improvement (Desta and Ortiz, 2014; Lipka et al., 2015; Spindel et al., 2016).

While there is a general tendency to move towards high-density genotyping as the cost per data point falls, low-coverage genotype data, which are often easier for rice breeding programmes to produce, can also be a useful tool for assessing rice diversity and implementing genomic-assisted breeding efforts. Genotyping-by-sequencing (GBS), for example, by using both 96-plex and 384-plex automation, has produced between 50 000 and 200 000 SNPs on biparental or multi-parent rice breeding populations, depending on the diversity and composition of the population (Bandillo et al., 2013; Spindel et al., 2013,

2015a). At lower levels of multiplexing (i.e. using skim sequencing strategies to generate 1x genome coverage using Illumina NGS platforms), similar protocols have produced medium-density genotype datasets consisting of ~1 million SNPs after imputation (Han and Huang, 2013; Huang et al., 2009).

For breeding applications, several fixed arrays have been developed and used for marker-assisted backcrossing, mapping of biparental populations or multi-parent crosses, or for assessing diversity in rice diversity panels. For these applications, fixed arrays comprising SNPs of 384 (Thomson et al., 2012), 1500 (Zhao et al., 2010), ~6000 (Thomson, 2014; Yu et al., 2014), 44 000 (Zhao et al., 2011), 50 000 (Singh et al., 2015) and ~700 000 (McCouch et al., 2016) have been reported in rice. Unlike the sequencing-derived approaches (i.e. GBS, NGS or PAC BIO), genotyping data from fixed arrays is subject to ascertainment bias; however, if the chips are well designed to serve the appropriate breeding applications, they provide many benefits when compared to sequence-based approaches, including rapid turnaround time, high-quality detection of heterozygosity, reliable SNP calls that are consistent across experiments and ease of data management in applied breeding programmes.

The public availability of genomic information for both wild and cultivated forms of rice provides the foundation for the application of marker-assisted breeding strategies to facilitate the use of diverse parents in crossing and to increase the efficiency of selection for new traits and alleles among the offspring. Examples from the literature include Dixit et al., 2014; Hittalmani et al., 2000; Nguyen et al., 2003; Septiningsih et al., 2009; Singh et al., 2001; Fukuoka et al., 2014, 2015 and Wang et al., 2014. Genomic information can also be leveraged to drive genomic selection, genome editing or transformation strategies, as demonstrated by Grenier et al., 2015; Onogi et al., 2015; Pinto et al., 1999; Shan et al., 2014; Singh et al., 2001; Spindel et al., 2015, 2016; Xu et al., 2014, 2015.

5 Pre-breeding programmes in rice

An important strategy for broadening the genetic base of breeding populations is the establishment of large, pre-breeding programmes. Pre-breeding refers to the development of new base populations or intermediate breeding populations derived from wide crosses between wild or exotic germplasm and elite breeding materials, where the donor germplasm contains desirable characteristics or genes to be introgressed into elite genetic backgrounds. The purpose of pre-breeding programmes is to facilitate the screening and selection of favourable recombinants in advanced breeding backgrounds without disrupting the normal breeding programme. As promising offspring are identified, they are subsequently introduced into the normal variety development pipeline, often after an additional generation of backcrossing to minimize linkage drag and increase favourable allele frequencies. Whether or not a breeder chooses to adopt a pre-breeding strategy depends on the resources available, the goals and objectives of the breeding programme, the need to look outside the adapted gene pool for trait variation to meet those goals and objectives, and the time and cost involved in developing and screening the pre-breeding populations.

As mentioned previously, the wild relatives of rice constitute an important reservoir of valuable genes, but the association of these species with weedy traits and incompatibility barriers limits the transfer of useful genes into cultivated species (Sano, 1993). The major goal of a wide cross programme is to selectively transfer agronomically and commercially useful alleles from wild species or unadapted material while simultaneously limiting linkage drag. This is generally accomplished by using a combination of conventional and molecular plant breeding strategies, sometimes involving tissue culture (Shakiba and Eizenga, 2014; Zhang and Xie, 2014; Brar and Khush, 1997). It is worth noting that many important genes and traits present in the wild relatives of rice or in low-yielding landraces may be masked and thus cannot be evaluated in the native background. Once they are transferred to a cultivated species, however, they may measurably enhance performance (Ying et al., 2014; McCouch et al., 2007; Xiao et al., 1998).

There are many ways by which standard breeding practices can be modified to make them more efficient and to increase the rate of genetic gain. Pedigree breeding is the most common method of rice improvement, and it has a relatively good track record, punctuated by the successes of the Green Revolution, the development of hybrid and super-hybrid rice, and breeding for the 'ideal plant type' (Yuan, 2014; Li et al., 2009; Xue et al., 2014). While traditional pedigree breeding is slow, requiring at least ten years for the release of a new variety (Collard and Mackill, 2008), it will be important to consider how breeding programmes can be modified in the future to incorporate new tools for selection, develop new strategies to shorten the generation time and make better use of natural variation to accelerate genetic gain.

One important consideration will be to understand the genetic architecture of the traits that one is trying to improve, as it will impact the breeding strategy. Genetic architecture refers to the number of quantitative trait loci (QTL), the size of QTL effects and the frequencies of alleles segregating in a population of interest for a given phenotype. It is a direct result of breeding and domestication history and thus varies widely by trait and population. For example, in elite rice breeding populations, most alleles that have a major effect on commercially important agronomic and quality traits such as flowering time or cooking quality of the grain are already fixed in the homozygous condition – as a result, the genetic architecture of these adaptive traits will appear 'infinitesimal', meaning that only small-effect loci contribute to the observed phenotypic variation. If a new allele from an exotic source is introduced for one of these traits, however, that allele will frequently appear to have a very large effect. In inbred rice, breeding lines that carry introgressions from distantly related subpopulations or specific, rare macro-mutations are often observed to carry large-effect QTL.

In rice, most traits are governed by a combination of a few genes of large effect and many genes of small effect. Historically, rice breeders have concentrated trait-based selection on favourable alleles of large effect, either via traditional pedigree breeding and phenotypic backcrossing or via Marker-assisted selection (MAS). These large-effect QTL also represent the majority of genes that have been cloned by plant biologists (Mickelbart et al., 2015; Wang et al., 2014; Xue et al., 2014). For example, WAXY, which encodes a granule-bound starch synthase on chromosome 6, and is the major-effect gene for grain starch content, was cloned in 1990 (Wang et al., 1990), and breeders have since incorporated numerous, sub-population-specific alleles related to starch quality into different market classes of rice as a way of adjusting the cooking quality of the grain (Teng et al., 2013; Fitzgerald et al., 2009).

Genes of small effect, by contrast, have been selected unconsciously for the most part, as part of 'black box' breeding schemes, while they have been largely ignored by molecular geneticists since they are difficult to measure and cannot be easily manipulated by using MAS. With the advent of NGS, high-resolution genotyping and big data analysis methods, the contribution of small-effect alleles to quantitatively inherited phenotypes is gaining recognition. Further, the potential to methodically recombine many alleles of small effect by using genomic selection (GS) (discussed in detail later in this chapter) is rapidly gaining traction as a way of enhancing the rate of genetic gain in many crops, including rice (Bernardo, 2008; Heffner et al., 2010; Lorenz et al., 2011; Meuwissen et al., 2001).

It is widely acknowledged that changes to current breeding practices will be essential to achieve sustainable rice cultivation in the future, but there is no consensus among breeders on what kinds of changes are needed or on how these will happen. Each breeding programme will have its own unique trajectory in this regard, depending on the traits of interest, the target population of environments in which the new varieties will be deployed, the availability of financial and human resources, access to diverse germplasm resources, the demands of the market or populations they serve, the flexibility of the organizational structure, and the unforeseen opportunities that may present themselves. In the next section, we explore some of the ways in which breeding programmes are evolving to take advantage of new technologies and new genetic variations, and some of the challenges they encounter as they move forward.

6 Using marker-assisted selection (MAS) and genomic selection (GS) to identify individuals with desirable traits or performance

Molecular markers have proved to be an essential tool for introgressing traits of interest in rice breeding programmes. MAS describes the use of molecular markers to select for one or a small number of genes of interest in breeding materials, once marker-trait associations have been discovered and validated. MAS can be used for forward selection, that is, to select *for* a desirable gene or QTL; for background selection, that is, to select *against* deleterious genes or unwanted background introgressions, such as from a wild relative; or for recombinant selection, that is, to select for particular recombination events (Collard and Mackill, 2008; Varshney et al., 2014). In pre-breeding programmes, all three MAS techniques have been invaluable for selecting recombinant individuals that retain only very small pieces of exotic donor chromosomes, thus vastly reducing linkage drag and the overall time required for pre-breeding.

MAS, targeting major genes and/or QTLs of large effect, has been used successfully for over two decades in rice to incorporate resistance to both biotic and abiotic stress. Examples include bacterial blight, rice blast, submergence, salinity and drought (Dixit et al., 2014; Gregorio et al., 2013; Ndjiondjop et al., 1999; Septiningsih et al., 2009; Singh et al., 2001; Venuprasad et al., 2012; Uga et al., 2013; Wang et al., 2014; Mickelbart et al., 2015; Shakiba and Eizenga, 2014). However, with the increased efficiencies offered by new sequencing and genotyping technologies, breeders have the potential to utilize a larger number of markers than was previously possible and to perform selection at a genome-wide level that more accurately accounts for the complex genetic architecture underlying many traits of interest.

GS, introduced in 2001 by Meuwissen and colleagues, offers an alternative to traditional MAS by allowing for selection based on an infinite number of loci across the genome (Meuwissen et al., 2001). In a GS breeding schema, genome-wide DNA markers are used to predict which individuals in a breeding population are most valuable as parents of the next generation of offspring. These estimated values, termed the genome estimated breeding values (GEBVs), are output from a model of the relationship between the genome-wide markers and phenotypes of the individuals undergoing selection. The GEBVs are then used to select the best parents for making new crosses.

The GS model is derived from a training population that is representative of the breeding population. The training population is both genotyped and phenotyped, while the testing population is genotyped only. The testing population genotypes are then entered into the model to calculate the GEBVs of all the individuals in the population, even those that have not been phenotyped. Thus, the key difference between GS and traditional MAS is that genotyping is not limited to a selected set of markers that tag putative genes, but rather breeding value is predicted on the basis of all available marker data. Including all markers in the model regardless of effect size makes it possible to track and select for small-effect genes/QTL in addition to large-effect genes/QTL. Statistical shrinkage, Bayesian and/or machine learning methods are used to fit the many thousands of effects (Lorenz et al., 2011; Meuwissen et al., 2001). GS can make breeding more efficient (see following section), but by itself it will not preserve or enhance genetic diversity. Ultimately, GS is a conservative breeding strategy – selection on the basis of GEBVs output from GS models will result only in the selection of alleles already demonstrated to be advantageous in the training population. Indeed, simulation studies have suggested that long-term GS programmes will decrease genetic diversity in a breeding population over time (Jannink, 2010).

Due to potential sterility barriers between highly divergent rice species/varieties, as well as the challenges posed by linkage drag, breeding strategies that do not rely on physical compatibility and ability to generate fertile crosses also hold promise for expanding the genetic base of deployed rice cultivars. These strategies include the use of transgenics and, more recently, genome editing by using CRISPR/Cas9 (Shan et al., 2014; Xu et al., 2014, 2015). Transgenic strategies can be used to introduce genes from entirely different species, such as the expression of the maize pro-vitamin A gene in rice endosperm to create 'golden rice' (Beyer et al., 2002), but are expensive due to extensive deregulation processes and are controversial among consumers.

Traditionally, genome editing could only be accomplished via homologous recombination by using targeted DNA repair of double-strand breaks – an expensive and laborious process. This has recently changed with the development of the CRISPR/Cas9 system, a relatively easy-to-use and inexpensive technique for inducing highly specific genome editing in a variety of species, including rice (Xu et al., 2015). By using genome editing, it is possible to substitute alleles at specific loci without any crossing or use of transgenes. Depending on how these genome-edited varieties are regulated in the future, genome editing could become a transformational strategy for introducing novel diversity into elite rice varieties while simultaneously bypassing the genetically modified organism (GMO) controversy.

The primary benefit of GS over the widely used traditional pedigree breeding method is one of breeding efficiency. Gain from selection during GS is proportional to GEBV accuracy. As a result, when GEBV accuracy is high enough, GS can reduce breeding time by increasing the proportion of high-performing offspring in a breeding population, thus accelerating gain from selection (Bernardo, 2010; Heffner et al., 2010). A number of recent studies suggest that it can work this way for rice, as well as for maize and wheat, the crops in which the strategy was pioneered (Grenier et al., 2015; Onogi et al., 2015; Spindel et al., 2015a,b). GS can further shorten generation time for temperate rice plants by enabling selection to take place in an off-season nursery (as phenotyping is not necessary, only genotyping), or for tropical rice, by combining GS with rapid generation advance (RGA) (Heffner et al., 2010). In RGA, seedlings are brought to maturity rapidly in the greenhouse, and in combination with GS, multiple rounds of selection could occur in the space of a single season (Hickey et al., 2009; Jongdee et al., 2006; Ohnishi et al., 2011).

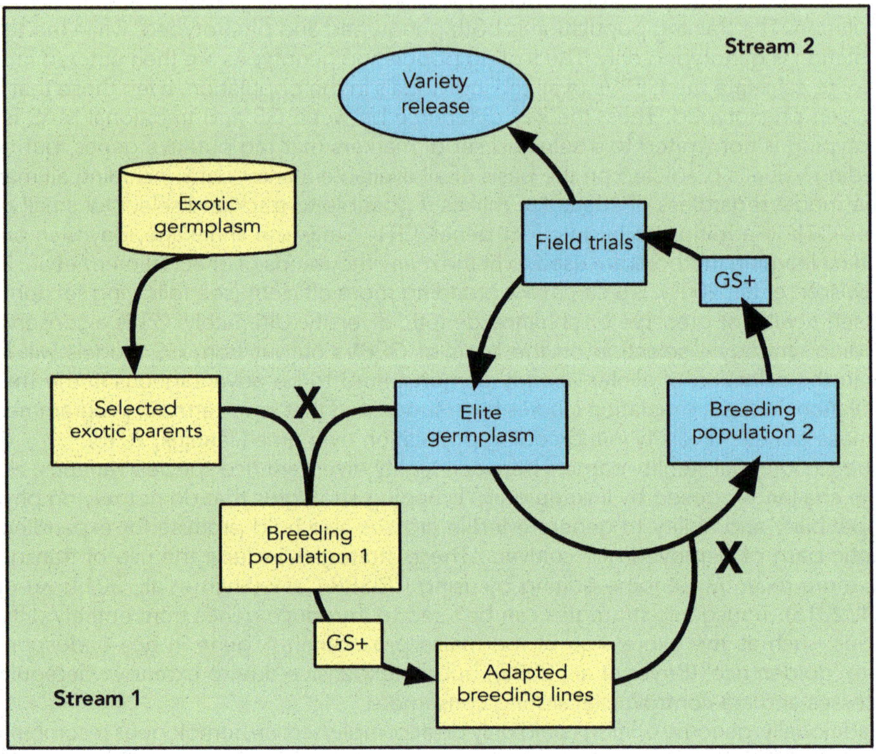

Figure 1 Diagram of proposed two-stream GS breeding program. Stream 1 (white boxes) consists of pre-breeding, in which favorable alleles from exotic germplasm are introduced into adapted germplasm via backcrossing. To identify the favorable alleles, exotic donor parents are backcrossed with elite recurrent parents to develop Breeding Population 1. Selection of individuals from Breeding Population 1 is performed using a combination of phenotypic selection and GS + *de novo* GWAS models (GS+), in which the exotic QTL are fit as fixed effects. The training population for GS would be a subset of breeding population 1, i.e., a fraction of breeding population 1 would be both genotyped and phenotyped, while the rest of breeding population 1 would be genotyped only. Individuals with the highest GEBVs from Breeding Population 1 are crossed into Breeding Population 2 (Stream 2, grey boxes) where they are further refined using GS + *de novo* GWAS models, where the fixed effects would include valuable QTL identified based on GWAS performed *de novo* in Breeding Population 2, the exotic QTL from Stream 1, or any other large-effect QTL a breeder might want to target for trait improvement. Output from Stream 2 can be advanced toward variety release or fed back into Stream 1 to serve as parents for further crossing and population development.

Reproduced with minor revisions by permission from: Spindel, J. E., Begum, H., Akdemir, D., Collard, B., Redoña, E., Jannink, J. L. and McCouch, S. R. (2016). Genome-wide prediction models that incorporate *de novo* GWAS are a powerful new tool for tropical rice improvement. Heredity. doi:10.1038/hdy.2015.113.

What many breeding programmes would ideally like to do is to combine the advantages of GS, which allows them to maintain valuable allele combinations present in elite germplasm, with the ability to selectively incorporate novel variation from exotic sources, similar to the way in which MAS has been used to facilitate interspecific crosses (Bernardo, 2014; Rutkoski et al., 2014; Owens et al., 2014; Bentley et al., 2014; Zhao et al., 2014).

A way to do this in rice has been proposed by Spindel et al. (2015b) by using a breeding schema termed 'GS + *de novo* GWAS'. This strategy consists of two parts or 'streams'. Stream 1 is a pre-breeding effort, in which new alleles sourced from diverse germplasm are sequentially introduced into a population of adapted material. The pre-breeding stream begins with a biparental cross between an exotic variety and an elite adapted variety. After several rounds of backcrossing and recombination, which are necessary to break linkages after the initial F_1 cross, GS would be performed on the backcrossed (BC) population along with GWAS to identify beneficial large-effect QTL to fit as fixed effects in the GS model. This GS + *de novo* GWAS approach is expected to increase the frequency of desirable exotic QTL while simultaneously selecting against alleles from unadapted material that conferred negative effects.

Stream 2 continues the process of refining and improving existing elite material, and by feeding the output of Stream 1 into Stream 2, the genetic base of modern varieties would be expanded. Valuable QTL in the Stream 2 breeding population would also be identified via *de novo* GWAS and fit as fixed effects. These fixed effects could include the exotic QTL from Stream 1 and any other large-effect QTL that a breeder might wish to target for either positive or negative selection (Fig. 1) (Spindel et al., 2015b).

7 The use of recombinant populations in pre-breeding programmes

All breeding strategies intended to recombine natural variation rely on making crosses between divergent genotypes (with the previously aforementioned exceptions of crops developed by using transgenic or genome editing approaches). Frequently, crosses are designed to disrupt population structure and open up new possibilities for recombination, as well as to maximize the number of polymorphisms segregating in the population. In this section, we discuss in greater detail the most common types of structured population crosses used for recombining natural variation, including biparental QTL mapping populations, interspecific advance backcross-QTL populations (BC-QTL/near isogenic lines/NIL populations), chromosome segment substitution line (CSSL) populations and multi-parent advanced generation inter-cross (MAGIC) populations.

7.1 QTL mapping populations

Biparental QTL mapping populations were initially generated from crosses between distantly related parents to increase the probability of detecting polymorphisms in an era when marker technology was in its infancy. In rice, this generally meant making *indica* x *japonica* crosses or *O. sativa* x *O. rufipogon* crosses, while mapping with more closely related crosses is feasible today. Regardless of the parents used, the resulting F_1 hybrids are often self-pollinated for six to seven generations to produce a genetically immortal population of recombinant inbred lines (RILs) that can be planted at numerous locations for replicated phenotyping. F_2, BC_1 and BC_1F_2 populations are also widely used for QTL mapping, but they have less resolution and power than RIL populations since there are fewer generations of meiosis and the lines are not fully homozygous, making replicated phenotyping difficult.

The purpose of a QTL mapping population is to identify regions of the genome containing loci that contribute to a phenotype of interest (Tanksley, 1993). The genetic distance between the parents of the biparental population determines the number of polymorphic markers segregating in the progeny, and ideally, the distribution of the phenotypic observations among the progeny will be normal (i.e. a bell curve). These two qualities of the mapping population contribute to the robustness of the statistical analysis and help maximize the probability that significant QTL will be identified. In general, QTL that are identified in such populations have medium to large effects for the phenotype of interest; smaller effects are detectable only in very large populations.

7.2 Interspecific advanced BC-QTL populations

Interspecific advanced BC-QTL populations (also known as near isogenic lines or NILs) are a specific type of biparental QTL population in which two different rice species, most commonly a cultivated *O. sativa* and a wild rice ancestor, are crossed to each other and the resulting progeny backcrossed to the cultivated parent over several generations to generate a population of individuals that contain many small introgressions derived from the wild parent against a cultivated genetic background (Tanksley and Nelson, 1996). Once again, the goal of QTL mapping in a BC-QTL population of NILs is to identify QTL segregating in the population for a trait of interest (McCouch et al., 2007; Tian et al., 2006). Using NILs, however, maximizes the chances of identifying beneficial 'wild' alleles that perform well in the genetic background of cultivated varieties, and the NILs may be immediately useful as breeding parents to expand the genetic base of the cultivated rice gene pool.

7.3 Chromosome segment substitution line (CSSL) populations

CSSL populations are similar to NIL populations, except that each CSSL in the population is selected to carry only a single introgression in a known region of the genome, and a set of CSSLs typically encompasses an entire genome (Ali et al., 2010). The advantage of CSSL populations is that the contribution of the donor parent alleles can be directly measured because the size and the location of the introgressions are known from marker data (Xu et al., 2010; Arbelaez et al., 2015). While CSSL populations take considerable time and effort to develop, they are among the most powerful tools for identifying and quantifying the effect of specific, exotic alleles in selected breeding backgrounds. CSSLs make a particularly powerful QTL mapping population, especially for characterizing the molecular basis of traits inherited from wild species (Jacquemin et al., 2013).

7.4 Multi-parent advanced generation inter-cross (MAGIC) populations

All of the previously discussed populations are derived from crosses of only two parents. MAGIC populations, by contrast, are developed by inter-crossing multiple inbred parents for several generations and then developing inbred lines from the resulting progeny. The two main advantages of MAGIC populations over the previously described biparental populations are that they (1) result in greater mixing of diverse alleles and thus increased genetic variation within the population and (2) undergo a greater number of recombination events, allowing finer resolution of QTLs during QTL mapping/GWAS. The number of parents and exact structure of crosses depends on the MAGIC population (Bandillo et al., 2013; Cavanagh et al., 2008; Huang et al., 2015). In rice, MAGIC populations are being developed at IRRI and elsewhere for both QTL mapping/GWAS and breeding purposes (Bandillo et al., 2013; Li et al., 2014). At IRRI, these include a MAGIC population developed by inter-crossing eight founder lines from the *indica* subpopulation (*indica* MAGIC population) and by inter-crossing eight founder lines from the japonica subpopulation (*japonica* MAGIC population), respectively. A third global MAGIC population was developed by inter-crossing the *indica* and *japonica* MAGIC populations, and a fourth involving the inter-crossing of multiple wild ancestors – these populations are being extensively phenotyped and have since been used for genome-wide association studies (Bandillo et al., 2013; Huang et al., 2015; Li et al., 2014). Because MAGIC populations mix genetic variation from a variety of sources, they are a good strategy for identifying potentially useful novel alleles that could be introduced into elite germplasm to ensure genetic diversity.

8 Summary and future trends

To meet the rising demand for food in the face of population and income growth, coupled with increasing climate volatility, steps must be taken to increase agricultural production by an estimated 50% without greatly increasing water usage or expanding the total land area dedicated to agriculture. These challenges require that we learn new ways of breeding staple food crops, adopting methods that allow for greater flexibility, reduced breeding cycle time and more efficient use of genetic variation than in the past. To make this possible, we must make transformative leaps in the way we use technology to increase the rate of genetic gain in plant breeding programmes.

In this chapter, we discussed a variety of approaches related to germplasm diversity and the use of genomics and molecular technology in breeding and selection, but we have not examined the computational and data management challenges confronting plant breeders who seek to use the massive quantity of genetics and genomics data that are being generated. The sheer volume of information about valuable genes, alleles and genetic variation in rice has made it extremely challenging for most breeding programmes to be able to routinely digest and utilize the information that is being produced. The availability of so many different genotyping and sequencing platforms further complicates the ability to integrate data generated at different times and in different places, and the fact that much of the digital information is not directly associated with seeds or other physical resources used by breeders creates additional translational hurdles.

To bridge the worlds of discovery research and applied plant breeding, the rice community urgently needs a flexible, open-source system of databases, tools and analytical pipelines that will allow them to routinely apply high-density genomic information in cultivar development. This will require the creation of large genomic databases to store high-density genotypic data in condensed format with links to tools that can convert genotype data into information that is immediately useful to breeders, that is, alleles, haplotypes, linkage blocks that are identical-by-descent (IBD), graphical representations of genomes, estimates of genetic distance, etc. Such tools are essential for breeders to be able to take advantage of the wealth of information and technology options that exist at their doorstep. By working together with people trained in computer science, software engineering, statistical genomics, quantitative and population genetics, and mathematics, rice breeders and pre-breeding specialists can begin developing the tools and the databases that they need to manage information in a robust and reliable way so that it can be harnessed to help improve the rate of genetic gain delivered to farmers and, in turn, help achieve sustainable rice cultivation for the future.

Gene banks currently conserve over seven million diverse accessions of wild and cultivated crop species and serve as one of the most accessible sources of untapped genetic diversity available for plant improvement (McCouch et al., 2012). While previously it was not feasible to screen large numbers of accessions either genotypically or phenotypically to identify alleles that might contribute useful variation, with the rapid decline in the cost of NGS and the establishment of high-throughput genotyping facilities capable of processing tens of thousands of breeding lines, it could soon be possible to genotype the majority of accessions used in variety improvement as well as those housed in gene banks. This information would provide critical datasets for implementing genome-wide prediction in breeding and linking breeding programmes more directly with the wider range of natural variation found in gene banks.

The rice community faces the grave challenge of increasing rice production by 25–50% over the next 30 years, in a world with diminishing natural resources and land available for rice cultivation and an unstable, changing climate. In order to ensure that crops continue to succeed in this context, it is more important than ever that breeders utilize the many new tools and strategies described here and elsewhere to integrate novel sources of variation into their routine variety development programmes, and that they work with others to focus on the many dimensions of diversity that are critical to sustainable rice cultivation across the agricultural landscape.

9 Where to look for further information

Review papers:

Garnett, T. and Godfray, C. (2012). *Sustainable Intensification in Agriculture. Navigating a Course through Competing Food System Priorities.* Food Climate Research Network and the Oxford Martin Programme on the Future of Food, University of Oxford, UK.

Jacquemin, J., Bhatia, D., Singh, K. and Wing, R. A. (2013). The international *Oryza* map alignment project: development of a genus-wide comparative genomics platform to help solve the 9 billion-people question. *Curr. Opin. Plant Biol.* 16(2): 147–56.

McCouch, S., Baute, G. J., Bradeen, J., Bramel, P., Bretting, P. K., Buckler, E., et al. (2013a). Agriculture: feeding the future. *Nature* 499: 23–4.

Seminal papers on population structure and domestication in rice:

Civáň, P., Craig, H., Cox, C. J. and Brown, T. A. (2015). Three geographically separate domestications of Asian rice. *Nature Plants* 1(11): 15164. doi:10.1038/nplants.2015.164.

Garris, A., Tai, T., Coburn, J., Kresovich, S. and McCouch, S. (2005). Genetic structure and diversity in *Oryza sativa* L. *Genetics* 169: 1631–58.

Gross, B. L. and Zhao, Z. (2014). Archaeological and genetic insights into the origins of domesticated rice. *PNAS* 111(17): 6190–7.

Huang, X., Kurata, N., Wei, X., Wang, Z. X., Wang, A., Zhao, Q., Zhao, Y., Liu, K., Lu, H., Li, W., Guo, Y., Lu, Y., Zhou, C., Fan, D., Weng, Q., Zhu, C., Huang, T., Zhang, L., Wang, Y., Feng, L., Furuumi, H., Kubo, T., Miyabayashi, T., Yuan, X., Xu, Q., Dong, G., Zhan, Q., Li, C., Fujiyama, A., Toyoda, A., Lu, T., Feng, Q., Qian, Q., Li, J. and Han, B. (2012). A map of rice genome variation reveals the origin of cultivated rice. *Nature* 490(7421): 497–501. doi: 10.1038/nature11532.

Kovach, M. J., Sweeney, M. T. and McCouch, S. R. (2007). New insights into the history of rice domestication. *Trends Genet* 23(11): 578–87.

Understanding genomic and marker assisted breeding:

Collard, B. C. and Mackill, D. J. (2008). Marker-assisted selection: an approach for precision plant breeding in the twenty-first century. *Philos. Trans. R. Soc. Lond. B. Biol. Sci.* 363(1491): 557–72.

Desta, Z. A. and Ortiz, R. (2014). Genomic selection: genome-wide prediction in plant improvement. *Trends Plant Sci.* 19(9): 592–601. doi: 10.1016/j.tplants.2014.05.006.

Spindel, J. E., Begum, H., Akdemir, D., Collard, B., Redoña, E., Jannink, J. L. and McCouch, S. (2016). Genome-wide prediction models that incorporate *de novo* GWAS are a powerful new tool for tropical rice improvement. *Heredity*. doi: 10.1038/hdy.2015.113.

Varshney, R., Terauchi, R. and McCouch, S. (2014). Harvesting the promising fruits of genomics: applying genome sequencing technologies to crop breeding. *PLoS Biol.* 12(6): e1001883. doi: 10.1371/journal.pbio.1001883.

Useful websites:

Rice Genome Annotation Project Database and Resource
http://rice.plantbiology.msu.edu/

International Rice Genome Sequencing Program (IRGSP)
http://rgp.dna.affrc.go.jp/IRGSP/index.html

The Oryza Map Alignment Project (OMAP)
http://www.omap.org/resources.html

The Oryza SNP project
http://oryzasnp.plantbiology.msu.edu/

National Science Foundation Rice Diversity Project
http://www.ricediversity.org/

The International Rice Informatics Consortium
http://iric.irri.org/

Recommended scientific journals/issues:

Heredity
Nature Plants
Nature
Trends in Plant Science
Crop Science
Theoretical and Applied Genetics
PNAS
PLoS Genetics
PLoS Biology
Current Opinion in Plant Biology
Current Opinion in Environmental Sustainability (particularly, special issue on Sustainability governance and transformation, edited by Paul C. Struik and Thom W. Kuyper, 2014: 8).

10 References

Alexandrov, N., Tai, S., Wang, W., Manuseto, L., Palis, K., Fuentes, R., Ulat, V., Chebotarov, D., Zhang, G., Li, A., Mauleon, R., Hamilton, R. and McNally, K. (2014). SNP-Seek database of NSPs derived from 3000 rice genomes. *Nucleic Acids Res.* doi: 10.1093/nar/gku1039.

Ali, M. L., Sanchez, P. L., Yu, S. B., Lorieux, M. and Eizenga, G. C. (2010). Chromosome segment substitution lines: a powerful tool for the introgression of valuable genes from *Oryza* wild species into cultivated rice (*O. sativa*). *Rice* 3: 218–34.

Araus, J. L. and Cairns, J. E. (2014). Field high-throughput phenotyping: the new crop breeding frontier. *Trends Plant Sci.* 19(1): 52–61. doi: 10.1016/j.tplants.2013.09.008.

Arbelaez, J., Moreno, L., Singh, N., Tung, C.-W., Maron, L., Ospina, Y., et al. (2015). Development and GBS-genotyping of introgression lines (ILs) using two wild species of rice, *O. meridionalis* and *O. rufipogon*, in a common recurrent parent, *O. sativa* cv. Curinga. *Mol. Breed* 35(2): 1–18.

Asch, F. and Wopereis, M. C. S. (2001). Responses of field-grown irrigated rice cultivars to varying levels of floodwater salinity in a semi-arid environment. *Field Crops Res.* 70(2): 127–37.

Atwell, B. J., Wang, H. and Scafaro, A. P. (2014). Could abiotic stress tolerance in wild relatives of rice be used to improve *Oryza sativa*? *Plant Sci.* 215–16: 48–58.

Bandillo, N., Raghavan, C., Muyco, P. A., Sevilla, M. A., Lobina, I. T., Dilla-Ermita, C. J., et al. (2013). Multi-parent advanced generation inter-cross (MAGIC) populations in rice: progress and potential for genetics research and breeding. *Rice* 6(1): 11. doi: 10.1186/1939-8433-6-11.

Bentley, A. R., Scutari, M., Gosman, N., Faure, S., Bedford, F., Howell, P., et al. (2014). Applying association mapping and genomic selection to the dissection of key traits in elite European wheat. *Theor. Appl. Genet.* 127(12): 2619–33.

Bernardo, R. (2008). Molecular markers and selection for complex traits in plants: learning from the last 20 years. *Crop Sci.* 48(5): 1649–64.

Bernardo, R. (2010). Genomewide selection with minimal crossing in self-pollinated crops. *Crop Sci.* 50(2): 624–7.

Bernardo, R. (2014). Genomewide selection when major genes are known. *Crop Sci.* 54(1): 68–75.

Beyer, P., Al-Babili, S., Ye, X., Lucca, P., Schaub, P., Welsch, R., et al. (2002). Golden rice: introducing the β-carotene biosynthesis pathway into rice endosperm by genetic engineering to defeat vitamin A deficiency. *J. Nutr.* 132(3): 506S–510S.

Bonman, J. M. (1992). Durable resistance to rice blast disease, Äî environmental influences. In R. Johnson and G. J. Jellis (eds), *Breeding for Disease Resistance*. Springer Netherlands. Vol. 1, pp. 115–23.

Brar, D. S. and Khush, G. S. (1997). Alien introgression in rice. *Plant Mo.l Biol.* 35(1): 35–47.

Cavanagh, C., Morell, M., Mackay, I. and Powell, W. (2008). From mutations to MAGIC: resources for gene discovery, validation and delivery in crop plants. *Curr. Opin. Plant Biol.* 11(2): 215–21.

Chandre Gowda, M. J. and Jayaramaiah, K. M. (1998). Comparative evaluation of rice production systems for their sustainability1. *Agric. Ecosyst. Environ.* 69(1): 1–9.

Civáň, P., Craig, H., Cox, C. J. and Brown, T. A. (2015). Three geographically separate domestications of Asian rice. *Nature Plants* 1(11): 15164. doi: 10.1038/nplants.2015.164.

Cobb, J. N., Declerck, G., Greenberg, A., Clark, R. and McCouch, S. (2013). Next-generation phenotyping: requirements and strategies for enhancing our understanding of genotype-phenotype relationships and its relevance to crop improvement. *Theor. Appl. Genet.* 126(4): 867–87. doi: 10.1007/s00122-013-2066-0.

Collard, B. C. and Mackill, D. J. (2008). Marker-assisted selection: an approach for precision plant breeding in the twenty-first century. *Philos. Trans. R. Soc. Lond. B. Biol. Sci.* 363(1491): 557–72.

Desta, Z. A. and Ortiz, R. (2014). Genomic selection: genome-wide prediction in plant improvement. *Trends Plant Sci.* 19(9): 592–601. doi: 10.1016/j.tplants.2014.05.006.

Dixit, S., Huang, B. E., Sta Cruz, M. T., Maturan, P. T., Ontoy, J. C. E. and Kumar, A. (2014). QTLs for tolerance of drought and breeding for tolerance of abiotic and biotic stress: an integrated approach. *PLoS ONE* 9(10): e109574. doi: 10.1371/journal.pone.0109574.

Duitama, J., Silva, A., Sanabria, Y., Cruz, D. F., Quintero, C., Ballen, C., et al. (2015). Whole genome sequencing of elite rice cultivars as a comprehensive information resource for marker assisted selection. *PLoS ONE* 10(4): e0124617. doi: 10.1371/journal.pone.0124617.

Famoso, A. N., Zhao, K., Clark, R. T., Tung, C. W., Wright, M. H., Bustamante, C., et al. (2011). Genetic architecture of aluminum tolerance in rice (*Oryza sativa*) determined through genome-wide association analysis and QTL mapping. *PLoS Genet.* 7(8): e1002221. doi: 10.1371/journal.pgen.1002221.

Fitzgerald, M. A., McCouch, S. R. and Hall, R. D. (2009). Not just a grain of rice: the quest for quality. *Trends Plant Sci.* 14(3): 133–9.

Ford-Lloyd, B. V., Brar, D., Khush, G. S., Jackson, M. T. and Virk, P. S. (2008). The association of multiple interacting genes with specific phenotypes in rice using gene coexpression networks. *Plant Physiol.* 154: 13–24.

Fukuoka, S., Yamamoto, S.-I., Mizobuchi, R., Yamanouchi, U., Ono, K., Kitazawa, N., Yasuda, N., Fujita, Y., Thi Thanh Nguyen, T., Koizumi, S., Sugimoto, K., Matsumoto, T. and Yano, M. (2014). Multiple functional polymorphisms in a single disease resistance gene in rice enhance durable resistance to blast. *Sci. Rep.* 4: 4550. doi: 10.1038/srep04550.

Fukuoka, S., Saka, N., Mizukami, Y., Koga, H., Yamanouchi, U., Yoshioka, Y., Hayashi, N., Ebana, K., Mizobuchi, R. and Yano, M. (2015). Gene pyramiding enhances durable blast disease resistance in rice. *Sci. Rep.* 5: 7773. doi: 10.1038/srep07773.

Furbank, R. T. and Tester, M. (2011). Phenomics–technologies to relieve the phenotyping bottleneck. *Trends Plant Sci.* 16(12): 635–44. doi: 10.1016/j.tplants.2011.09.005.

Gamuyao, R., et al. (2012). The protein kinase Pstol1 from traditional rice confers tolerance of phosphorus deficiency. *Nature* 488: 535–9.

Garnett, T. and Godfray, C. (2012). *Sustainable Intensification in Agriculture. Navigating a Course through Competing Food System Priorities.* Food Climate Research Network and the Oxford Martin Programme on the Future of Food, University of Oxford, UK.

Garris, A., Tai, T., Coburn, J., Kresovich, S. and McCouch, S. (2005). Genetic structure and diversity in *Oryza sativa* L. *Genetics* 169: 1631–58.

Ghanem, M. E., Marrou, H. and Sinclair, T. R. (2015). Physiological phenotyping of plants for crop improvement. *Trends Plant Sci.* 20(3): 139–44. doi: 10.1016/j.tplants.2014.11.006.

Glaszmann, J. C. (1987). Isozymes and classification of Asian rice varieties. *Theor. Appl. Genet.* 74(1): 21–30.

Glemin, S. and Bataillon, T. (2009). A comparative view of the evolution of grasses under domestication. *New Phytol.* 183(2): 273–90. doi: 10.1111/j.1469-8137.2009.02884.x.

Gregorio, G. B. I. M., Vergara, G. V. and Thirumeni, S. (2013). Recent advances in rice science to design salinity and other abiotic stress tolerant rice varieties. *SABRAO J. Breed.Genet.* 45(1): 31–41.

Grenier, C., Cao, T.-V., Ospina, Y., Quintero, C., Châtel, M. H., Tohme, J., et al. (2015). Accuracy of genomic selection in a rice synthetic population developed for recurrent selection breeding. *PLoS ONE* 10(8): e0136594. doi: 10.1371/journal.pone.0136594.

Han, B. and Huang, X. (2013). Sequencing-based genome-wide association study in rice. *Curr. Opin. Plant Biol.* 16(2): 133–8.

Heffner, E. L., Lorenz, A. J., Jannink, J. L. and Sorrells, M. E. (2010). Plant breeding with genomic selection: gain per unit time and cost. *Crop Sci.* 50(5): 1681–90.

Hickey, L. T., Dieters, M. J., DeLacy, I. H., Kravchuk, O. Y., Mares, D. J. and Banks, P. M. (2009). Grain dormancy in fixed lines of white-grained wheat (Triticum aestivum L.) grown under controlled environmental conditions. *Euphytica* 168(3): 303–10.

Hittalmani, S., Parco, A., Mew, T. V., Zeigler R. S. and Huang N. (2000). Fine mapping and DNA marker-assisted pyramiding of the three major genes for blast resistance in rice. *Theor. Appl. Genet.* 100(7): 1121–8.

Huang, B. E., Verbyla, K., Verbyla, A., Raghavan, C., Singh, V., Gaur, P., et al. (2015). MAGIC populations in crops: current status and future prospects. *Theor. Appl. Genet.* 128(6): 999–1017.

Huang, X., Feng, Q., Qian, Q., Zhao, Q., Wang, L., Wang, A., et al. (2009). High-throughput genotyping by whole-genome resequencing. *Genome Res.* 19(6): 1068–76.

Huang, X., Zhao, Y., Wei, X., Li, C., Wang, A., et al. (2012a). Genome-wide association study of flowering time and grain yield traits in a worldwide collection of rice germplasm. *Nat. Genet.* 44(1): 32–41.

Huang, X., Kurata, N., Wei, X., Wang, Z. X., Wang, A., Zhao, Q., et al. (2012b). A map of rice genome variation reveals the origin of cultivated rice. *Nature* 490(7421): 497–501.

IRRI (1984). *Terminology for Rice Growing Environments.* International Rice Research Institute.

Jacquemin, J., Bhatia, D., Singh, K. and Wing, R. A. (2013). The international oryza map alignment project: development of a genus-wide comparative genomics platform to help solve the 9 billion-people question. *Curr. Opin. Plant Biol.* 16(2): 147–56.

Jannink, J. L. (2010). Dynamics of long-term genomic selection. *Genet. Sel. Evol.* 42: 35. doi: 10.1186/1297-9686-42-35.

Jena, K. K. (2010). The species of the genus *Oryza* and transfer of useful genes from wild species into cultivated rice, *O. sativa*. *Breed Sci.* 60(5): 518–23.

Jones, M. P., Dingkuhn, M., Aluko, G. K. and Semon, M. (1997). Interspecific *Oryza sativa* L. x *O. glaberrima* Steud. progenies in upland rice improvement. *Euphytica* 92: 237–46.

Jongdee, B., Pantuwan, G., Fukai, S. and Fischer, K. (2006). Improving drought tolerance in rainfed lowland rice: an example from Thailand. *Agric. Water Manage* 80(1–3): 225–40.

Khush, G. (1997). Origin, dispersal, cultivation and variation of rice. In T. Sasaki and G. Moore (eds), *Oryza: From Molecule to Plant.* Springer Netherlands, pp. 25–34.

Klee, M., Zach, B. and Neumann, K. (2000). Four thousand years of plant exploitation in the Chad Basin of northeast Nigeria I: the archaeobotany of Kursakata. *Vegetation History and Archaeobotany* 9(4): 223–37.

Kovach, M. J., Sweeney, M. T. and McCouch, S. R. (2007). New insights into the history of rice domestication. *Trends Genet.* 23(11): 578–87.

Kuyper, T. W. and Struik, P. C. (2014). Epilogue: global food security, rhetoric, and the sustainable intensification debate. *Curr. Opin. Environ. Sustain.* 8: 71–9.

Li, J., Xin, Y. and Yuan, L.P. (2009). Hybrid Rice Technology Development. IFPRI Discussion Paper 00918. 2020 Vision Initiative.

Li, Z., Ye, G., Yang, M., Liu, Z., Lu, D., Mao, X., et al. (2014). Genetic characterization of a multiparent recombinant inbred line of rice population. *Research on Crops* 15(1): 28–37.

Li, Z.-M., Zheng, X.-M. and Ge, S. (2011). Genetic diversity and domestication history of African rice (*Oryza glaberrima*) as inferred from multiple gene sequences. *Theor. Appl. Genet.* 123: 21–31. doi: 10.1007/s00122-011-1563-2.

Linares, O. F. (2002). African rice (*Oryza glaberrima*): history and future potential. *Proc. Natl. Acad. Sci. U.S.A.* 99(25): 16360–5.

Lipka, A. E., Kandianis, C. B., Hudson, M. E., Yu, J., Drnevich, J., Bradbury, P. J. and Gore, M. A. (2015). From association to prediction: statistical methods for the dissection and selection of complex traits in plants. *Curr. Opin. Plant Biol.* 24: 110–18. doi: http://dx.doi.org/10.1016/j.pbi.2015.02.010.

Lorenz, A. J., Chao, S., Asoro, F. G., Heffner, E. L., Hayashi, T., Iwata, H., et al. (2011). Genomic selection in plant breeding: knowledge and prospects. *Adv. Agron.* 110: 77–123. doi: 10.1016/B978-0-12-385531-2.00002-5.

Lu, H., Liu, Z., Wu, N., BernÉ, S., Saito, Y., Liu, B., et al. (2002). Rice domestication and climatic change: phytolith evidence from East China. *Boreas* 31(4): 378–85.

Matson, P. A., Parton, W. J., Power, A. G. and Swift, M. J. (1997). Agricultural intensification and ecosystem properties. *Science* 277: 504–9.

Matsuoka, Y., Vigouroux, Y., Goodman, M., Sanchez, G. J., Buckler, E. and Doebley, J. (2002). A single domestication for maize shown by multilocus microsatellite genotyping. *Proc. Natl. Acad. Sci. U.S.A.* 99: 6080–4.

McCouch, S. R., McNally, K. L., Wang, W. and Sackville Hamilton, R. (2012). Genomics of gene banks: A case study in rice. *Am. J. Bot.* 99(2): 407–23.

McCouch, S. R., Baute, G. J., Bradeen, J., Bramel, P., Bretting, P. K., Buckler, E., et al. (2013a). Agriculture: feeding the future. *Nature* 499: 23–4.

McCouch, S. R., Wright, M. H., Tung, C.-W., Maron, L. G., McNally, K. L., Fitzgerald, M., et al. (2016). Open access resources for genome wide association mapping in rice. *Nature Communications* 7: 10532. doi: 10.1038/ncomms10532.

McCouch, S. R., Sweeney, M., Li, J., Jiang, H., Thomson, M., Septinginsih, E., Edwards, J., Moncada, P., Xiao, J., Garris, A., Tai, T., Martinez, C., Tohme, J., Sugiono, M., McClung, A., Yuan, L. P. and Ahn, S. N. (2007). Through the genetic bottleneck: *O. rufipogon* as a source of trait-enhancing alleles for *O. sativa. Euphytica* 154(3): 317–39.

Meuwissen, T. H. E., Hayes, B. J. and Goddard, M. E. (2001). Prediction of total genetic value using genome-wide dense marker maps. *Genetics* 157(4): 1819–29.

Mickelbart, M. V., Hasegawa, P. M. and Bailey-Serres, J. (2015). Genetic mechanisms of abiotic stress tolerance that translate to crop yield stability. *Nat. Reviews Genet.* 16(4): 237–51.

Ndjiondjop, M. N., Albar, L., Fargette, D., Fauquet, C. and Ghesquiére, A. (1999). The genetic basis of high resistance to Rice Yellow Mottle Virus (RYMV) in cultivars of two cultivated rice species. *Plant Disease* 83(10): 931–5.

Nguyen, B. D., Brar, D. S., Bui, B. C., Nguyen, T. V., Pham, L. N. and Nguyen, H. T. (2003). Identification and mapping of the QTL for aluminum tolerance introgressed from the new source, Oryza rufipogon Griff., into indica rice (*Oryza sativa* L.). *Theor. Appl. Genet.* 106(4): 583–93.

Ohnishi, T., Yoshino, M., Yamakawa, H. and Kinoshita, T. (2011). The biotron breeding system: a rapid and reliable procedure for genetic studies and breeding in rice. *Plant Cell Physiol.* 52(7): 1249–57.

Oka, H. (1988). *The Origin of Cultivated Rice*. Japan Scientific Society Press, Tokyo.

Onogi, A., Ideta, O., Inoshita, Y., Ebana, K., Yoshioka, T., Yamasaki, M. and Iwata, H. (2015). Exploring the areas of applicability of whole-genome prediction methods for Asian rice (*Oryza sativa* L.). *Theor. Appl. Genet.* 128(1): 41–53.

Owens, B. F., Lipka, A. E., Magallanes-Lundback, M., Tiede, T., Diepenbrock, C. H., Kandianis, C. B., et al. (2014). A foundation for provitamin A biofortification of maize: genome-wide association and genomic prediction models of carotenoid levels. *Genetics* 198(4): 1699–716.

Paterson, A. H., Bowers, J. E. and Chapman, B. A. (2004). Ancient polyploidization predating divergence of the cereals, and its consequences for comparative genomics. *Proc. Natl. Acad. Sci. U.S.A.* 101(26): 9903–8.

Pinto, Y. M., Kok, R. A. and Baulcombe, D. C. (1999). Resistance to rice yellow mottle virus (RYMV) in cultivated African rice varieties containing RYMV transgenes. *Nat. Biotechnol.* 17: 702–7.

Plucknett, D. L., Smith, N. J. H., Williams, J. T. and Anishetty, N. M. (1987). *Gene Banks and the World's Food.* Princeton University Press.

Ren, Z.-H., Gao, J.-P., Li, L.-G., Cai, X.-L., Huang, W., Chao, D.-Y., Zhu, M.-Z., Wang, Z.-Y., Luan, S. and Lin, H.-X. (2005). A rice quantitative trait locus for salt tolerance encodes a sodium transporter. *Nat. Genet.* 37: 1141–6.

Rutkoski, J. E., Poland, J. A., Singh, R. P., Huerta-Espino, J., Bhavani, S., Barbier, H., et al. (2014). Genomic selection for quantitative adult plant stem rust resistance in wheat. *Plant Genome* 7(3). doi: 10.3835/plantgenome2014.02.0006.

Sakai, H., Ikawa, H., Tanaka, T., Numa, H., Minami, H., Fujisawa, M., et al. (2011). Distinct evolutionary patterns of Oryza glaberrima deciphered by genome sequencing and comparative analysis. *Plant J.* 66(5): 796–805.

Sano, Y. (1993). Constraints in using wild relatives in breeding: lack of basic knowledge on crop gene pools. In D. R. Buxton (ed.), *International Crop Science*. Crop Science Society of America, Madison, WI, Vol. I, pp. 437–43.

Sardesai, N., Kumar, A., Rajyashri, K., Nair, S. and Mohan, M. (2002). Identification and mapping of an AFLP marker linked to Gm7, a gall midge resistance gene and its conversion to a SCAR marker for its utility in marker aided selection in rice. *Theor. Appl. Genet.* 105(5): 691–8.

Sarla, N. and Swamy, B. P. M. (2005). Oryza glaberrima: a source for the improvement of *Oryza sativa. Curr. Sci.* 89(6): 955–63.

Second, G. (1982). Origin of the genic diversity of cultivated rice (*Oryza*): study of the polymorphism scored at 40 isozyme loci. *Jpn. J. Genet.* 57(1): 25–57.

Semon, M., Nielsen, R., Jones, M. P. and McCouch, S. R. (2005). The population structure of African cultivated rice Oryza glaberrima (Steud.): evidence for elevated levels of linkage disequilibrium caused by admixture with *O. sativa* and ecological adaptation. *Genetics* 169(3): 1639–47.

Septiningsih, E. M., Pamplona, A. M., Sanchez, D. L., Neeraja, C. N., Vergara, G. V., Heuer, S., et al. (2009). Development of submergence-tolerant rice cultivars: the Sub1 locus and beyond. *Ann. Bot.* 103(2): 151–60.

Shakiba, E. and Eizenga, G. (2014). Unraveling the secrets of rice wild species. In W. Yan and J. Bao (eds), *Rice – Germplasm, Genetics and Improvement*, ISBN 978-953-51-1240-2, Chapter 1. doi: 10.5772/58393.

Shan, Q. W., Wang, Y. P., Li, J. and Gao, C. (2014). Genome editing in rice and wheat using the CRISPR/Cas system. *Nat. Protoc.* 9(10): 2395–410.

Singh, N., Jayaswal, P. K., Panda, K., Mandal, P., Kumar, V., Singh, B., et al. (2015). Single-copy gene based 50,ÄâK SNP chip for genetic studies and molecular breeding in rice. *Sci. Rep.* 5(11600). doi: 10.1038/srep11600.

Singh, S., Sidhu, J. S., Huang, N., Vikal, Y., Li, Z., Brar, D. S., et al. (2001). Pyramiding three bacterial blight resistance genes (xa5, xa13 and Xa21) using marker-assisted selection into indica rice cultivar PR106. *Theor. Appl. Genet.* 102(6–7): 1011–15.

Soreng, R. J., Peterson, P. M., Romaschenko, K., Davidse, G., Zuloaga, F. O., Judziewicz, E. J., Filgueiras, T. S., Davis, J. I. and Morrone, O. (2015). A worldwide phylogenetic classification of the Poaceae (Gramineae). *J. Syst. Evol.* 53(2): 117–37.

Spindel, J. E., Begum, H., Akdemir, D., Virk, P., Collard, B., Redoña, E., et al. (2015). Genomic selection and association mapping in rice (*Oryza sativa*): effect of trait genetic architecture, training population composition, marker number and statistical model on accuracy of rice genomic selection in elite, tropical rice breeding lines. *PLoS Genet.* 11(2): e1004982. doi: 10.1371/journal.pgen.1004982.

Spindel, J. E., Wright, M., Chen, C., Cobb, J., Gage, J., Harrington, S., et al. (2013). Bridging the genotyping gap: using genotyping by sequencing (GBS) to add high-density SNP markers and new value to traditional bi-parental mapping and breeding populations. *Theor. Appl. Genet.* 126(11): 2699–716.

Spindel, J. E., Begum, H., Akdemir, D., Collard, B., Redoña, E., Jannink, J. L. and McCouch, S. (2016). Genome-wide prediction models that incorporate *de novo* GWAS are a powerful new tool for tropical rice improvement. *Heredity*. doi: 10.1038/hdy.2015.113.

Struik, P. C., Kuyper, T. W., Brussaard, L. and Leeuwis, C. (2014). Deconstructing and unpacking scientific controversies in intensification and sustainability: why the tensions in concepts and values? *Curr. Opin. Environ. Sustain.* 8: 80–8.

Tanksley, S. D. (1993). Mapping polygenes. *Ann. Rev. Genet.* 27: 205–33.

Tanksley, S. D. and Nelson, J. C. (1996). Advanced backcross QTL analysis: a method for the simultaneous discovery and transfer of valuable QTLs from unadapted germplasm into elite breeding lines. *Theor. Appl. Genet.* 92: 191–203.

Tanksley, S. D. and McCouch, S. R. (1997). Seed banks and molecular maps: unlocking genetic potential from the wild. *Science* 277(5329): 1063–6.

Teng, B., Zhang, Y., Wu, J. D., Cong, X. H., Wang, R. Y., Han, Y. H., et al. (2013). Association between allelic variation at the Waxy locus and starch physicochemical properties using single-segment substitution lines in rice (*Oryza sativa* L.). *Starch-Starke* 65(11–12): 1069–77.

The rice genome project (2014). The 3000 rice genomes project. *GigaScience* 3(7). doi: 10.1186/2047-217X-3-7.

Thomson, M. J. (2014). High-throughput SNP genotyping to accelerate crop improvement. *Plant Breed Biotech* 2(3): 195–212.

Thomson, M. J., Zhao, K., Wright, M., McNally, M. L., Rey, J., Tung, C.-W., Reynolds, A., Scheffler, B., Eizenga, G., McClung, A., Kim, H., Ismail, A. M., de Ocampo, M., Ma, C. M., Reveche, Y.,

Dilla-Ermita, C. J., Mauleon, R., Leung, H., Bustamante, C. and McCouch, S. R. (2012). High-throughput single nucleotide polymorphism genotyping for breeding applications in rice using the BeadXpress platform. *Mol Breeding* 29(4): 875–86.

Tian, F., Li, D. J., Fu, Q., Zhu, Z. F., Fu, Y. C., Wang, X. K. and Sun, C. Q. (2006). Construction of introgression lines carrying wild rice (*Oryza rufipogon* Griff.) segments in cultivated rice (*Oryza sativa* L.) background and characterization of introgressed segments associated with yield-related traits. *Theor. Appl. Genet.* 112: 570–80.

Tilman, D. (1998). The greening of the green revolution. *Nature* 396(6708): 211–12.

Uga, Y., Sugimoto, K., Ogawa, S., Rane, J., Ishitani, M., Hara, N., Kitomi, Y., Inukai, Y., Ono, K., Kanno, N., Inoue, H., Takehisa, H., Motoyama, R., Nagamura, Y., Wu, J., Matsumoto, T., Takai, T., Okuno, K. and Yano, M. (2013). Control of root system architecture by *DEEPER ROOTING 1* increases rice yield under drought conditions. *Nat. Genet.* 45: 1097–102.

van der Ploeg, J. D. and Ventura, F. (2014). Heterogeneity reconsidered. *Curr. Opin. Environ. Sustain.* 8: 23–8.

Vanlauwe, B., Coyne, D., Gockowski, J., Hauser, S., Huising, J., Masso, C., Nziguheba, G., Schut, M. and Van Asten, P. (2014). Sustainable intensification and the African smallholder farmer. *Curr. Opin. Environ. Sustain.* 8: 15–22.

Varshney, R., Terauchi, R. and McCouch, S. (2014). Harvesting the promising fruits of genomics: applying genome sequencing technologies to crop breeding. *PLoS Biol.* 12(6): e1001883. doi: 10.1371/journal.pbio.1001883.

Vaughan, D. A., Lu, B.-R. and Tomooka, N. (2008b). The evolving story of rice evolution. *Plant Sci.* 174(4): 394–408.

Vaughan, D. A., Morishima, H. and Kadowaki, K. (2003). Diversity in the Oryza genus. *Curr. Opin. Plant Biol.* 6(2): 139–46.

Vaughan, D. A., Ge, S., Kaga, A. and Tomooka, N. (2008a). On the phylogeny and biogeography of the genus *Oryza*. *Rice Biology in the Genomics Era*, pp. 219–34.

Vaughan, D. A., Kadowaki, K.-I., Kaga, A. and Tomooka, N. (2005). On the phylogeny and biogeography of the genus *Oryza*. *Breed Sci.* 55(2): 113–22.

Venuprasad, R., Bool, M. E., Quiatchon, L., Cruz, M. T. S., Amante, M. and Atlin, G. N. (2012). A large-effect QTL for rice grain yield under upland drought stress on chromosome 1. *Mol. Breed* 30(1): 535–47.

Wang, M., Yu, Y., Haberer, G., Marri, P. R., Fan, C., Goicoechea, J. L., et al. (2014). The genome sequence of African rice (*Oryza glaberrima*) and evidence for independent domestication. *Nat. Genet.* 46(9): 982–8.

Wang, X., Lee, S., Wang, J., Ma, J., Bianco, T. and Jia, Y. (2014). Current advances on genetic resistance to rice blast disease. In W. Yan and J. Bao (eds), *Rice - Germplasm, Genetics and Improvement,* ISBN 978-953-51-1240-2, Chapter 7. doi: 10.5772/56824.

Wang, Z. Y., Wu, Z. L., Xing, Y. Y., Zheng, F. G., Guo, X. L., Zhang, W. G., et al. (1990). Nucleotide sequence of rice *waxy* gene. *Nucleic Acids Res.* 18(19): 5898.

Xiao, J., Li, J., Grandillo, S., Ahn, S. N., Yuan, L., Tanksley, S. D., et al. (1998). Identification of trait-improving quantitative trait loci alleles from a wild rice relative, *Oryza rufipogon*. *Genetics* 150(2): 899–909.

Xu, J., Zhao, Q., Du, P., Xu, C., Wang, B., Feng, Q., Liu, Q., Tang, S., Gu, M., Han, B. and Liang, G. (2010). Developing high throughput genotyped chromosome segment substitution lines based on population whole-genome re-sequencing in rice (*Oryza sativa* L.). *BMC Genomis* 11(656). doi: 10.1186/1471-2164-11-656.

Xu, K., Xu, X., Fukao, T., Canlas, P., Maghirang-Rodriguez, R., Heuer, S., Ismail, A. M., Bailey-Serres, J., Ronald, P. C. and Mackill, D. J. (2006). *Sub1A* is an ethylene-response factor-like gene that confers submergence tolerance to rice. *Nature* 442: 705–8.

Xu, R. F., Li, H., Qin, R. Y., Li, J., Qiu, C. H., Yang, Y. C., et al. (2015). Generation of inheritable and 'transgene clean' targeted genome-modified rice in later generations using the CRISPR/Cas9 system. *Sci. Rep.* 5(11491). doi: 10.1038/srep11491.

Xu, R. F., Li, H., Qin, R. Y., Wang, L., Li, L., Wei, P. C. and Yang, J. B. (2014). Gene targeting using the *Agrobacterium tumefaciens*-mediated CRISPR-Cas system in rice. *Rice* 7(5). doi: 10.1186/s12284-014-0005-6.

Xu, X., Liu, X., Ge, S., Jensen, J. D., Hu, F. Y., Li, X., et al. (2012). Resequencing 50 accessions of cultivated and wild rice yields markers for identifying agronomically important genes. *Nat. Biotechnol.* 30(1): U105–U157.

Xue, D., Qian, Q. and Teng, S. (2014). Identification and utilization of elite genes from elite germplasms for yield improvement. In W. Yan and J. Bao (eds), *Rice - Germplasm, Genetics and Improvement*, ISBN 978-953-51-1240-2, Chapter 5. doi: 10.5772/56390.

Ying, J. Z., Chen, Y. Y. and Zhang, H. W. (2014). Functional characterization of genes/QTLs for increasing rice yield potential. In W. Yan and J. Bao (eds), *Rice - Germplasm, Genetics and Improvement*, ISBN 978-953-51-1240-2, Chapter 6. doi: 10.5772/56820.

Yu, H. H., Xie, W. B., Li, J., Zhou, F. S. and Zhang, Q. F. (2014). A whole-genome SNP array (RICE6K) for genomic breeding in rice. *Plant Biotechnol. J.* 12(1): 28–37.

Yuan, L. P. (2014). Development of hybrid rice to ensure food security. *Rice Sci.* 21(1): 1–2.

Zhang, F. and Xie, J. (2014). Genes and QTLs resistant to biotic and abiotic stresses from wild rice and their applications in cultivar improvements. In W. Yan and J. Bao (eds), *Rice - Germplasm, Genetics and Improvement*, ISBN 978-953-51-1240-2, Chapter 7. doi: 10.5772/56825.

Zhao, Y., Mette, M. F., Gowda, M., Longin, C. F. H. and Reif, J. C. (2014). Bridging the gap between marker-assisted and genomic selection of heading time and plant height in hybrid wheat. *Heredity* 112(6): 638–45.

Zhao, K., Tung, C.-W., Eizenga, G. C., Wright, M. H., Ali, M. L., Price, A. H., et al. (2011). Genome-wide association mapping reveals a rich genetic architecture of complex traits in *Oryza sativa*. *Nat Commun* 2(467). doi: 10.1038/ncomms1467.

Zhao, K., Wright, M., Kimball, J., Eizenga, G., McClung, A. and Kovach, M., et al. (2010). Genomic diversity and introgression in *O. sativa* reveal the impact of domestication and breeding on the rice genome. *PLoS ONE* 5(5): e10780. doi: 10.1371/journal.pone.0010780.

Zhao, Z. and Piperno, D. R. (2000). Late pleistocene/holocene environments in the middle Yangtze river valley, China and rice (*Oryza sativa* L.) domestication: the phytolith evidence. *Geoarchaeology* 15(2): 203–22.

Zhu, Y., Chen, H., Fan, J., Wang, Y., Li, Y., Chen, J., et al. (2000). Genetic diversity and disease control in rice. *Nature* 406(6797): 718–22.

Zong, Y., Chen, Z., Innes, J. B., Chen, C., Wang, Z. and Wang, H. (2007). Fire and flood management of coastal swamp enabled first rice paddy cultivation in east China. *Nature* 449(7161): 459–62.

Chapter 2

Advances in molecular breeding techniques for rice

R. B. Angeles-Shim and M. Ashikari, Nagoya University, Japan

1 Introduction

Rice (*Oryza sativa* L.) is a principal food for 3 billion people, providing up to 23% of the caloric needs worldwide. Although rice is cultivated in more than 100 countries across the world, Asia still accounts for 91% of rice production and 87% of rice consumption (Khush, 1997; http://apps.fao.org; www.irri.org; GRiSP, 2013). In sub-Saharan Africa, the high rate of population growth, urbanization, changes in employment patterns, rising income levels and changes in consumer preferences make rice the fastest growing staple crop in the region (Seck et al., 2012; GRiSP, 2013). A similar trend in population growth, coupled with the steady increase in per capita consumption is continuously spurring rice demands in Latin America and Middle East. In the United States and countries in the European Union, shifts from protein to high-fibre diet, as well as the increase in the number of Asian and Hispanic immigrants has resulted in a growing per capita consumption of rice (GRiSP, 2013).

Between the 1960s and 1980s, a remarkable increase in rice yields was achieved by introducing the semi-dwarf (*sd1*) gene in rice varieties using conventional breeding methods. The resulting short stature of the rice plant made it responsive to fertilizers without succumbing to lodging. The release of these high-yielding, semi-dwarf rice varieties, combined with mechanization, implementation of modern irrigation technology and application of high inputs (i.e. fertilizers and pesticides) ushered in the Green Revolution that saved the world from the brink of mass starvation (Khush, 1999). In recent years, however, the yield gains made during this landmark period in agricultural history

http://dx.doi.org/10.19103/AS.2016.0003.03

has begun widespread deceleration. In fact, since 2000, global rice consumption has far exceeded that of production (Grassini et al., 2013).

By 2050, the current world population will have surpassed 9 billion. To feed this 9 billion people, agriculture production has to increase by 77%. Global cereal production alone has to increase sustainably by 38% over the next 34 years (Tester and Langridge, 2010; http://faostat.fao.org/default.aspx). Sustainable yield increases of this scale would require significant improvements in both crop breeding methods and management strategies that takes into serious consideration decreasing water, land and labour resources; climate change (i.e. global warming, changes in rainfall patterns, increasing sea level and desertification); and emergence of new pests and pathogens (Godfray et al., 2010; http://apps.fao.org).

To this end, plant breeders and agricultural scientists are finding new and more efficient strategies to overcome bottlenecks in plant breeding. For the last two decades, traditional marker-assisted selection (MAS) has reshaped breeding programmes and facilitated gains from selection. Recent, rapid developments in genomic technologies including the advent of high-throughput DNA sequencing technologies (i.e. next-generation sequencing; NGS) and cost-effective genotyping platforms (i.e. genotyping by sequencing (GBS), KASPar) (Smith and Maughan, 2015) are effecting new shifts in the prevailing framework of plant breeding towards a more precise utilization of genetic variation for crop improvement.

In this chapter, we first describe the importance of selection and hybridization in crop improvement, citing as example the utilization of this tandem approach in realizing the breeding targets of the Green Revolution. We then describe the current trends in rice molecular breeding, including quantitative trait loci (QTL) mapping and MAS, and presented the case of *Sub1A* to illustrate how these techniques increased breeding efficiency and precision for target traits. Finally, we present the emerging tools in genomics-based breeding that promises to increase the efficiency of modern rice breeding towards a more rapid varietal development and release.

2 Selection and hybridization tools

The cultivated Asian rice species, *O. sativa*, is the oldest domesticated crop species, having fed more people than any other plant in human history (Caicedo et al., 2007). Since its domestication 8000–9000 years ago, rice has been subjected to the most primitive form of plant breeding: selection. Ancient farmers and gatherers selected naturally occurring wild rice variants with desirable phenotypes (e.g. non-shattering, high yield and environmental adaptability) and used them as progenitors for subsequent generations. Throughout this process, individuals with less desirable characteristics are eliminated. Generations of artificial selection for random mutations and gene combinations that give better yield, grain size and traits of agronomic value have led to the establishment of several rice lineages with distinct identity, genotype variability and local adaptation to specific agricultural ecosystems (Villa et al., 2006; Ray et al., 2013; Breseghello, 2013).

By the turn of the twentieth century, the importance of hybridization in crop improvement started gaining recognition through early breeding studies aimed at increasing yields in corn. Crossing of two, genetically diverse, inbred corn strains resulted in hybrids that gave higher yields relative to either parent (Shull, 1909). Results of this experiment demonstrated the power of hybridization in shuffling alleles in several loci to create numerous gene combinations that could give rise to favourable phenotypes. The vast array of possible genotypes that can be generated by crossing diverse parents expands the potential genetic gain that can be realized with optimal selection (Moose and Mumm, 2008; Breseghello, 2013).

Success in the development of modern rice cultivars during the 1960s has relied primarily on the tandem process of hybridization to create useful genetic variation and selection of individuals with superior phenotypes for cultivar improvement. IR8, possibly the most noteworthy achievement in rice breeding during the mid-twentieth century, was developed by this traditional approach. Based on the already established concept of dwarfism in wheat and sorghum, a breeding strategy for rice was devised to address the problems of low yields due to lodging. Hybridization experiments between the tall, Indonesian rice variety Peta and the short, Taiwanese rice variety Deegeewoogen generated tall F_1 plants. Selection of short, early maturing plants from thousands of F_2 plants, followed by generations of self-pollination of selected individuals finally yielded the semi-dwarf variety IR8, the miracle rice that increased yields from 1 to 10 t/ha with the application of 120 kg N/ha (Khush et al., 2001; Guimaraes, 2009).

With the successful release of IR8 as a variety, breeding targets shifted to improving grain quality and milling recovery, as well as to the introduction of disease and insect resistance in rice. IR36, the first improved variety with multiple resistance to rice diseases (blast, bacterial blight, grassy stunt virus and tungro virus) and pests (green leafhopper, brown leafhopper, stem borer and gall midge), was developed by crossing IR8 with 12 other varieties and the wild rice species, *Oryza nivara*. IR64, on the other hand, the first variety with a desirable combination of intermediate amylose content, soft gel consistency, intermediate gelatinization temperature, as well as long, slender and translucent grains, was bred from 20 landraces from eight countries (Khush et al., 2001; Guimaraes, 2009).

During this golden age of crop improvement, hybridization and selection served as powerful tools in realizing the breeding targets that drove the first agricultural revolution in rice. However, the efficiency of this conventional approach is largely based on the complementarity of traits from both parents and is therefore most applicable in breeding for qualitative traits such as plant stature and simple disease resistance (Breseghello, 2013). Selection for superior individuals was also based on observable phenotypes alone; hence, little or nothing is known about the proportion of the genome that came from either parent. With the advent of molecular markers, it became possible to dissect the genetic basis of more complex traits and use functional DNA markers to monitor the introgression of the gene of interest into target cultivars for improvement.

3 From conventional to molecular breeding approaches

3.1 QTL/gene discovery and mapping using molecular marker technology

Unlike monogenic traits, most agronomically important characteristics including high yield, biotic stress resistance and abiotic stress tolerance are quantitatively inherited, and thus exhibit a wide range of phenotypic variation. Such traits are influenced by multiple QTLs, most of which have only small effects on the trait. Phenotypic expression of these traits are also influenced by epistatic interactions, genetic background, environmental effects and QTL x environment (Zhu et al., 2008; Spindel et al., 2015). For breeding applications, prior information on the QTLs controlling the trait of interest, including the number of loci, approximate position of the QTLs in the genome, and the magnitude of QTL effects and the stability of these effects across environments, are necessary (Moose and Mumm, 2008; Varshney et al., 2014).

The advent of molecular marker systems brought QTL analysis into the mainstream and made mapping of genetic loci controlling complex traits possible. A set of plants derived from a single cross that differed genetically for a trait can be phenotyped for any trait of interest and genotyped at the same time using molecular markers. Based on the frequency of genetic recombination during meiosis, the order and relative distance of each marker in the chromosome can be inferred to create a genetic map. QTLs are then identified using statistical tools that can then establish the association between the genotype and the observed phenotype. Markers that are genetically linked to a QTL controlling a trait of interest will segregate more frequently with values observed for that particular trait, whereas unlinked markers will not show significant association with the phenotype (Tanksley, 1993; Breseghello, 2013).

Restriction fragment length polymorphisms (RFLPs) were the first DNA markers ever used to develop a linkage map for rice (McCouch et al., 1998; Saito et al., 1991). RFLPs are co-dominant, phenotype-neutral and highly locus-specific. Although originally assayed using Southern hybridization, RFLPs were converted later on to polymerase chain reaction (PCR)-based markers known as sequence-tagged sites for technical simplicity. Despite improvements in the assay, the limitations of RFLP in terms of high-quality template DNA requirement and low-throughput and amenability to automation led to the development of PCR-based markers that include random amplified polymorphic DNA (RAPD), amplified fragment length polymorphism (AFLP), cleaved amplified polymorphic sequences and simple sequence repeats (SSRs) or microsatellites. Collectively, these markers significantly contributed in saturating linkage maps for rice (Causse et al., 1994; Mackill et al., 1996).

Microsatellites are simple, tandemly repeated DNA sequences that can detect high levels of polymorphism in the genome. These markers are robust, co-dominant, and easily and economically assayed using PCR (Panaud et al., 1996). The wealth of SSR information (approximately 19 000) (International Rice Genome Sequencing Project, 2005; http://www.gramene.org/) identified in the rice genome has accelerated QTL analysis, mapping and positional cloning in rice. Although SSRs are currently being superseded by single-nucleotide polymorphisms (SNPs), which are more abundant and amenable to high-throughput genotyping, many laboratories are still equipped for SSR analysis and it is very likely that SSRs will remain a marker of choice for genetic and breeding applications in many years to come (Collard et al., 2008).

To date, thousands of QTLs underlying traits of agronomic importance have been identified by various research groups (http://www.gramene.org/) and some of them have been isolated and cloned as corresponding genes. A few examples of genes/QTLs that have been cloned include *Grain number 1a* (*Gn1a*) (Ashikari et al., 2005), *ABERRANT PANICLE ORGANIZATION1* (*APO1*) (Ikeda-Kawakatsu et al., 2009), *SPIKELET NUMBER* (*SPIKE*)/*NARROW LEAF1* (*NAL1*) (Fujita et al., 2013) and *Ghd7* (Xue et al., 2008) controlling grain number; *GS3* (Fan et al., 2006), *GW2* (Song et al., 2007) and *qSW5* (Shomura et al., 2008) for grain size; *DENSE AND ERECT PANICLE* (*DEP1*) (Huang et al., 2009); and *WEALTHY FARMERS PANICLE* (*WFP*) (Miura et al., 2010) conditioning panicle architecture and *STRONG CULM 1* and *2* (*SCM1* and *SCM2*) (Ookawa et al., 2010) for strong stem/lodging tolerance. Cloned QTLs for abiotic stress tolerance include *SUB1A* (Xu et al., 2006) and *SNORKEL 1* and *2* (*SK1* and *SK2*) (Hattori et al., 2009) for submergence tolerance and *DEEP ROOTING* (*DRO1*) (Uga et al., 2013) for drought tolerance. The availability of a wide array of molecular marker systems for rice provides an opportunity for the fine mapping and positional cloning of these genes/QTLs for their utilization in breeding programmes.

3.2 Marker-assisted selection

MAS refers to the indirect selection of target traits using molecular markers that are tightly linked to genes responsible for an agricultural trait. The utilization of MAS in breeding programmes requires the mapping of QTLs underlying important agronomic traits. In some cases, it has been possible to directly use markers associated with large-effect QTLs in MAS. Most of the time, however, QTL confirmation, validation, fine mapping and even positional cloning are necessary to develop a set of useful DNA markers that can be used for efficient selection (Collard and Mackill, 2008; Breseghello, 2013). The applicability of DNA markers in MAS depends on the reliability of the marker, template DNA requirement, means of marker assay, level of polymorphism and costs (Collard and Mackill, 2008).

Given the advantages of MAS compared to phenotypic screening, it has found several genetics and breeding applications in rice including genotype identity testing, genetic diversity analysis of breeding materials, gene surveys in parental material, marker-evaluated selection, marker-assisted backcrossing (MABC) and marker-assisted pyramiding (MAP) (Collard et al., 2008).

Backcrossing is an approach in plant breeding that has been widely used to incorporate one or few genes into an elite or adapted rice variety that is deficient only in a few characteristics (Allard, 1999). MABC, in particular, is a fast and precise approach of introgressing target traits and recovering the genetic background of the recurrent parent via three levels of selection namely foreground, recombinant and background selection. Foreground selection is used to screen for target gene/QTL introgressions in backcross progenies. It is particularly valuable when trait expression depends on specific environmental conditions (e.g. abiotic stress tolerance) or when phenotypic screening is difficult, expensive or destructive in nature. For traits that are expressed at later stages of plant development (e.g. grain characteristics) or are controlled by recessive genes, foreground selection allows the early identification of the best plants for backcrossing (Moose and Mumm, 2008; Collard and Mackill, 2008).

Recombinant selection, on the other hand, identifies rare recombinants in the proximate vicinity of the introduced gene and allows trimming of the introgressed chromosome segment as close as possible to the target gene. Linkage of many undesirable genes or loci to the target gene from a donor line is common. The transfer of these undesirable genes or loci together with the target gene that results in reduced agronomic performance of the recurrent parent is known as linkage drag. The use of flanking markers at less than 5 cM upstream and downstream of the target gene minimizes the deleterious effects of linkage drag. Using recombinant selection, the size of the donor chromosome segment can be reduced only after two backcrosses as opposed to up to ten backcrosses using conventional breeding methods (Frisch et al., 1996; Collard and Mackill, 2008; Breseghello, 2013).

In contrast to both foreground and recombinant selection, background selection uses markers that are not linked to the gene/QTL of interest and targets loci across all chromosomes. It is useful in selecting against the donor genome and in the identification of backcross progenies with the greatest proportion of the recurrent genome. Using background selection, more than 95% of the genetic background of the recurrent parent can be recovered after only four backcrosses as opposed to up to six backcrosses using the conventional approach (Frisch et al., 1996). The use of background selection during MAB to recover the genetic background of the recurrent parent having the additional gene is also referred to as 'complete line conversion' (Ribaut et al., 2002).

Aside from single-gene introgression, MAS can also be used to simultaneously monitor several gene introgressions following a breeding scheme known as MAP. The most widespread application of MAP in rice is in breeding for multiple disease (e.g. bacterial blight and blast) and pest (e.g. brown planthopper) resistance (Collard and Mackill, 2008; Breseghello, 2013; Hu et al., 2016). The introgression of disease resistance genes in existing rice cultivars remains the most effective and economical means of disease control. Over time, however, resistance to the disease breaks down due to the occurrence of new pathotypes (Rao et al., 2002). The use of MAS to pyramid in a single target variety multiple disease resistance genes that will provide a more durable and broad-spectrum resistance against pathogens addresses this problem (Yasuda et al., 2015; Fukuoka et al., 2015). Using MAS, genotypes with multiple resistance genes can be differentiated in a set of plants with generally the same phenotype. It can also be used to replace screening of single plants for multiple diseases or pathogen races (Collard et al., 2008; Collard and Mackill, 2008; Breseghello, 2013).

To date, MAS has been successfully used in rice breeding to selectively transfer major genes/QTLs controlling abiotic stress tolerance including salt, drought and submergence tolerance (Gregorio et al., 2013), pest and disease resistance (Myint et al., 2012; Yasuda et al., 2015; Fukuoka et al., 2015), yield (Peng et al., 2008) and grain quality (Brar et al., 2012).

3.3 The case of the *Sub1A* gene

Flooding is a major challenge for rice production in South and Southeast Asia, where more than 90% of rice is produced. The annual occurrence of flash floods and typhoons, as well as the overflowing of rivers nearby paddy fields usually results in heavy production losses in rice. Submergence of 20 million hectares of flood-prone paddy fields in the region results in losses of up to 1 billion USD per year (Septiningsih et al., 2015). In India and Bangladesh alone, 4 million tons of rice is lost annually due to flooding. These figures are expected to considerably increase in the future given the increasing frequency and intensity of flooding brought about by extreme weather patterns (Bates et al., 2008).

Although rice thrives under shallow flooding, it cannot survive prolonged submergence in water. Flood-prone paddy fields in Asia are commonly affected by two types of flooding: (1) stagnant and (2) flash flooding. Stagnant flooding is characterized by prolonged partial flooding that does not completely submerge the plants (Mackill et al., 2010; Septiningsih et al., 2015). Traditionally, deep-water rice is grown in areas that are prone to stagnant flooding. This type of rice exhibits an escape mechanism via extreme internode elongation during submergence. Elongation of internodes pushes the plants above the water surface to gain access to oxygen and light necessary for respiration and photosynthesis. The low yields and late maturity led to the gradual replacement of deep-water rice with early maturing varieties that can be grown before and after the peak of the flooding season (Vergara and Mazaredo, 1975; Vergara et al., 1976; Catling, 1992; Hattori et al., 2009).

In contrast to stagnant flooding, flash flooding completely submerges rice fields for up to 2 weeks at any time during the planting season. This type of flooding is more common and can occur more than once during a cropping season in South and Southeast Asia. As early as the 1970s, landraces such as FR13A and FR43B have been identified to have tolerance to transient complete submergence (Vergara and Mazaredo, 1975; Mackill et al., 1996). In particular, FR13A from Odisha, India, has been determined to survive up to 2 weeks of complete submergence. However, like several other landraces, FR13A has poor agronomic characteristics including tall stature, photoperiod sensitivity, low yields and poor grain quality. Nevertheless, the identification of a source of submergence tolerance provided a starting point for breeders to develop rice that can survive flash flooding (Bailey-Serres et al., 2010).

Conventional breeding to transfer submergence tolerance from FR13A to advance breeding lines began in the 1980s. However, it was not until the early 1990s that submergence tolerance was successfully introgressed from FR13A to a semi-dwarf, high-yielding rice line, IR493830-7-1-2-2 (Mackill et al., 1993). Until this time, the genetic control of submergence tolerance remains unclear, with only a few reports on the quantitative nature of the trait (Mohanty and Khush, 1985; Haque, 1989).

With the development of DNA markers, linkage mapping using RAPDs and RFLPs identified a locus for submergence tolerance, designated as *Sub1*, near the centromeric region of chromosome 9 of FR13A. This locus accounted for 70% of the phenotypic variation observed for the trait (Xu and Mackill, 1995). High-resolution mapping using a bigger F_3 population that is segregating for the trait, as well as additional RFLP and AFLP markers further narrowed down the *Sub1* locus into a 0.16 cM region (Xu et al., 2000). Fine mapping of the *Sub1* QTL finally identified a 182 kb candidate region that includes annotations for three ethylene response factor-like genes designated as *Sub1A*, *Sub1B* and *Sub1C* (Xu et al., 2006).

Sub1A has two recognized alleles, that is, *Sub1A-1* and *Sub1A-2* that were identified in submergence-tolerant *indica* and *aus*, and -intolerant *indica* accessions, respectively. Overexpression of either allele in a submergence-intolerant *O. sativa* subsp. *japonica* identified *Sub1A-1* as the primary determinant of submergence tolerance. In contrast to the escape mechanism of deep-water rice (Hattori et al. 2009), *Sub1A-1* induces a quiescence strategy that limits ethylene-induced, gibberellic acid promoted elongation in plants under submergence. Plants carrying *Sub1A-1* consume carbohydrate energy reserves more slowly when inundated (Xu et al., 2006).

The identification and cloning of the large-effect QTL, *Sub1*, provided not only an opportunity to elucidate the molecular mechanisms underlying submergence tolerance in rice but also allowed the development of functional markers that will facilitate the precise introduction of *Sub1A* into target varieties by marker-assisted breeding. Efforts were soon underway to convert rice mega varieties into submergence-tolerant varieties. *Sub1A* was first successfully introduced into Swarna, a popular, high-yielding rice variety in India, within a 2-year time frame by MABC (Neeraja et al., 2007). This was followed by the introduction of *Sub1* into Samba Mahsuri and CR1009 from India, IR64 from the Philippines (IRRI), Thadokkham 1 (TDK1) from Laos and BR11 from Bangladesh. All mega varieties carrying the *Sub1* locus exhibited higher survival rate during inundation compared to the recurrent parents only after 2–3 rounds of MABC (Septiningsih et al., 2009). Incorporation of *Sub1A* into these varieties also gave a yield advantage of 1–3 tons following flooding for 10–15 days. At present, flood-tolerant varieties that have been released and are now being planted include Swarna-*Sub1A* in India, Sambha Mahsuri-*Sub1A* in Bangladesh and IR64-*Sub1A* in the Philippines (www.irri.org).

The success story of *Sub1A* clearly underscores the importance of molecular markers in increasing the efficiency and precision of breeding strategies to improve trait performance in rice. Before the identification of the *Sub1* locus, conventional breeding for submergence tolerance from FR13A to target varieties have been slow (up to 10 years) and progress with only limited success (Mohanty et al. 2000; Mackill et al., 1993). In some cases, donor loci with deleterious effects or that reduce the fitness of the target cultivar are transferred along with the *Sub1* locus (Mackill et al., 1993). The cloning of *Sub1A* provides a precise target for selection for MABC. This made identification of lines carrying the target trait possible even at early generations, thereby effectively reducing the number of plants for backcrossing. The use of molecular markers to select against donor introgressions across the recurrent genome that might have undesirable phenotypic effects on the target varieties also sped up the selection process, thereby increasing genetic gain per unit time.

4 The genomics era in rice breeding

4.1 Reference genome sequence for rice

Rice is not only one of the most important crops in the world but also a model species for monocotyledonous plants. For centuries, rice breeders and geneticists have accumulated a large number of resources for this crop including wild relatives, mutants, cultivated landraces (Shimamoto and Kyozuka, 2002; Paterson et al., 2005), monosomic alien addition lines (Jena and Khush, 1990; Jena et al., 1991; Multani et al., 1994) chromosome segment substitution lines (Furuta et al., 2014, Shim et al., 2010; Ando et al., 2008; Takai et al., 2007; Ebitani et al., 2005; Doi et al., 1997) and multi-parent advanced generation intercrosses (Bandillo, 2013). Compared to the other major cereal crops, the rice genome is relatively compact (400–430 Mb), with dense, well-characterized linkage maps. It shares a high degree of synteny and colinearity of genes with maize, wheat, barley, oats, rye and sorghum, and has a well-established transformation and regeneration system. For these reasons, rice has become an excellent model for crop genetics research and the obvious choice for the first whole-genome sequencing in cereals (Goff et al., 2002; International Rice Genome Sequencing Project, 2005).

The release in 2001 of the draft sequences of *indica* rice cv. 93–11 (Yu et al., 2002) and of *japonica* rice cv. Nippponbare (Goff et al., 2002), as well as the completion of the map-based reference genome of cv. Nipponbare (International Rice Genome Sequencing Project, 2005) marked milestone achievements in rice genomics. With a publicly available reference sequence, the stage was set for the functional characterization of the rice genome. In the years that followed, the Institute of Genome Research in the United States and the Rice Genome Research Program in Japan led intensive efforts dedicated to annotating the map-based sequence of cv. Nipponbare. These concerted undertakings led to the identification of major structural features of the rice genome including genes, centromeres, transposons, repetitive elements and regulatory sequences. Low and high stringency alignments of gene indices between rice and other plant species, as well as syntenic alignments between maize and rice that provides alternative views of sequence conservation across several plant species were generated. Sequence variations that represent new targets for the development of new molecular markers were identified. These includes an abundance of tandemly repeated DNAs that were used to develop close to 19 000 class 1, bi-, tri- and tetra-nucleotide SSRs (Yuan et al., 2003; Zhang et al.,

2004; Wu et al., 2004; Ouyang et al., 2007; http://rapdb.dna.affrc.go.jp/; http://rice.plantbiology.msu.edu/). Systematic, full-length comparison of the draft genomic sequence of rice cultivars 93–11 and Nipponbare also identified thousands of insertions/deletions (indels) as well as SNPs (Shen et al., 2004; Feltus et al., 2004).

Integration of the advances in biotechnology, genetics and molecular marker applications with conventional breeding has transformed the practice of trait improvement in rice into the interdisciplinary science of molecular breeding in the twentieth century (Moose and Mumm, 2008). With the sequencing of the rice genome and the wealth of information that comes with it, rice geneticists and breeders alike are presented with an array of tools to systematically unlock the biological functions of more than 50 000 predicted genes in rice that can be used to revolutionize rice breeding in the twenty-first century.

4.2 Single-nucleotide polymorphisms

SNPs represent a single base variation in specific sites in the genome. They occur in both coding and non-coding regions of nuclear and plastid DNA and represent the bulk of natural genetic variations in the eukaryotic genome (Kwok et al., 1996). The study of Nasu et al. in 2002 discovered 2800 SNPs in approximately 250 000 aligned bases in 417 regions distributed throughout the genome of three *japonica* and two *indica* rice cultivars, and the wild rice species *O. rufipogon*. Comparative sequence alignment of the draft sequences of rice cultivars 93–11 and Nipponbare identified 1.7 million SNPs by Shen et al. (2004) and a set of 384 431 high-quality SNPs by Feltus et al. (2004). Alignment of BAC libraries generated by the Oryza map alignment project also identified a wide array of SNPs in a panel of 12 wild rice species despite the variable quality of the obtained sequence data (Ammiraju et al., 2006; Wing et al., 2007; Ammiraju et al., 2010). In McNally et al. (2009), reported the distribution of 160 000 non-redundant SNP across 20 varieties of rice. More recently, re-sequencing projects have discovered several million SNP along the entire rice genome (Chen et al., 2014).

Through the years, SNPs have been increasingly used in genetics and breeding applications in rice. Mapping based on high-density SNP markers have been shown to locate the proximate position of QTLs for grain weight (*GS3*) and (*GW5/qSW5*) with more precision and accuracy compared to mapping using the more traditional RFLP and SSR markers (Yu et al., 2011). The utility of SNPs in fine mapping and cloning has also been demonstrated in studies that identified a functional SNP in the 5′ regulatory region of the *qSH1* gene that is responsible for the loss of seed shattering in rice. In this case, the SNP was identified to be functionally associated with natural variation in the trait (Konishi et al., 2006). Aside from QTL mapping and positional cloning, the applicability of SNPs for MAS has also been explored. Analysis of sequences from a blast-susceptible Japanese cultivar and two cultivars having rice blast resistance genes revealed the abundance of SNPs in the loci containing the genes, *Piz* and *Piz-t*. The dense distribution of SNPs generated numerous SNP markers in the *Piz* and *Piz-t* loci that can be used in MAS for blast resistance breeding (Hayashi et al., 2004).

Realizing the applications of SNPs in various biological researches led to more SNP discovery efforts to expand the available SNP data resources that can be used to explore and utilize natural variation in rice. These efforts spurred rapid synergistic developments in sequencing technology and genotyping platforms that allow the more efficient integration of SNPs as molecular markers in breeding applications. As the amount of both sequence and biological information grows exponentially, there is also a parallel growth in the development of computational tools and methodologies for the systematic storage, organization, indexing and analysis of such data.

4.3 Sequencing technologies and genotyping platforms

Advances in molecular biology, cloning technology, computational tools and experimentation have allowed the rapid sequencing of large portions as well as whole genomes of several species. From 1995, the full genomes of more than 200 organisms have been completely sequenced. In plants, the number has reached more than 50 since the publication of the full genomic sequence of *Arabidopsis thaliana* in 2000 (Michael and Jackson, 2013; Arabidopsis Genome Initiative, 2000). This includes the working draft sequences for *O. sativa* subsp. *indica* cv. 93–11 and *O. sativa* subsp. *japonica* cv. Nippponbare.

Sequencing for most of these plant species has been carried out entirely or partially by Sanger method. Classical Sanger sequencing technology provides long (approx. 1000 bp), high-quality reads but requires a DNA cloning step that is both very expensive and time consuming. The high demand for low-cost sequence data has driven the development of a wide range of high-throughput, NGS technologies that allow parallel sequencing of millions to billions of short DNA strands. These techniques included sequencing by hybridization, parallel signature sequencing and pyrosequencing. It was the pyrosequencing-based, 454 platform by Roche, however, that was to become the first stand-alone next-generation sequencer. Within a short period of time following the release of the Roche 454, more next-generation sequencers such as Roche 454 FLX Titanium, Illumina Solexa, and HiSeq2500, Ion Torerent and Helicos entered the market (Llaca, 2012, Perez-de Castro, 2012). More recently, revolutionary improvements in sequencing technologies have spurred the emergence of third-generation sequencers (TGS) that are based on a single DNA molecule, real-time sequencing such as the one developed by Pacific BioSciences (PACBio). Unlike the NGS technologies that offer short read lengths from an ensemble of DNA molecules following a 'wash-and-scan' technique, PACBio sequencing generates long reads without the need to pause in between read steps. It is currently limited, however, by low-throughput, higher error rate and higher costs compared to NGS. To overcome the limitations of TGS, 'hybrid sequencing' strategies have been developed that allow the utilization of the short reads of NGS in conjunction with the long reads produced by TGS (Rhoads and Au, 2015; Schadt et al., 2010).

In rice, the first re-sequencing efforts generated the high-resolution, shotgun sequences that made up the draft sequences of the *indica* cultivar 93–11 (Yu et al., 2002) and the *japonica* cultivar Nipponbare (Goff et al., 2002). Since then, the advent of a wide array of NGS technologies for *de novo* sequencing, whole-genome sequencing and whole-genome re-sequencing has provided high-throughput sequences for applications in both functional genomics studies and breeding in rice. The rapid advances in NGS technologies continuously reduce costs of DNA sequencing, which in turn drives the development of other NGS application such as GBS. GBS is a simple, cost-effective application of NGS protocols for SNP discovery and genotyping that does not require a reference genome sequence (He et al., 2014). To date, GBS has been shown to be a powerful tool for genetic diversity analysis, linkage mapping, association studies and genomic selection (GS) in rice (Spindel et al., 2013; Jannik, 2010).

Although thousands of non-redundant and high-quality SNPs (i.e. SNPs that were called correctly and with high accuracy by algorithms designed to identify DNA polymorphisms) have been discovered in the rice genome, SNPs are generally informative only for a particular set of germplasm or populations. Due to this, SNP sets have to be validated and optimized first for the population to be studied before it can be used for any biological application (McCouch et al., 2010). The most popular genotyping platforms that are currently being used for SNP validation include the Illumina BeadArray technology-based Golden Gate and Infinium assays (Steemers and Gunderson, 2007; Thomson et al., 2012), Life Technologies TaqMan assay coupled with OpenArray platform (Livak, 1995) and KBiosciences Competitive Allele Specific PCR (KASPar) (Smith and Maughan, 2015). Based on these genotyping platforms, custom-made assays have been made available for a suite of biological applications in rice. Examples of these SNP assay include the seven GoldenGate VeraCode oligo pool assay sets designed for the Illumina BeadXpress reader Six 384-SNP Illumina BeadXpress assays for QTL mapping and background selection for crosses between different germplasm pools in rice (Thomson et al., 2012). More recently, a breeders chip based on Golden Gate platform was developed for MAP of genes/QTLs underlying yield-related traits as well as disease resistance in a selected set of rice varieties (Kurokawa et al., 2016). The synergistic evolution of sequencing technologies and genotyping platforms has provided not only a means for a more efficient SNP detection but also for a fast and cost-effective method of MAS for breeding.

5 Applying genomics for more rapid varietal development and release

5.1 Genetic diversity and genomics breeding

The unprecedented increase in food production that offset the Malthusian prediction (Fedoroff, 2010) of population collapse during the Green Revolution of the 1960s truly represents an extraordinary moment in agricultural history. Through the integration of the semi-dwarf, high-yielding varieties of rice in farming systems and the mechanization of cultural management practices, significant increases in food production was achieved at a critical time when global population was also increasing by a factor of 2.3 (i.e. from 3.1 to 7 billion people from 1961 to 2010) (Khush, 1999). Building on the success of IR8, the international breeding efforts that followed were mostly based on a small selection of high-yielding varieties (HYVs) with the *sd1* gene. By mid-1980s, one-third to one-half of the rice growing areas in developing countries in Asia were growing HYVs (Guimaraes, 2009; Khush et al., 2001).

Without a doubt, the Green Revolution has benefited humankind with the groundbreaking increase in rice production that secured food supply for millions of people. However, the extensive adaptation by farmers of only a few semi-dwarf, HYVs led to the gradual loss of traditional varieties and landraces, and eventually, to genetic erosion in rice. In addition, the use of limited germplasm source of *sd1* has caused a bottleneck in the genetic background of rice. With narrow genetic variation, rice production became vulnerable not only to disease and pest outbreaks but also to declining agro-environmental conditions (Hargrove and Cabanilla, 1979; Pingali, 2012).

Rice holds a tremendous amount of genetic diversity in the collection of cultivated accessions, landraces and wild relatives maintained at various international genebanks such as those in the International Rice Research Institute (http://irri.org/our-work/research/

genetic-diversity/international-rice-genebank) and AfricaRice (http://www.africarice.org/warda/genebank.asp), as well as in the national germplasm repositories of Japan (https://www.gene.affrc.go.jp/databases-core_collections_jr_en.php), China (Han and Huang, 2013), India, Thailand and Malaysia (http://www.fao.org/docrep/006/y4751e/y4751e0b.htmref). To mitigate the effects of genetic erosion from years of utilizing only a small genetic base in improvement programmes, rice breeders would have to tap on the existing genetic diversity of rice to create and select novel gene combinations that gives the highest productivity under a wide range of agricultural ecosystems. The initial step to efficiently utilize this natural variation is to unlock the genetic basis of phenotypic variations observed across the rice germplasm. To this end, the complete genome sequence of rice provides a starting point to evaluate genetic diversity at the DNA level and link this variation to observable phenotypic performances. Understanding this relationship will enable breeders to make informed decisions in selecting germplasm for crosses, resulting in the faster development and deployment of rice varieties.

Technological advances in genotyping and sequencing, coupled with refinements in computational methodologies have spurred breeding strategies based on the genomics of genotype–phenotype relations in rice. Two of these strategies viz., gene/QTL discovery and mapping by genome-wide association studies (GWAS) and GS are discussed here.

5.2 QTL/gene discovery and mapping via GWAS

The elucidation of the genetic basis of agronomic traits has remained one of the major challenges in crop breeding. Conventionally, mapping of complex traits has been carried out by linkage analysis. Genotyping of a few hundred lines in a segregating population derived from a bi-parental cross using approximately 100–200 markers (e.g. SSRs) has been sufficient in detecting QTLs controlling various agronomic traits. However, the restricted number of meiotic events that is captured in bi-parental populations, as well as the relatively sparse marker coverage usually result in a low-resolution genetic map where the estimated QTL intervals extend to 10–30 cM. This genetic distance translates into genomic regions covering a considerable number of candidate genes (Holland, 2007). To obtain sufficient genetic resolution to clone the gene/QTL of interest, large populations have to be genotyped using markers with a denser coverage (Zhao et al., 2011; Pasam et al., 2012).

With the advances in sequencing technology and genotyping platforms, GWAS has emerged as an alternative to traditional linkage analysis. Unlike linkage mapping, GWAS utilizes a set of diverse, unrelated germplasm (i.e. diversity panel) to identify associations between markers and traits of interest. The historical and evolutionary recombination events that differentiate the diversity panel from their last common ancestor allows for higher genetic resolution due to the faster linkage disequilibrium decay in unrelated lines (Zhu et al., 2008; Varshney et al., 2014). To date, GWAS has been widely used in dissecting the genetic basis of complex agronomic traits in various crops including maize (Tian et al., 2011; Li et al., 2013), wheat (Reif et al., 2011; Sukumaran et al., 2015), sorghum (Morris et al., 2013), barley (Pasam et al., 2012) and sugarcane (Debibakas et al., 2014).

In rice, the availability of a high-quality reference genome, abundance of germplasm and self-fertilizing nature of the crop make it an ideal candidate for GWAS applications. Association mapping for 14 agronomic traits using 517 landraces genotyped with approximately 3.6 million SNPs identified loci for drought tolerance, spikelet number and 12 other agronomic traits. Six loci that were identified for grain width and length, gelatinization temperature, amylose content, apiculus and pericarp, colour and heading date coincided with previously reported genes (Huang et al., 2010). A related GWAS for flowering time and grain yields in a diverse sample of 950 rice varieties also successfully identified 32 new loci associated with flowering time and 10 loci responsible for grain-related traits. From these newly identified loci, 18 were assigned candidate genes through detailed annotation (Huang et al., 2011). In another study, association mapping for 19 agronomic traits using an elite breeding population of irrigated rice identified 52 QTLs for 11 agronomic traits, including large-effects QTLs for flowering time, grain length:grain width:grain length/breadth ratio. Haplotypes that can be used to directly select for short stature, early flowering and high yield in the studied population were also identified (Begum et al., 2015). More recently, GWAS using 176 *japonica* rice varieties developed in Japan detected 26 loci associated with heading date. From the five loci the authors concentrated their study on two coincided with positions of the known heading date genes *Hd6* and *Hd2*, whereas one corresponded with *Hd1* based on gene-based association analysis. Two other unreported loci in chromosome 1 and 11 controlling heading date were confirmed to affect the trait by transgenic complementation, with the chromosome 11 locus exhibiting pleiotropic effects on plant height and panicle length. In addition, genes associated with panicle number per plant, spikelet number per panicle, leaf blade width and awn length were also identified using the same panel of rice varieties (Yano et al., 2016).

With GWAS application becoming commonplace in gene/QTL discovery and mapping, the challenge for breeders now is in determining how information derived from such studies can be directly utilized for MAS and/or GS in applied breeding programmes.

5.3 Genomic selection

The application of MAS in rice breeding has been mostly effective in the incorporation of traits that are controlled by major genes or QTLs with a known association to a marker (Yano et al., 2000; Ashikari et al., 2002, 2005; Xu et al., 2006; Miura et al., 2010). However, not all traits of agronomic importance are controlled by major genes/QTLs but by many genes with small, additive effects or a combination of genes with major and minor effects. Several others are controlled by genes/QTLs that are influenced by epistatic and/or genotype x environment interactions. In breeding for such traits, the applicability of MAS is limited (Crosbie et al., 2003; Rutkoski, 2011; Lorenz et al., 2011; Varshney et al., 2014).

GS is an alternative breeding strategy that was originally developed by Meuwissen and colleagues in 2001 for livestock breeding. This approach capitalizes on genotyping technologies that provide low-cost, high-throughput, genome-wide marker coverage. Unlike MAS which utilizes a defined set of markers associated with a trait for selection, GS uses all available marker (i.e. SNP) data across the genome and feeds it into a prediction model that would then estimate the genome estimated breeding value (GEBV) of an individual in a breeding population. Prediction models are generated from the integration of genotype and phenotype data of a training population (TP), which is usually representative of the selection candidates in a breeding programme. Essentially, GEBV serves as a predictor of the value of individuals as parents for crossing based on the similarity of its genomic profile to other individuals in the TP that are known to have good performance in a target environment. Therefore, even without phenotype data, individuals can be selected for advancement in a breeding cycle based on GEBVs alone (Meuwissen, 2001; Heffner et al., 2010; Jannik et al., 2010; Lorenz et al., 2011; Breseghello, 2013; Varshney et al., 2014).

Success in the use of GS in animal breeding (Hayes et al., 2009; Duchemin et al., 2012; Gaspa et al., 2015) has generated a lot of interest in the application of this approach to leverage breeding for the improvement of crops. To date, GS simulation and empirical studies have been applied to a wide range of crops including maize, wheat, barley, cassava and sugar beet. In all cases, GS outperformed traditional MAS in terms of breeding efficiency (Bernardo and Yu, 2007; Guo et al., 2012; Crossa et al., 2014; Poland et al., 2012; Rutkoski et al., 2011; Wurschum et al., 2013; De Oliveira et al., 2012).

In rice, limited simulation studies indicate that the successful application of GS in rice breeding would depend on the relationship between the training and the breeding population, the genetic architecture of the trait (i.e. small- or large-effect QTL), the type and number of markers used for genotyping, the statistical method used to build the GS prediction model and the accuracy of phenotyping (Varshney et al., 2014; Spindel et al., 2015; Xu et al., 2014; Zhong et al., 2009).

To date, results of GS simulation and cross-validation studies have not been evaluated in actual rice breeding programmes. A shift in the use of GS from theory to practice is necessary to realize the potential of GS in increasing the efficiency in rice breeding via increasing rate of genetic gain per unit time. To do this, however, significant investments in time, labour and organizational resources are required to set up both genotyping and phenotyping pipelines (Breseghello, 2013; Eathington et al., 2007) that are both requirements for genome-wide selection.

6 Future trends and conclusion

The development of molecular marker systems (RFLP, AFLP and SSRs) for rice has greatly transformed plant breeding from a simple activity of visual selection for desirable traits into an interdisciplinary science of shuffling alleles to create favourable recombinations and selecting variants with desirable phenotypes. For the past few decades, the use of molecular markers have greatly contributed in accelerating the pace of varietal development by moving target genes underlying important agronomic traits across genomes with better precision and accuracy. The use of molecular markers also facilitated gene pyramiding for durable resistance to pests and diseases, a feat that was almost impossible using conventional approaches. Using the same pyramiding technique, combining multiple abiotic stress tolerance in mega varieties was also made easier.

Despite genetic gains from selection via the application of molecular markers, emerging challenges such as population explosion and adverse climatic conditions are negatively affecting world food security. To address these problems, new strategies that will further accelerate breeding cycles have to be developed. Breeding targets also have to be shifted to focus not only on increasing yields but also on improving the tolerance of the crop to abiotic stresses. Last but not least, breeding in this century has to ensure that a high degree of genetic diversity exists in breeding populations to ascertain the resilience of the crop to emerging pest and pathogens.

Rapid developments in sequencing technology have allowed whole-genome sequencing at a faster rate and at significantly lower costs. This has spurred the development of sequencing applications for SNP discovery and genotyping that have generated large-scale sequence data for rice. While such data sets have been critical in enhancing our understanding of genome architecture and dynamics in rice, the bigger challenge now for rice geneticists and breeders alike is in translating these genomics information into improved breeding products that is acceptable to both farmers and consumers.

To this end, GWAS and GS are two of the emerging genomics-based breeding strategies that promise to complement existing paradigms in rice breeding. Both strategies capitalize on the abundance of SNP markers in the rice genome, as well as on the availability of a vast germplasm for the crop. Both methods also have the potential to deliver genetic gains by speeding the whole process of varietal improvement and dissemination to target regions. Although several challenges remain to be addressed before these methods can be fully integrated in actual breeding programmes, simulation and empirical evidences are available that show the efficient utilization of these methods in breeding applications in rice.

7 Where to look for further information

A good review on the basic aspects of GS and the value of GS in achieving long-term genetic gains in crops are detailed in Lorenz et al. (2011). Results of recent GS studies on tomato by Yamamoto et al. (2016, 2017) provide empirical basis for the use of GS in breeding programmes. The results of such studies can be used to leverage GS in rice.

8 References

Allard, R. W. (1999), *Principles of Plant Breeding*, 2nd edition, Wiley.

Ammiraju, J. S. S., Luo, M., Goicoechea, J. L., Wang, W., Kudrna, D., Mueller, C., Talag, J., Kim, H., Sisneros, N. B., Blackmon, B., Fang, E., Tomkins, J. B., Brar, D. S., Mackill, D., McCouch, S., Kurata, N., Lambert, G., Galbraith, D. W., Arumuganathan, K., Rao, K., Walling, J. G., Gill, N., Yu, Y., SanMiguel, P., Soderlund, C., Jackson, S. and Wing, R. A. (2006), 'The *Oryza* bacterial artificial chromosome library resource: Construction and analysis of 12 deep-coverage large-insert BAC libraries that represent the 10 genome types of the genus *Oryza*', *Genome Res.*, 16(1), 140–7.

Ammiraju, J. S. S., Song, X., Luo, M., Sisneros, N., Angelova, A., Kudrna, D., Kim, H., Yu, Y., Goicoechea, J. L., Lorieux, M., Kurata, N., Brar, D. S., Ware, D., Jackson, S. and Wing, R. A. (2010), 'The *Oryza* BAC resource: a genus-wide and genome scale tool for exploring rice genome evolution and leveraging useful genetic diversity from wild relatives', *Breed. Sci.*, 60(5), 536–43.

Ando, T., Yamamoto, T., Shimizu, T., Ma, X. M., Shomura, A., Takeuchi, Y., Lin, S. Y. and Yano, M. (2008), 'Genetic dissection and pyramiding of quantitative traits for panicle architecture by using chromosomal segment substitution lines rice', *Theor. Appl. Genet.*, 116, 881–90.

Arabidopsis Genome Initiative (2000), 'Analysis of the genome sequence of the flowering plant *Arabidopsis thaliana*', *Nature*, 408, 796–815.

Ashikari, M., Sakakibara, H., Lin, S., Yamamoto, T., Takashi, T., Nishimura, A., Angeles, E. R., Qian, Q., Kitano, H. and Matsuoka, M. (2005), 'Cytokinin oxidase regulates rice grain production', *Science*, 309(5735), 741–5.

Ashikari, M., Sasaki, A., Ueguchi-Tanaka, M., Itoh, H., Nishimura, A., Datta, S., Ishiyama, K., Saito, T., Kobayashi, M., Khush, G. S., Kitano, H. and Matsuoka, M. (2002), 'Loss-of-function of a rice gibberellin biosynthetic gene, GA20 oxidase (GA20ox-2), led to the rice Green Revolution', *Breed. Sci.*, 52, 143–50.

Bailey-Serres, J., Fukao, T., Ronald, P. Ismail, A., Heuer, S. and Mackill, D. (2010), 'Submergence tolerance rice: *Sub1*'s journey from landrace to cultivar', *Rice*, 3, 138–47.

Bandillo, N., Raghavan, C., Muyco, P. A., Sevilla, M. A. L., Lobina, I. T., Dilla-Ermita, C. J., Tung, C., McCouch, S., Thomson, M., Mauleon, R., Singh, R. K., Gregorio, G., Redona, E. and Leung, H.

(2013), 'Multi-parent advanced generation inter-cross (MAGIC) populations in rice: progress and potential for genetics research and breeding', *Rice*, 6, 11.

Bates, B. C., Kundzewicz, Z. W., Wu, S. and Palutikof, J. P., eds, (2008), 'Climate change and water. Technical paper of the intergovernmental panel on climate Change, IPCC Secretariat, Geneva', www.ipcc.ch/ipccreports/tpclimate-change-water.htm.

Begum, H., Spindel, J. E., Lalusin, A., Borromeo, T., Gregorio, G., Hernandez, J., Virk, P. Collard, B. C. Y. and McCouch, S. R. (2015), 'Genome-wide association mapping for yield and other agronomic traits in an elite breeding population of tropical Rice (*Oryza sativa*)', *PloS ONE*, 10(3), e0119873.

Bernardo, R. and Yu, J. (2007), 'Prospects for genome-wide selection for quantitative traits in maize', *Crop Sci.*, 47(3), 1082–90.

Brar, D. S., Virk, P. S., Grewal, D., Slamet-Loedin, I., Fitzgerald, M. and Khush, G. S. (2012), 'Breeding rice varieties with improved grain and nutritional quality', *Quality Assurance and Safety of Crops and Foods*, 4(3), 137.

Breseghello, F. (2013), 'Traditional and modern plant breeding methods with examples in rice (*Oryza sativa* L.)', *J. Agric. Food. Chem.*, 61(35), 8277–86.

Caicedo, A. L., Williamson, S. H., Hernandez, R. D., Boyko, A., Fledel-Alon, A., York, T. L., Polato, N. R., Olsen, K. M., Neilsen, R., McCouch, S. R., Bustamante, C. D. and Purugganan, M. D. (2007), 'Genome-wide patterns of nucleotide polymorphism in domesticated rice', *PloS Genet.*, 3(9), e163.

Catling, D. (1992), *Rice in Deep Water*, Macmillan.

Causse, M. A., Fulton, T. M., Cho, Y. G., Ahn, S. N., Chunwongse, J., Wu, K., Xiao, J., Yu, Z., Ronald, P. C., Harrington, S. E., Second, G., McCouch, S. R. and Tanksley, S. D. (1994), 'Saturated molecular map of the rice genome based on interspecific backcross population', *Genetics*, 138(4), 1251–74.

Chen, H., Xie, W., He, H., Yu, H., Li, J., Yu, R., Yao, Y., Zhang, W., He, Y., Tang, X., Zhou, F., Deng, X. W. and Zhang, Q. (2014), 'A high-density SNP genotyping array for rice biology and molecular breeding', *Mol. Plant*, 7(3), 541–53.

Collard, B. C. Y, Vera Cruz, C. M., McNally, K. L., Virk, P. S. and Mackill, D. J. (2008), 'Rice molecular breeding laboratories in the genomics era: current status and future considerations', *Int. J. Plant Genom.*, 524847.

Collard, B. Y. C. and Mackill, D. J. (2008), 'Marker-assisted selection: an approach for precision plant breeding in the twenty-first century', *Philos. Trans. R. Soc. Lond. B. Biol. Sci.*, 363, 557–72.

Crosbie, T. M., Eathington, S. R., Johnson, G. R., Edwards, M., Reiter, R., Stark, S., Mohanty, R. G., Oyervides, M., Buehler, R. E., Walker, A. K., Dobert, R., Dalannay, X., Pershing, J. C., Hall, M. A. and Lamkey, K. (2006), 'Plant breeding: past, present, and future', in Lamkey, K. R. and Lee, M., (Eds), *Plant Breeding*. The Arnel R. Hallauer International Symposium, Blackwell.

Crossa, J., Perez, P., Hickey, J., Burgueno, J., Ornella, L., Ceron-Rojas, J., Zhang, X., Dreisigacker, S., Babu, R., Lil, Y., Bonnett, D. and Mathews, K. (2014), 'Genomic prediction in CIMMYT maize and wheat breeding programs', *Heredity*, 112, 48–60.

De Oliveira, E. D., De Resende, M. D. V., Santos, V. D. S., Ferreira, C. F., Oliveira, G. A. F., Da Silva, M. S., De Oliveira, L. A. and Aguilar-Vildoso, C. A. (2012), 'Genome-wide selection in cassava', *Euphytica*, 187(2), 263–76.

Debibakas, S., Rocher, S., Garsmeur, O., Toubi, L., Roques, D., D'Hont, A., Hoarau, J. Y. and Daugrois, J. H. (2014), 'Prospecting sugarcane resistance to sugarcane yellow leaf virus by genome-wide association', *Theor. Appl. Genet.*, 127(8), 1719–32.

Doi, K., Iwata, N. and Yoshimura, A. (1997), 'The construction of chromosome substitution lines of African rice (*Oryza glaberrima* Steud.) in the background of japonica (*O. sativa* L.)', *Rice Genet. Newsl.*, 14, 395–9.

Duchemin, S. I., Colombani, C., Legarra, A., Baloche, G., Larroque, H., Astruc, J. M., Barillet, F., Robert-Granié, C. and Manfredi, E. (2012), 'Genomic selection in the French Lacaune dairy sheep breed', *J. Dairy Sci.*, 95(5), 2723–33.

Eathington, S. R., Crosbie, T. M., Edwards, M. D., Reiter, R. S. and Bull, J. K. (2007), 'Molecular markers in a commercial breeding program', *Crop Sci.*, 47, S154–S163.

Ebitani, T., Takeuchi, Y., Nonue, Y., Yamamoto, T. Takeuchi, K. and Yano, M. (2005), 'Construction and evaluation of chromosome segment substitution lines carrying overlapping chromosome segments of indica rice cultivar Kasalath in a genetic background of japonica elite cultivar Koshihikari', *Breed. Sci.*, 55, 65–73.

Fan, C., Xing, Y., Mao, H., Lu, T., Han, B., Xu, C., Li, X. and Zhang, Q. (2006), '*GS3*, a major QTL for grain length and weight and minor QTL for grain width and thickness in rice, encodes a putative transmembrane protein', *Theor. Appl. Genet.*, 112(6), 1164–71.

Fedoroff, N. V. (2010), 'The past, present and future of crop genetic modification', *N. Biotechnol.*, 27(5), 461–5.

Feltus, F. A., Wan, J., Schulze, S. R., Estill, J. C., Jiang, N. and Paterson, A. H. (2004), 'A SNP resource for rice genetics and breeding based on subspecies indica and japonica genome alignments', *Genome Res.*, 14(9), 1812–19.

Frisch, M., Bohn, M. and Melchinger, A. E. (1996), 'Minimum sample size and optimal positioning of flanking markers in marker-assisted backcrossing for transfer of a target gene', *Crop Sci.*, 39, 967–75.

Fujita, D., Trijatmiko, K. R., Tagle, A. G., Sapasap, M. V., Koide, Y., Sasaki, K., Tsakirpaloglou, N., Gannaban, R. B., Nishimura, T., Yanagihara, S., Fukuta, Y., Koshiba,T., Slamet-Loedin, I. H., Ishimaru, T. and Kobayashi, N. (2013), '*NAL1* allele from a rice landrace greatly increases yield in modern indica cultivars', *Proc. Natl.Acad. Soc.*, 110(51), 20431–6.

Fukuoka, S., Saka, N., Mizukami, Y., Koga, H., Yamanouchi, U., Yoshioka, Y., Hayashi, N., Ebana, K. Mizobuchi, R. and Yano, M. (2015), 'Gene pyramiding enhances durable blast resistance in rice', *Sci. Rep.*, 5, 7773.

Furuta, T., Uehara, K., Angeles-Shim, R. B., Shim, J., Ashikari, M. and Takashi, T. (2014), 'Development and evaluation of chromosome segment substitution lines (CSSLs) carrying chromosome segments derived from *Oryza rufipogon* in the genetic background of *O. sativa* L.', *Breed. Sci.*, 63, 468–57.

Gaspa, G., Veerkamp, R. F., Calus, M. P. L. and Windig, J. J. (2015), 'Assessment of genomic selection for introgression of polledness into Holstein Friesian cattle by simulation', *Livestock Sci.*, 179, 86–95.

Godfray, H. C. J., Beddington, J. R., Crute, I. R., Haddad, L., Lawrence, D., Muir, J. F., Pretty, J., Robinson, S., Thomas, S. M. and Toulmin, C. (2010), 'Food Security: The challenge of feeding 9 billion people', *Science*, 327, 812–17.

Goff, S. A., Ricke, D., Lan, T. H., Presting, G., Wang, R., Dunn, M., Glazebrook, J. Sessions, A., Oeller, P., Varma, H., Hadley, D., Hutchison, D., Martin, C., Katagiri, F., Lange, B. M., Moughamer, T., Xia, Y., Budworth, P., Zhong, J., Miguel, T., Paszkowski, U., Zhang, S., Colbert, M., Sun, W., Chen, L., Cooper, B., Park, S., Wood, T. C., Mao, L., Quail, P., Wing, R., Dean, R., Yu, Y., Zharkikh, A., Shen, R., Sahasrabudhe, S., Thomas, A., Cannings, R., Gutin, A., Pruss, D., Reid, J., Tavtigian, S., Mitchell, J., Eldredge, G., Scholl, T., Miller, R. M., Bhatnagar, S., Adey, N., Rubano, F, Tusneem, N., Robinson, R., Feldhaus, J., Macalma,T., Oliphant, A. and Briggs, S. (2002), 'A draft sequence of the rice genome (*Oryza sativa* L. ssp. *Japonica*)', *Science*, 296, 99–100.

Grassini, P., Eskridge, K. M. and Cassman, K. G. (2013), 'Distinguishing between the yield advances and yield plateaus in historical crop production trends', *Nat. Comm.*, 4, 2918.

Gregorio, G. B., Vergara, G. V. and Thirumeni, S. (2013), 'Recent advance in rice science to design salinity and other abiotic stress tolerant rice varieties', *SABRAO J. Breed. Genet.*, 45, 31–41.

GRiSP (Global Rice Science Partnership) (2013), *Rice Almanac*, 4th edition, International Rice Research Institute, Philippines.

Guimaraes, E. P. (2009), 'Rice breeding', in Carena, M. J. (Ed.), *Cereals*. Springer.

Guo, Z., Tucker, D. M., Lu, J., Kishire, V. and Gay, G. (2012), 'Evaluation of genome-wide selection efficiency in maize nested association mapping populations', *Theor. Appl. Genet.*, 124(2), 261.275.

Han, B. and Huang, X. (2013), 'Sequencing-based genome-wide association study in rice', *Curr. Opin. Plant Biol.*, 16, 133–8.

Haque, Q. A., HilleRisLambers, D., Tepora, N. M. and Dela Cruz, Q. D. (1989), 'Inheritance of submergence tolerance in rice', *Euphytica*, 41, 247–51.

Hargrove, T. R. and Cabanilla, V. L. (1979), 'The impact of semi-dwarf varieties on Asian rice breeding programs', *Bioscience*, 29, 731–5.

Hattori, Y., Nagai, K., Furukawa, S., Song, X. J., Kawano, R., Sakakibara, H., Wu, J., Matsumoto, T., Yoshimura, A., Kitano, H., Matsuoka, M., Mori, H. and Ashikari, M. (2009), 'The ethylene response factors *SNORKEL1* and *SNORKEL2* allow rice to adapt to deep water', *Nature*, 460(7258), 1026–30.

Hayashi, K., Hashimoto, N., Daigen, M. and Ashikawa, I. (2004), 'Development of PCR-based SNP markers for rice blast resistance genes at the *Piz* locus', *Theor. Appl. Genet.*, 108(7), 1212–20.

Hayes, B. J., Bowman, P. J., Chamberlain, A. J. and Goddard, M. E. (2009), 'Invited review: Genomic selection in dairy cattle: progress and challenges', *J. Dairy Sci.*, 92, 433–43.

He, J., Zhao, X., Loraoche, A., Lu, Z., Liu, H. and Li, Z. (2014), 'Genotpying-by-sequencing, an ultimate marker-assisted selection (MAS) tool to accelerate breeding', *Front. Plant Sci.*, 5, 484.

Heffner, E. L., Lorenz, A. J., Jannink, J. L. and Sorrells, M. E. (2010), 'Plant breeding with genomic selection: gain per unit time and cost', *Crop Sci.*, 50, 1681–90.

Holland, J. B. (2007), 'Genetic architecture of complex traits in plants', *Curr. Opin. Plant Biol.*, 10, 156–61.

http://apps.fao.org. (accessed 22 December 2015)

http://faostat.fao.org/default.aspx (accessed 22 December 2015)

http://irri.org/our-work/research/genetic-diversity/international-rice-genebank (accessed 12 February 2016)

http://rapdb.dna.affrc.go.jp/ (accessed 8 January 2016)

http://rice.plantbiology.msu.edu/ (accessed 8 January 2016)

http://www.africarice.org/warda/genebank.asp (accessed 12 February 2016)

http://www.fao.org/docrep/006/y4751e/y4751e0b.htmref (accessed12 February 2016)

http://www.gramene.org/ (accessed 1 March 2016)

https://www/gene.affrc.go.jp/databases-core_collections_jr_en.php (accessed 12 February 2016)

Huang, X., Lu, T. and Han, B. (2013). 'Resequencing rice genomes: an emerging new era of rice genomics', *Trends in Genet.*, 29(4), 225–32.

Huang, X., Qian, Q., Liu, Z., Sun, H., He, S., Luo, D., Xia, G., Chu, C., Li, J. and Fu, X. (2009), 'Natural variation at the *DEP1* locus enhances grain yield in rice', *Nat. Genet.*, 41, 494–7.

Huang, X., Wei, X., Sang, T., Zhao, Q., Feng, Q., Zhao, Y., Li1, C., Zhu, C., Lu, T., Zhang, Z., Li, M., Fan, D., Guo, Y., Wang, A., Wang, L., Deng, L., Li1, W., Lu, Y., Weng, Q., Liu, K., Huang, T., Zhou, T., Jing, Y., Li, W., Lin, Z., Buckler, E. S., Qian, Q., Zhang, Q., Li, J. and Han, B. (2010), 'Genome-wide association studies of 14 agronomic traits in rice landraces', *Nat. Genet.*, 42(11), 961–9.

Huang, X., Zhao, Y., Wei, X., Li, C., Wang, A., Zhao, Q., Li, W., Guo, Y., Deng, L., Zhu, C., Fan, D., Lu, Y., Weng, Q., Liu, K., Zhou, T., Jing, Y., Si, L., Dong, G., Huang, T., Lu, T., Feng, Q., Qian, Q., Li, J. and Han, B. (2011), 'Genome-wide association study of flowering time and grain yield traits in a worldwide collection of rice germplasm', *Nat. Genet.*, 44(1), 32–41.

Hu, J., Xiao, C and He, Y. (2016), 'Recent progress on the genetics and molecular breeding of brown planthopper resistance in rice', *Rice*, 9(30).

Ikeda-Kawakatsu, K., Yasuno, N., Oikawa, T., Iida, S., Nagato, Y., Maekawa, M. and Kyozuka, J. (2009), 'Expression level of *ABERRANT PANICLE ORGANIZATION1* determines rice inflorescence form through control of cell proliferation in the meristem', *Plant Physiol.*, 150(2), 736–47.

International Rice Genome Sequencing Project (2005), 'The map-based sequence of the rice genome', *Nature*, 436, 793–800.

Jannik, J., Lorenz, A. J. and Iwata, H. (2010), 'Genomic selection in plant breeding: from theory to practice', *Brief. Funct. Genomics*, 9(2), 166–77.

Jena, K. K., Multani , D. S. and Khush, G. S. (1991), 'Monosomic alien addition lines of *Oryza australiensis* and alien gene transfer In Rice Genetics II', International Rice Research Institute, Philippines.

Jena, K. K. and Khush, G. S. (1990), 'Introgression of genes from *Oryza officinalis* Well ex Watt to cultivated rice, *O. sativa* L.', *Theor. Appl. Genet.*, 80, 737–45.

Khush, G. S. (1999), 'Green revolution: the way forward', *Nat. Rev. Genet.*, 2, 815–22.

Khush, G. S. (1997), 'Origin, dispersal, cultivation and variation of rice', *Plant Mol. Biol.*, 35, 25–34.

Khush, G. S., Coffman, W. R. and Beachell, H. M. (2001), 'The history of rice breeding: IRRI's contribution', in Rockwood, W. G. (Ed.), *Proceedings Of Rice Research And Production In The 21st Century*, Symposium honoring Robert, F. Chandler Jr., International Rice Research Institute, Philippines.

Konishi, S., Izawa, T., Lin, S. Y., Ebana, K., Fukuta, Y., Sasaki, T. and Yano, M. (2006), 'An SNP caused loss of seed shattering during rice domestication', *Science*, 312(5778), 1392–6.

Kurokawa, Y., Noda, T., Yamagata, Y., Angeles-Shim, R. B., Sunohara, H., Uehara, K., Furuta, T., Nagai, K., Jena, K. K., Yasui, H., Yoshimura, A., Ashikari, M. and Doi, K. (2016), 'Construction of versatile SNP array for pyramiding useful genes of rice', *Plant Sci.*, 242, 131–9.

Kwok, P. Y., Deng, Q., Zakeri, H., Taylor, S. L. and Nickerson, D. (1996), 'Increasing the information content of STS-based genome maps: identifying polymorphism in mapped STS', *Genomics*, 31, 123–6.

Li, H., Peng, Z., Yang, X., Wang, W., Fu, J., Wang, J., Han, Y, Chai, Y, Guo, T., Yang, N., Liu, J., Warburton, M. L., Cheng, Y., Hao, X., Zhang, P., Zhao, J., Liu, Y., Wang, G., Li, J. and Yan, J. (2013), 'Genome-wide association study dissects the genetic architecture of oil biosynthesis in maize kernels', *Nat. Genet.*, 45, 43–50.

Livak, K. J., Flood, S. J. A., Marmaro, J., Giusti, W. and Deetz, K. (1995), 'Oligonucleotides with fluorescent dyes at opposite ends provide a quenched probe system useful for detecting PCR product and nucleic acid hybridization', *Genome Res.*, 4(6), 357–62.

Llaca, V. (2012), 'Sequencing technologies and their use in plant biotechnology and breeding', in Munshi, A. (Ed.), *DNA Sequencing-Methods And Applications*. Intech, Croatia.

Lorenz, A. J., Chao, S., Asoro, F. G., Heffner, E. L., Hayashi, T. Iwata, H, Smith, K. P., Sorrells, M. E. and Jannink, J. (2011), 'Genomic selection in plant breeding: Knowledge and prospects', *Adv. Agr.*, 110, 77–123.

Mackill, D. J, Coffman, W. R. and Garrity, D. P. (1996), *Rainfed Lowland Rice Improvement*, International Rice Research Institute, Philippines.

Mackill, D. J., Amante, M. M., Vergara, B. S. and Sarkarung, S. (1993), 'Improved semi-dwarf lines with tolerance to submergence of seedlings', *Crop Sci.*, 33, 749–53.

Mackill, D. J., Ismail, A. M., Pamplona, A. M., Sanchez, D. L., Carandang, J. J. and Septiningsih, E. M. (2010), 'Stress-tolerant rice varieties for adaptation to a changing climate', *Crop Environ. Bioinformatics*, 7, 250–9.

Mackill, D. J., Zhang, Z., Redona, E. D. and Colowit, P. M. (1996), 'Level of polymorphism and genetic mapping of AFLP markers in rice', *Genome*, 39(5), 967–11.

McCouch, S. R., Kochert, G., Yun, Z. H., Wang, Z. Y., Khush, G. S., Coffman, W. R. and Tanksley, S. D. (1998), 'Molecular mapping of rice chromosomes', *Theor. Appl. Genet.*, 76, 815–29.

McCouch, S. R., Zhao, K., Wright, M., Tung, C. W., Ebana, K., Thomson, M., Reynolds, A., Wang, D., DeClerck, G., Ali, Md. L., McClung, A., Eizenga, G. and Bustamante, C. (2010), 'Development of genome-wide SNP assays for rice', *Breed. Sci.*, 60, 524–35.

McNally, K. L., Childs, K. L., Bohnert, R., Davidson, R. M., Zhao, K., Ulat, V. J., Zeller, G., Clark, R. M., Hoen, D. R., Bureau, T. E., Stokowski, R., Ballinger, D. G., Frazer, K. A., Cox, D. R., Padhukasahasram, B., Bustamante, C. D., Weigel, D., Mackill, D. J., Bruskiewich, R. M., Rätsch, G., Buell, C. R., Leung, H. and Leach, J. E. (2009), 'Genomewide SNP variation reveals relationships among landraces and modern varieties of rice', *Proc. Natl. Acad. Sci.*, 106, 2273–8.

Meuwissen, T. H. E., Hayes, B. J. and Goddard, M. E. (2001), 'Prediction of total genetic value using genome-wide dense marker maps', *Genetics*, 157, 1819–29.

Michael, T. P and Jackson, S. (2013), 'The first 50 plant genomes', *Plant Genome*, 6(2).

Miura, K., Ikeda, M., Matsubara, A., Song, X. J., Ito, M., Asano, K., Matsuoka, M., Kitano, H. and Ashikari, M. (2010), '*OsSPL14* promotes panicle branching and higher grain productivity in rice', *Nat. Genet.*, 42(6), 545–9.

Mohanty, H. K. and Khush, G. S. (1985), 'Diallel analysis of submergence tolerance in rice, *Oryza sativa* L.' *Theor. Appl. Genet.*, 70' 467–73.

Mohanty, H. K., Mallik, S. and Grover, A. (2000), 'Prospects for improving flooding tolerance in lowland rice varieties by conventional breeding and genetic engineering', *Curr. Sci.*, 78(2), 132–7

Moose, S. P. and Mumm, R. H. (2008), 'Molecular plant breeding as the foundation for 21st century crop improvement', *Plant Physiol.*, 147, 969–77.

Morris, G. P., Ramu, P., Deshpande, S. P., Hash, C. T., Shah, T., Upadhyaya, H. D., Riera-Lizarazu, O., Brown, P. J., Acharya, C. B., Mitchell, S. E., Harriman, J., Glaubitz, J. C., Buckler, E. S. and Kresovich, S. (2013), 'Population genomic and genome-wide association studies of agroclimatic traits in sorghum', *Proc. Natl. Acad. Sci.*, 110(2), 453–8.

Multani, D. S., Jena, K. K., Brar, D. S., De Los Reyes, B. G., Angeles, E. R. and Khush, G. S. (1994), 'Development of monosomic alien addition lines and introgression of genes from *Oryza australiensis* Domin. to cultivated rice *O. sativa* L.', *Theor. Appl. Genet.*, 88, 102–9.

Myint, K. K. M., Fujita, D., Matsumura, M., Sonoda, T., Yoshimura, A. and Yasui, H. (2012), 'Mapping and pyramiding of two major genes for resistance to the brown planthopper (*Nilaparvata lugens* [Stål]) in the rice cultivar ADR52', *Theor. Appl. Genet.*, 124(3), 495–504.

Nasu, S., Suzuki, J., Ohta, R., Hasegawa, K., Yui, R., Kitazawa, N., Monna, L. and Minobe, Y. (2002), 'Search for and analysis of single nucleotide polymorphisms (SNPS) in rice (*Oryza sativa, Oryza rufipogon*) and establishment of SNP markers', *DNA Res.*, 9(5), 163–71.

Neeraja, C. N., Maghirang-Rodriguez, R., Pamplona, A., Heuer, S., Collard, B. C. Y., Septiningsih, E. M., Vergara, G., Sanchez, D., Xu, K., Ismail, A. M and Mackill, D. J. (2007), 'A marker-assisted backcross approach for developing submergence-tolerant rice cultivars', *Theor. Appl. Genet.*, 115, 767–76.

Ookawa, T., Hobo, T., Yano, M., Murata, K., Ando, T., Miura, H., Asano. K, Ochiai, Y., Ikeda, M., Nishitani, R., Ebitani, T., Ozaki, H., Angeles, E. R., Hirasawa, T. and Matsuoka, M. (2010), 'New approach for rice improvement using a pleiotropic QTL gene for lodging resistance and yield', *Nat. Comm.*, 1, 132.

Ouyang, S., Zhu, W., Hamilton, J., Lin, H., Campbell, M., Childs, K., Thibaud-Nissen, F., Malek, R. L., Lee, Y., Zheng, L., Orvis, J., Haas, B., Wortman, J. and Buell, C. R. (2007), 'The TIGR Rice Genome Annotation Resource: improvements and new features', *Nucleic Acids Res.* 35 (Database Issue), D846–51.

Panaud, O., Chen, X. and McCouch, S. R. (1996), 'Development of microsatellite markers and characterization of simple sequence length polymorphism (SSLP) in rice (*Oryza sativa* L.)', *Mol. Gen. Genet.*, 252, 597–607.

Pasam, R. K., Sharma, R., Malosetti, M., Van Euwijk, F. A., Haseneyer, G., Kilian, B. and Graner, A. (2012), 'Genome-wide association studies for agronomical traits in a world spring barley collection', *BMC Plant Biol.*, 12, 16.

Paterson, A. H., Freeling, M. and Sasaki, T. (2005), 'Grains of knowledge: genomics of model cereals', *Genome Res.*, 15, 1643–50.

Peng, S., Khush, G. S., Virk, P., Tang, Q. and Zou, Y. (2008), 'Progress in ideotype breeding to increase yield potential', *Field Crops Res.*, 108(1), 32–8.

Perez-De Castro, A. M., Vilanova, S., Canizares, J., Pascual, L., Blanca, J. M., Diez, M. J., Prohens, J. and Pico, B. (2012), 'Application of genomic tools in plant breeding', *Curr. Genomics*, 13, 179–95.

Pingali, P. L. (2012), 'Green Revolution: Impacts, limits, and the path ahead', *Proc. Natl. Acad. Sci.*, 109(31), 12302–8.

Poland, J., Endelman, J., Dawson, J., Rutkoski, J., Wu, S., Manes, Y., Dreisigacker, S., Crossa, J., Sánchez-Villeda, H., Sorrells, M. and Jannink, J. L. (2012), 'Genomic selection in wheat breeding using genotyping-by-sequencing', *Plant Genome*, 5, 103–13.

Rao, K. K., Lakshminarasu, M. and Jena, K. K. (2002), 'DNA markers and marker-assisted breeding for durable resistance to bacterial blight disease in rice', *Biotechnol. Adv.*, 20, 33–47.

Ray, A., Deb, D., Ray, R. and Chattopadhayay, B. (2013), 'Phenotypic characters of rice landraces reveal independent lineages of short-grain aromatic indica rice', *AoB Plants*, 5, plt032.

Reif, J. C., Gowda, M., Maurer, H. P., Loning, C. F. H., Korzum, V., Ebmeyer, E., Bothe, R., Pietsch, C. and Wurschum, T. (2011), 'Association mapping for quality traits in soft winter wheat', *Theor. Appl. Genet.*, 122(5), 961–70.

Rhoads, A. and Au, K. F. (2015), 'PACBiosequencing and its applications', *Genomics, Proteomics, Bioinformatics*, 13, 278–89.

Ribaut, J. M., Jiang, C. and Hoisington, D. (2002), 'Simulation experiments on efficiencies of gene introgression by backcrossing', *Crop Sci.*, 42, 557–65.

Rutkoski, J. E., Heffner, E. L. and Sorrells, M. E. (2011), 'Genomic selection for durable stem rust resistance in wheat', *Euphytica*, 179, 161–73.

Saito, A., Yano, M., Kishimoto, N., Nakagahra, M., Yoshimura, A., Saito, K., Kuhara, S., Ukai, Y., Kawase, M., Nagamine, T., Yoshimura, S., Ideta, O., Ohsawa, R., Hayano, Y., Iwata, N. and Sigiura, M. (1991), 'Linkage map of restriction fragment length polymorphism loci in rice', *Japan J. Breed.*, 41, 665–70.

Schadt, E. E., Turner, S. and Kasarskis, A. (2010), 'A window into third generation sequencing', *Hum. Mol. Genet.*, 19(R2), R227–R240.

Seck, P. A., Diagne, A., Mohanty, S. and Wopereis, M. C. S. (2012), 'Crops that feed the world 7: Rice', *Food Security*, 4, 7–24.

Septiningsih, E. M., Hidayatun, N., Sanchez, D. L., Nugraha, Y., Carandang, J., Pamplona, A. M., Collard, B. C. Y., Ismail, A. M., and Mackill, D. J. (2015), 'Accelerating the development of new submergence tolerant rice varieties: the case of Ciherang-Sub1 and PSB RC18-Sub1', *Euphytica*, 202(2), 259–68.

Septiningsih, E. M., Pamplona, A. M., Sanchez, D. L., Neeraja, C. N., Vergara, G. V., Heuer, S., Ismail, A. M and Mackill, D. J. (2009), 'Development of submergence-tolerant rice cultivars: the *Sub1* locus and beyond', *Ann. Bot.*, 103, 151–60.

Shen, Y. J., Jiang, H., Jin, J. P., Zhang, Z. B., Xi, B., He, Y. Y., Wang, G., Wang, C., Qian, L., Li, X., Yu, Q. B., Liu, H. J., Chen, D. H., Gao, J. H., Huang, H., Shi, T. L. and Yang, Z. N. (2004), 'Development of genome-wide DNA polymorphism database for map-based cloning of rice genes', *Plant Physiol.*, 135, 1198–205.

Shim, R. A., Angeles, E. R., Ashikari, M. and Takashi, T. (2010), 'Development and evaluation of *Oryza glaberrima* Steud., chromosome segment substitution lines (CSSLs) in the background of *O. sativa* L. cv. Koshihikari', *Breed. Sci.*, 60, 613–19.

Shimamoto, K. and Kyozuka, J. (2002), 'Rice as a model comparative genomics of plants', *Annu. Rev. Plant Biol.*, 53, 339–419.

Shomura, A., Izawa, T., Ebana, K., Ebitani, T., Kanegae, H., Konishi, S. and Yano, M. (2008), 'Deletion in a gene associated with grain size increased yields during rice domestication', *Nat. Genet.*, 40, 1023–8.

Shull, G. H. (1909), 'A pure-line method in corn breeding', *Am. Breeders Assoc. Rep.*, 5, 51–9.

Smith, S. M. and Maughan, P. J. (2015), 'SNP genotyping using KASPar assays', *Plant Genotyping*, 1245, 243–55.

Song, X. J., Huang, W., Shi, M., Zhu, M. Z. and Lin, H. X. (2007), 'A QTL for rice grain width and weight encodes a previously unknown RING-type E3 ubiquitin ligase', *Nat. Genet.*, 39, 623–30.

Spindel, J., Begum, H., Akdemir, D., Virk, P., Collard, B., Redona, E., Atlin, G., Jannik, J. and McCouch, S. R. (2015), 'Genomic selection and association mapping in rice (*Oryza sativa*): Effect of trait genetic architecture, training population composition, marker number and statistical model on accuracy of rice genomic selection in elite, tropical rice breeding lines', *PLoS Genet.*, 11(2), e1004982.

Spindel, J., Wright, M., Chen, C., Cobb, J., Gage, J., Harrington, S., Lorieux, M., Ahmadi, N. and McCouch, S. (2013), 'Bridging the genotyping gap: using genotyping by sequencing (GBS) to add high-density SNP markers and new value to traditional bi-parental mapping and breeding populations', *Theor. Appl. Genet.*, 126(11), 2699–716.

Steemers, F. J and Gunderson, K. L. (2007), 'Whole genome genotyping technologies on the BeadArray™ platform', *Biotech. J.*, 2(1), 41–9.

Sukumaran, S., Dreisigacker, S., Lopes, M., Chavez, P. and Reynolds, M. P. (2015), 'Genome-wide association study for grain yield and related traits in an elite spring wheat population grown in temperate irrigated environments', *Theor. Appl. Genet.*, 128(2), 353–63.

Takai, T., Nonue, Y., Yamamoto, S., Yamanouchi, U., Matsubara, K., Liang, Z., Hin, L., Ono, N., Uga, Y. and Yano, M. (2007), 'Development of chromosome segment substitution lines derived from backcross between indica donor rice cultivar Nona Bokra and japonica recipient cultivar "Koshihikari"', *Breed. Sci.*, 57, 257–61.

Tanksley, S. (1993), 'Mapping polygenes', *Annu Rev. Genet.*, 27, 205–33.

Tester, M. and Langridge, P. (2010), 'Breeding technologies to increase crop production in a changing world', *Science*, 327, 818–22.

Thomson, M. J., Zhao, K., Wright, M., McNally, K. L., Rey, J., Tung, C. W., Reynolds, A., Scheffler, B., Eizenga, G., McClung, A., Kim, H., Ismail, A. M., De Ocampo, M., Mojica, C., Reveche, M. Y., Dilla-Ermita, C. J., Mauleon, R., Leung, H., Bustamante, C. and McCouch, S. R. (2012), 'High-throughput single nucleotide polymorphism genotyping for breeding applications in rice using BeadXpress platform', *Mol. Breed.*, 29, 875–86.

Tian, F., Bradbury, P. J., Brown, P. J., Hung, H., Sun, Q., Flint-Garcia, S., Rocheford, T. R., McMullen, M. D., Holland, J. B. and Buckler, E. S. (2011), 'Genome-wide association study of leaf architecture in the maize nested association mapping population', *Nat. Genet.*, 43, 159–62.

Uga, Y., Sugimoto, K., Ogawa, S., Rane, J., Ishitani, M., Hara, N., Kitomi, Y., Inukai, Y., Ono, K., Kanno, N, Inoue, H., Tekehisa, H., Motoyama, R., Nagamura, Y., Wu, J., Matsumoto, T., Takai, T., Okuno, K. and Yano, M. (2013), 'Control of root system architecture by *DEEPER ROOTING 1* increase rice yield under drought condition', *Nat. Genet.*, 45, 1097–102.

Varshney, R. K., Terauchi, R. and McCouch, S. (2014), 'Harvesting the promising fruits of genomics: applying genome sequencing technologies to crop breeding', *PLoS Biol.*, 12(6), 1001883.

Vergara, B. S. and Mazaredo, A. (1975), 'Screening for resistance to submergence under greenhouse conditions', In *Proceedings of the International Seminar on Deepwater Rice*, Bangladesh Rice Research Institute, Bangladesh.

Vergara, B. S., Jackson, B. and De Datta, S. K. (1976), 'Deepwater rice and its response to deep water stress', in *Climate and Rice*, International Rice Research Institute, Philippines.

Villa, T. C. C., Maxted, N., Scholten, M. and Ford-Lloyd, B. (2006), 'Defining and identifying crop landraces', *Plant Genet. Res.*, 3(3), 373–84.

Wing, R. A., Kim, H., Goicoechea, J. L., Yu, Y., Kudrna, D., Zuccolo, A., Ammiraju, J. S. S., Luo, M., Jinaxin, M., SanMiguel, P., Hurwit, B., Ware, D., Brar, D. S., Mackil, D. J., Soderlund, C., Stein, L. and Jackson, S. (2007), 'The *Oryza* map alignment project (OMAP): A new resource for comparative genome studies within *Oryza*', in Upadhyaya, N. M. (Ed.), *Rice Functional Genomics*, Springer, New York.

Wu, J., Yamagata, H., Hayashi-Tsugane, M., Hijishita, S., Fujisawa, M., Shibata, M., Ito, Y., Nakamura, M., Sakaguchi, M., Yoshihara, R., Kobayashi, H., Ito, K., Karasawa, W., Yamamoto, M., Saji, S., Katagiri, S., Kanamori, H., Namiki, N., Katayose, Y., Matsumoto, T. and Sasaki, T. (2004), 'Composition and structure of the centromeric region of rice chromosome 8', *Plant Cell*, 16(4), 967–76.

Würschum, T., Reif, J. C., Kraft, T., Janssen, G. and Zhao, Y. (2013), 'Genomic selection in sugar beet breeding populations', *BMC Genet.*, 14, 85–93.

www.irri.org (accessed 22 December 2015)

Xu, K., Xu, X., Ronald, P. C. and Mackil, D. J. (2000), 'A high-resolution linkage map in the vicinity of the rice submergence locus, *Sub1*', *Mol. Gen. Genet.*, 263, 681–9.

Xu, K. and Mackill, D. J. (1996), 'A major locus for submergence tolerance mapped on rice chromosome 9', *Mol. Breed.*, 2, 219–24.

Xu, K., Xu, X., Fukao, T., Canlas, P., Maghirang-Rodriguez, R., Heuer, S., Ismail, A. M, Bailey-Serres, J., Ronald, P. C. and Mackill, D. J. (2006), '*Sub1A* is an ethylene-response factor-like gene that confers submergence tolerance to rice', *Nature*, 442(10), 705–8.

Xu, S., Zhu, D. and Zhang, Q. (2014), 'Predicting hybrid performance in rice using genomic best linear unbiased prediction', *Proc. Natl. Acad. Soc.*, 111(34), 12456–61.

Xue, W., Weng, X., Zhao, Y., Tang, W., Zhou, H., Yu, S., Xu, C., Li, X. and Zhang, Q. (2008), 'Natural variation in *Ghd7* is an important regulator of heading date and yield potential in rice', *Nat. Genet.*, 40(6), 761–7.

Yamamoto, E., Matsunaga, H., Onogi, A., Ohyama, A., Miyatake, K., Yamaguchi, H., Iwata, H. And Fukuoka, H. (2017), 'Efficiency of genomic selection for breeding population design and phenotype prediction in tomato', *Heredity*, (118), 202-209.

Yamamoto, E., Matsunaga, H., Onogi, A., Kajiya-Kanegae, H., Minamikawa, M., Suzuki, A., Shirasawa, K., Hirakawa, H., Nunome, T., Yamaguchi, H., Miyatake, K., Ohyama, A., Iwata, H and Fukuoka, H. (2016), 'A simulation-based breeding design that uses whole-genome prediction in tomato', *Sci Reports*, 6: doi:10.1038/srep19454.

Yano, M., Katayose, Y., Ashikari, M., Yamabouchi, U., Monna, L., Fuse, T., Baba, T., Yamamoto, K., Umehara, Y., Nagamura, Y. and Sasaki, T. (2000), '*Hd1*, a major photoperiod sensitivity quantitative trait locus in rice, is closely related to the *Arabidopsis* flowering time gene *CONSTANS*', *Plant Cell*, 12(12), 2473–84.

Yano, K., Yamamoto, E., Aya, K., Takeuchi, H., Lo, P., Hu, L., Yamasaki, M., Yoshida, S., Kitano, H., Hirano, K. and Matsuoka, M. (2016), 'Genome-wide association study using whole-genome sequencing rapidly identifies new genes influencing agronomic traits in rice', *Nat. Genet.* doi:10.1038/ng.3596.

Yasuda, N., Mitsunaga, T., Hayashi, K., Koizumi, S., Fujita, Y. and Kanagawa, F. (2015), 'Effects of pyramiding quantitative resistance genes *pi21*, *Pi34*, and *Pi35* on rice leaf blast disease', *Plant Dis.*, 99(7), 904–9.

Yu, H., Xie, W., Wang, J., Xing, Y., Xu, C., Li, X., Xiao, J. and Zhang, Q. (2011), 'Gains in QTL detection using an ultra-high density SNP map based on population sequencing relative to traditional RFLP/SSR markers', *PloS ONE*, 6(3), e17595.

Yu, J., Hu, S., Wang, J., Wong, G. K., Li, S., Liu, B., et al. (2002), 'A draft sequence of the rice genome (*Oryza sativa* L. ssp. *indica*)', *Science*, 296(5565), 79–92.

Yuan, Q., Ouyang, S., Liu, J., Suh, B., Cheung, F., Sultana, R., Lee, D., Quackenbush, J. and Buell, C. R. (2003), 'The TIGR rice genome annotation resource: annotating the rice genome and creating resources for plant biologists', *Nucleic Acid Res.*, 31(1), 229–33.

Zhang, Y., Huang, Y., Zhang, L., Li, Y., Lu, T., Lu, Y., Feng, Q., Zhao, Q., Cheng, Z., Xue, Y., Wing, R. A. and Han, B. (2004), 'Structural features of the rice chromosome 4 centromere', *Nucleic Acids Res.*, 32(6), 2023–30.

Zhao, K., Tung, C. W., Eizenga, G. C., Wright, M. H., Ali, M. L., Price, A. H., Norton, G. J., Islam, M. R., Reynolds, R., Mezey, J., McClung, A. M., Bustamante, C. D. and McCouch, S. R. (2011), 'Genome-wide association mapping reveals a rich genetic architecture of complex traits in *Oryza sativa* L.', *Nat. Comm.*, 2, 467.

Zhong, S., Dekkers, J. C. M., Fernando, R. L. and Jannink, J. L. (2009), 'Factors affecting accuracy from genomic selection in populations derived from multiple inbred lines: a barley case study', *Genetics*, 182, 355–64.

Zhu, C., Gore, M., Buckler, E. S. and Yu, J. (2008), 'Status and prospects of association mapping in plants', *Plant Genome*, 1, 5–20.

Chapter 3

Breeding strategies to improve rice yields: an overview

K. K. Jena and G. Ramkumar, International Rice Research Institute, The Philippines

1 Introduction

Cereals are a major food source for most of the human population as measured in calories (Xu et al. 2005), and account for 173 kg/person/year of food intake in developing countries. They provide more than 50% of the calories consumed directly by the human populations and make up 44% of animal feed as an indirect source of energy (http://www.fao.org/docrep/006/Y4683E/y4683e06.htm#TopOfPage). Among the cereals, rice is one of the most important crops that feeds more than half of the world population (Khush 1997). Rice production is also a major source of livelihood as it is one of the most valuable and economically viable crops in Asia. It is planted on about 154 million hectares annually, which is equal to 11% of the world's cultivable land (Khush 2005). Rice is not only an economically valuable crop, but also an important model cereal for understanding crop

http://dx.doi.org/10.19103/AS.2016.0003.20

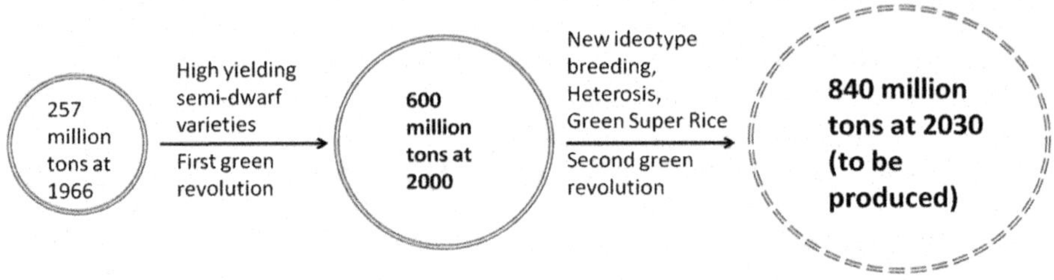

Figure 1 Rice production quantity in the past and requirement in future.

science on account of its relatively small genome size (~430Mb), availability of high-quality genome sequences, annotated genes and their mRNA, and protein products. Available rice genome information (DNA sequences and expression data) is used as a reference to improve other cereals such as barley, maize and sorghum (Xu et al. 2005; Cai et al. 2012).

In the early 1960s, the world desperately needed to increase grain yield to cope up with a steep acceleration in population growth and to maintain the balance between food production and population growth. Fortunately, by introducing the semi-dwarf1 (*sd1*) gene into a traditional indica variety, IR8 was developed by the International Rice Research Institute (IRRI), which made possible the 'green revolution' (Khush 1999, 2005) in which food production was more than doubled, while the population of low-income countries grew by 80% (Khush 1999). Due to an efficient adoption strategy, many social and economic upheavals were avoided and food security ensured. The first green revolution gene, *Sd1*, codes for gibberellin (plant growth hormone) biosynthesis, and the plants harbouring *sd1* have gibberellin deficiency, which results in reduced plant height (Spielmeyer et al. 2002; Monna et al. 2002a), which leads in turn to reduced lodging. Replacement of traditional varieties with high-yielding cultivars, coupled with farm management practices has led to (un-milled) rice production of 5 to 7 tons per hectare in comparison with traditional varieties producing 1 to 3 t/h (Khush 1999). However, at present, the increase in rice production rate (~1.0% per year) is lower than the increase in population of 1.5% per year (Jeon et al. 2011). It is estimated that 23% more rice must be produced by 2035 (Khush and Jena 2009). If the present trend of rice production is not increased, the world may face a large-scale food deficit. Moreover, the available land for agriculture may dwindle, as land is being utilized for industrialization and other purposes; water sources are being depleted and the working labour force is shifting to urban areas. Increasing rice yield per hectare with limited land, water and labour is the primary and most important task of agricultural scientists. Hence, a second green revolution in rice is imperative to increase yield to maintain the balance between food quantity and population size, especially in the developing countries (Fig. 1).

2 Modern breeding techniques to increase rice yield: ideotype breeding, heterosis and green super rice

Irrespective of the progress made in rice production, 3 billion people are affected by micronutrient deficiencies; deaths due to chronic hunger is ~2,400 lives every day (Khush 2005). Hence, it is necessary to deploy different and novel strategies to improve the grain yield.

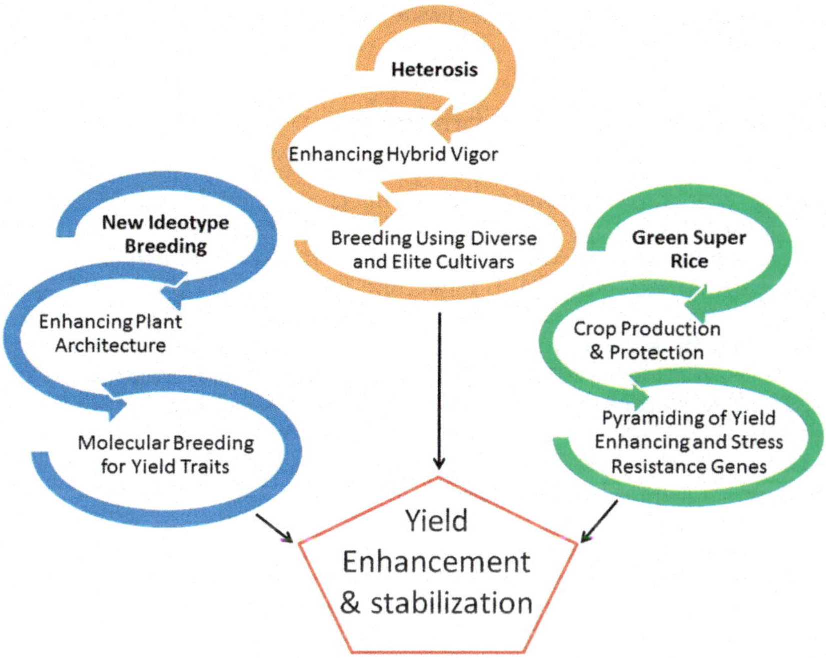

Figure 2 Key molecular breeding strategies to increase rice crop yield.

This chapter will discuss three major strategies that have a higher potential to improve crop yields than conventional techniques (Fig. 2).

2.1 Ideotype breeding

Some evidence indicates that yield gain can be significantly enhanced by improving yield component traits and fine-tuning the plant architecture, such as plant height, productive tiller number and main culm diameter/strength. Complex agronomic traits such as yield can be influenced by modifying simple phenotypic traits. Although the modern rice varieties have a high-yield capacity, the yield can be enhanced further by incorporating yield-enhancing genes such as *SPL14* for primary and secondary branching of the panicle and panicle length; *SCM2* for increasing stem diameter for lodging resistance (Ookawa et al. 2010); *Gn1a*, *DEP1* and *DEP3* genes for grain number and panicle size; and so on. By incorporating a particular gene or combination of genes, scientists can achieve the desired plant architecture and hence maximize the yield potential. This can be realized by applying conventional breeding along with modern molecular biology tools such as gene-specific DNA markers.

Desirable plant architecture may have the following characteristics: 350 to 500 grains per panicle, dense and lengthy panicles, each tiller should have a panicle (~10 panicles/tillers per plant), a strong culm to avoid lodging, dark green and erect leaves for photosynthetic efficiency, a vigorous root system, growth duration of 90 to 110 days and a harvest index of 0.5 to 0.6. The development of plants combining yield-enhancing traits using a genomic approach through molecular markers and wide hybridization methods will be necessary to achieve the desired rice production.

2.2 Heterosis in hybrid rice breeding

In general, the F_1 hybrid seeds show superiority over both their parents, owing to dominance or over dominance and epigenetic effects (Hochholdinger and Hoecker 2007). The yield can be achieved by increasing heterosis in hybrid rice breeding (Zhou et al. 2012), which is popular in sorghum, wheat, pearl millet and recently in rice as well. Sufficient molecular markers were reported to three- and two-line hybridization methods, which can be exploited to produce F_1 hybrid seeds (Rajendran et al. 2007; Rajendrakumar et al. 2007). It may be possible to increase heterosis if more diverse parents are crossed (Breseghello Flavio Coelho 2013). Hybrids developed from indica and japonica crosses showed greater yield potential than intra-sub-species crosses (Gao et al. 2005). However, in those sub-species inter-crosses, partial hybrid sterility is often observed (Kato et al. 1928). This problem can be tackled by using wide compatible varieties (discussed in Section 3), which can produce the seeds normally, without any hybrid sterility (Ikehashi and Araki 1984).

There are a few challenges associated with hybrid seed production: (1). Hybrid vigour will be reduced in future generations and, hence, fresh F_1 seeds have to be produced every season to maintain the high level of heterosis. (2). Hybrid seeds are usually produced by a two- or three-line hybrid system, which means that two or three separate populations have to be maintained to produce hybrid seeds. (3). If the desired trait is governed by a recessive gene or semi-dominant gene, that gene should be incorporated in both the parents to get proper function/expression of the gene. However, if dominant genes are selected which can function in heterozygous status in the same way as in homozygous status, it is enough to incorporate them in either of the parents. Despite the abovementioned challenges, achieving high yield through heterosis can help to overcome the cost associated with crossings and other practices to produce hybrid seeds.

2.3 Green super rice

Due to advancements in genetics and genomics, hundreds of genes have been identified and isolated for yield and resistance to biotic and abiotic stresses. On the other hand, most consumer-preferred rice varieties are either not high yielding or are susceptible to biotic and abiotic stresses. It is possible to avert this problem by incorporating the genes for yield, resistance to biotic and abiotic stresses into elite rice varieties (Fig. 3). Although the rice yield can be enhanced through various difficult strategies, the yield should be well protected from diseases and pests and other abiotic stresses. The concept of green super rice is that rice should give higher yield as well as showing resistance to biotic stresses such as bacterial blight, rice blast and insect pests such as brown planthopper (BPH) and abiotic stresses such as drought and salinity. Furthermore, green super rice should show high nutrient efficiency, fertilizer use efficiency, and pesticides and water use efficiency. Many reviews have reported green super rice development strategy and probable genes to achieve green super rice (Zhang 2007, Li chapter 5 of this book). The major challenge in creating green super rice is incorporating many favourable genes in a single elite cultivar. However, this can be managed by producing hybrid plants from each

Figure 3 Functional genes of rice to increase production.

parent having different sets of favourable genes, which ensures proper functioning and expression of targeted genes.

3 Genetic diversity and utilization of wild *Oryza* species

The *Oryza* genus contains two cultivated (*O. sativa* and *O. glaberrima*) and 22 different wild species (Jena 2010). *O. sativa* is grown throughout the world, while *O. glaberrima* is grown in specific regions of West Africa. *O. sativa* has two sub-species, indica and japonica, which diverged 0.44 million years ago (McCouch et al. 2007). The sub-species of *O. sativa* were domesticated from *O. rufipogon*, a perennial species; however, there is another view that *O. sativa* might have been domesticated from *O. nivara*, an annual species (Kovach et al. 2007; Sweeney and McCouch 2007). Because of restricted crossings (due to incompatibility) within the sub-population, the allelic diversity is very much limited and the beneficial alleles were trapped within sub-species groups. It is estimated that the modern cultivars have retained only 10 to 20% of the genome diversity that was present in the progenitor species *O. rufipogon* (Kovach and McCouch 2008). Traditional breeders have been trying several combinations of crosses within *O. sativa* sub-populations to increase the crop yield. However, since the last decade, the yield increase has been nearly stagnant and fails to match the level of human population growth. Hence, to break the yield barrier, it is necessary to enhance the genetic base of the available genetic diversity. Increase in genetic variation in the population for enhancing the crop yield will lead to increased genetic gain (Rutkoski et al. 2015). The presence of rare alleles of wild *Oryza* species, which were left behind during evolution and manual selection will also lead to enhanced genetic gain in the breeding programmes targeting crop yield. In order to bring back the novel and superior yield-related alleles from wild *Oryza* species, exploration of wild *Oryza* species is crucial. The wild ancestors of our cultivated rice species are valuable reservoirs of yield-related genes and traits.

Crosses between high-yielding elite cultivars and low-yielding wild *Oryza* species often lead to superior offspring, which indicates that wild cultivars contribute beneficial alleles for yield and yield component traits (Kovach and McCouch 2008). Phenotypic evaluation may not correlate with breeding value in the case of wild *Oryza* accessions, as they may look phenotypically inferior, which may be due to a complex ancestral genome. Crosses involving wild *Oryza* accessions usually lead to transgressive segregations, due to the

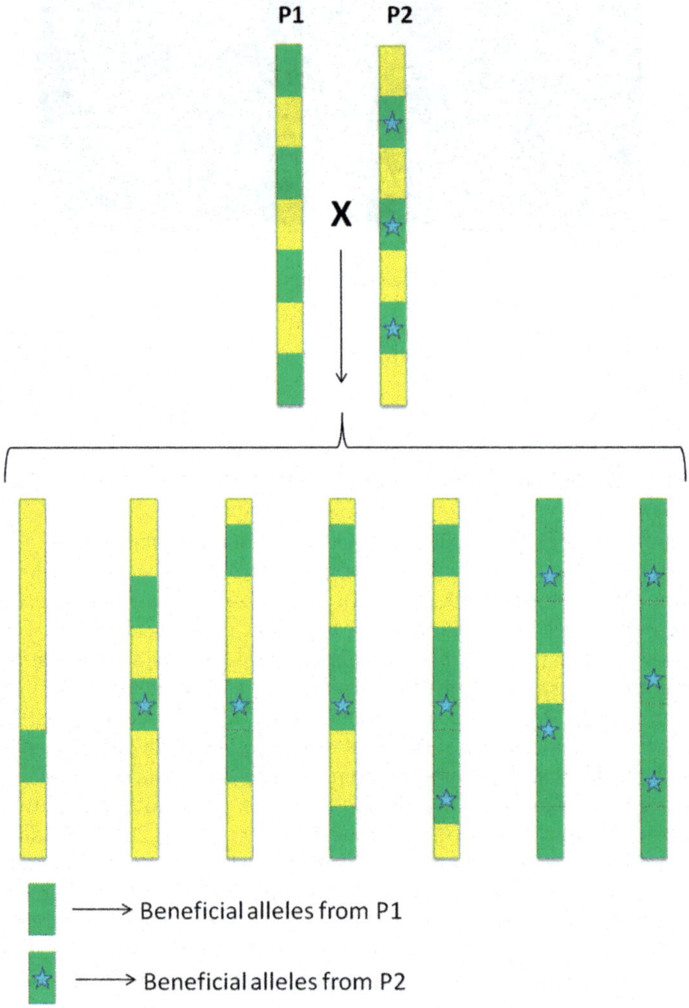

Figure 4 Selection of superior alleles of wild *Oryza* species through transgressive segregations: P1 – elite *O. sativa*; P2 – wild *Oryza* species.

introgression of 'wild' alleles in the elite rice genome background (Fig. 4). The idea that wild *Oryza* species possess novel and beneficial alleles has been proven in many studies. Yoon et al. (2006) clearly demonstrated that the wild *Oryza* species *O. grandiglumis* ($2n = 48$; CCDD genome) harbours beneficial alleles. The study used *O. grandiglumis* as donor, with *O. sativa* japonica cultivar Hwaseongbyeo as recipient parent. In the advanced backcross progenies they found 39 quantitative trait loci (QTLs), and among them, 18 (46.2%) QTLs were from *O. grandiglumis* that contributed to a desirable phenotype. Although *O. grandiglumis* had alleles that had a negative effect on a few traits, it also had positive alleles which led to preferable target traits. For example, two QTLs (gl6 and gl11) were found for grain length; gl6 increased, while the gl11 decreased the grain length. This study identified at least one positive QTL for many yield-related traits, such as days to heading, culm length, panicle length, panicle number, spikelets per panicle, grain thickness, thousand-grain weight and grain length/width ratio. In another similar study, despite the late heading date and maturity of *O. rufipogon* in comparison with *O. sativa* indica elite variety, IR64, four *O. rufipogon*-derived QTLs were identified to be associated with early heading date and maturity (Septiningsih et al. 2003). Moreover, regardless of the inferior performance of *O. rufipogon*, 33% of the total 42 QTLs identified were associated with a desirable phenotypic effect on yield and yield parameters such as panicle length, panicles per plant, percentage of seed set, grains per plant, grain weight and yield per plant (Septiningsih et al. 2003). Xie et al. (2006) identified different beneficial alleles from *O. rufipogon*, which increased grain length, grain width and grain thickness, even though the *O. rufipogon* had an inferior phenotype in the targeted grain size traits. Yuan et al. (2009) also identified favourable alleles from *O. rufipogon* accession for panicle length, panicle number, secondary branches, spikelets per panicle, days to heading and thousand-grain weight.

Wild *Oryza* species are reservoirs of beneficial alleles not only for yield traits but also for biotic and abiotic stress resistance. For instance, one of the major blast resistance genes, *Pi40*, was isolated from a wild *Oryza* species, *O. australiensis*, which belongs to the EE genome (Jeung et al. 2007). Another blast resistance gene, *Pi9*, was also identified and characterized from wild *Oryza* species, *O. minuta* (Qu et al. 2006). A bacterial blight resistance gene, *Xa23*, was identified from *O. rufipogon* (Wang et al. 2014). A bacterial blight broad-spectrum resistance gene, *Xa33*, was fine mapped from an advanced backcross line derived from *O. nivara* (Kumar et al. 2012). Many insect resistance genes were also isolated and characterized from wild *Oryza* species. One of the major BPH resistance genes, *BPH18*, was fine mapped and cloned from an introgression line derived from *O. australiensis* (Jena et al. 2006; Ji et al. 2016). A BPH resistance gene, *BPH10*, was also fine mapped from *O. australiensis*, while *BPH11*, *BPH12*, *BPH13*, *BPH14* and *BPH15* were fine mapped from *O. officinalis* and *BPH12* was fine mapped from *O. latifolia*. *BPH20* and *BPH21* were fine mapped from *O. minuta* (Ishii et al. 1994; Renganayaki et al. 2002; Ren et al. 2004; Yang et al. 2002; Rahman et al. 2009). Many beneficial alleles of different wild *Oryza* species and their introgression in elite cultivars were discussed in a related review (Brar and Khush 1997). Hence, these studies strongly suggest that wild *Oryza* species are important sources for many beneficial alleles, which can be used to improve yield-related traits as well as for other traits. Identification of the untapped beneficial genes from wild *Oryza* relatives will enhance yield significantly, which may lead to a second green revolution.

However, there are a few obstacles for using the wild *Oryza* species or distant relative crops in normal breeding programmes. The major problem associated with wild *Oryza* breeding is hybrid sterility. Many reasons for hybrid sterility were discussed elsewhere (Ouyang et al. 2010). However, developments in molecular breeding approaches, identification of wide compatibility (WC) alleles and established embryo rescue techniques assist the success of wild *Oryza* breeding programmes. The *O. glaberrima* is a rich resource for beneficial alleles, which could be transferred to *O. sativa* (Deng et al. 2010). Gene *S1* is responsible for a reproductive barrier that inhibits the cross of *O. sativa* with its other AA genome relatives, such as *O. longistaminata* and *O. glaberrima*. However, there are ways to overcome this genetic barrier; by incorporating WC genes, such as the *S1-g* allele from *O. glaberrima*, the *O. sativa* can be converted to a more crossable (wide compatible) variety and can act as a genetic bridge to cross with *O. glaberrima* (Deng et al. 2010). In the case of indica and japonica intra-sub-species crosses, the deleterious allelic effect is referred to as the Dobzhansky–Muller model, responsible for hybrid sterility, and is controlled by more than 28 genes (Sundaram et al. 2010). Among them, a major WC gene, *S5*, has been cloned through map-based cloning strategy (Chen et al. 2008). Cultivars possessing *S5* allele could produce hybrid seeds with normal fertility and these cultivars are called wide compatible varieties (Chen et al. 2008). Another gene, *Sa*, responsible for male sterility in indica–japonica crosses, was also characterized at molecular level (Long et al. 2008), which helps to overcome the hybrid sterility problems. Cultivars that possess these WC alleles can be selected and utilized in inter-sub-population (indica x japonica) or inter-*Oryza* species (e.g. *O. sativa* x *O. rufipogon*) crosses. These WC alleles will be useful for transferring or bringing back the beneficial alleles left behind. Linked simple sequence repeat (SSR) markers and gene-specific functional markers have already been reported for effective use of these WC genes, such as *S5* (Sundaram et al. 2010; Singh et al. 2006). The applicability of the *S5* allele and the corresponding gene-specific marker were validated in 150 different indica x japonica crosses and revealed the high utility of the allele (Revathi 2015). Other techniques such as embryo rescue are also helpful to achieve the successful hybridization and transfer of genes from wild *Oryza* species to *O. sativa* (Sharma et al. 1996).

4 The use of Multi-parent Advanced Generation Inter-Cross (MAGIC) populations in breeding

Multi-parent Advanced Generation Inter-Cross (MAGIC) is a novel strategy to increase genetic variation and diversity through shuffling a wide range of genomes to promote identification of beneficial alleles. Bandillo et al. (2013) developed four different kind of MAGIC populations that include indica MAGIC (crosses among 8 different indica parent), MAGIC plus (8 indica parent with two additional rounds of eight-way F_1 inter-crossing), japonica MAGIC (crosses among 8 *japonica* parent) and Global MAGIC (crosses among 8 indica and 8 japonica parents). As the parents of MAGIC populations showed resistance to many abiotic and biotic stresses, the study claimed that individuals of the population will have many favourable alleles (Bandillo et al. 2013). The strategy and parameters to be considered in preparation of MAGIC population and its applications were discussed in detail in a recent review (Huang et al. 2015). This MAGIC population serves as a permanent mapping population for varietal improvement or development breeding programmes. Unlike natural germplasm with genomic variation, these populations are artificially tailor-made with maximum beneficial alleles for breeders. Hence, these materials may have many novel genes or superior alleles of known and unknown genes. Following the same strategy, a wild MAGIC concept was also employed in which cultivated species (indica and japonica) were crossed with wild *Oryza* species. This may bring out many beneficial genes/alleles that were untapped during evolution (Shakiba and Eizenga 2014). Care was taken while inter-crossing and selecting progenies; the plants with beneficial genes were selected using phenotypic as well as genotypic analysis. The pool of beneficial alleles from the novel MAGIC strategy may prove immensely helpful to protect and enhance the rice crop yield.

5 Molecular marker techniques

Crop yield can be increased by utilizing beneficial alleles and eliminating unfavourable negative genes. Molecular markers play a significant role for this purpose and for transferring the beneficial genes with precision from any related or distant (wild) species, without inefficiency in time, effort and cost. Markers are reliable and free from environmental influence, in comparison with phenotype-based plant evaluation.

SSRs are very useful markers due to their high information on different alleles, relatively high abundance in genome and co-dominance pattern. Several thousands of SSR markers are reported in the rice genome and have been used in many marker-assisted breeding programmes as well as diversity analysis studies (McCouch et al. 2002; Kalia et al. 2011). Single-nucleotide polymorphism (SNP) is a recently developed marker concept targeting single-nucleotide changes in the genome. Due to its applicability in different modern molecular marker applications, the SNP marker is rapidly replacing the SSR markers in the molecular breeding research field. SNP is preferred due to its greater abundance than SSR, di-allelic pattern that will provide flexibility to adapt for modern techniques such as bead expression (Thomson et al. 2012) and easier adaptation for high-throughput technologies (Thomson 2014). Next-generation sequence technology, which is becoming cheaper, provides a huge amount of SNP data. More than a million SNPs are reported for rice genomes, which are available publically (http://oryzasnp.org/iric-portal/). Using the available SNP data, many fixed array SNP platforms have been developed such as BeadArray (Steemers and Gunderson 2007), GoldenGate (Lin et al. 2009) and GeneChip Array (Matsuzaki et al. 2004), which can be applied for whole-genome analysis/selection. Recently, Infinium 6K chips were developed, which reduced cost per sample significantly and can be applied to high-density SNP-based genome-wide analysis. Recent developments in marker field, such as SNP chips, DarT and genotype by sequencing technologies, facilitate precise transfer of the genes with greater efficiency in less time (Boualaphanh et al. 2011). Application of SNP markers and its recent developments were reviewed in detail by Thomson et al. (2012) and Thomson (2014).

Although general markers are important for whole-genome analysis, diversity analysis and many other applications, gene-specific markers targeting a candidate gene or set of genes are also very important in any breeding programme. Gene-specific markers are applied to track the favourable alleles in segregating the population, and to identify the favourable alleles in a given set of diverse plants (Wang et al. 2008a; Ramkumar et al. 2015; Singh et al. 2015). Among the gene-specific markers, functional markers are preferred over linked or tightly linked markers, as they are located within the gene or targeting the functional polymorphic site of the gene, which co-segregate perfectly with the targeted allele without any re-combination between the allele and marker. Numbers of genes that influence crop yield have been reported with their respective gene-specific markers. Crop yield can be enhanced significantly by utilizing the yield genes through application of the molecular markers in any well-designed breeding programmes. This chapter further discusses the utility of agronomic trait-related genes and application of molecular markers in breeding programmes with special reference to crop yield.

6 Understanding and enhancing yield traits

6.1 Grain size and grain weight

Understanding yield and yield-related traits may lead to the development of better strategies to increase crop yield. In general, rice yield is governed by three major traits:

- grain size and weight
- number of grains per panicle
- number of panicles per plant

Grain weight is highly dependent on grain width, length and filling rate. The grains per panicle depend on the number of spikelets in the panicle and fertility percentage, panicle branching, and panicle length and density. The panicle number depends on the plant's ability to produce productive primary and secondary tillers. These yield traits are influenced by many QTLs, which are usually complex in nature. However, modern molecular concepts facilitate greater understanding of the nature and mechanism of the yield QTLs. Moreover, available rice cultivars, genome sequences and advanced molecular breeding technology such as SNP chips (which cover the entire genome) and efficient breeding programmes make it possible to identify the genes underpinning QTLs.

Grain weight, which influences grain size, is a major determining factor for crop yield. As it positively correlates with grain width, length, width and length ratio, it is considered an important quality trait as well. Hence, grain weight and size are the more focused-upon breeding traits. Many genes have been cloned or fine mapped which influence grain size. By incorporating those appropriate genes in elite recipient lines, it is possible to achieve the desired grain length or width. One of the major genes that influence the grain length is *GS3*, which was cloned and characterized at molecular level by Fan et al. (2006). The QTL were located on the peri-centromeric region of rice chromosome 3, which explained 80 to 90% of the grain length and weight variations, and acts as a minor QTL for grain width as well (Fan et al. 2006). The gene *GS3* consists of five exons and codes for 232 amino acids; this gene regulates grain length and weight negatively, as all wild *Oryza* species favour small grains for better dispersal (Fan et al. 2006). The gene's second exon has functional SNP, which differentiate the long and short grain type. The dominant *GS3* allele codes for complete protein, which leads to short grain, while the recessive allele codes for prematured protein, due to adenine replacement with cytosine at the second exon, leading to long grain (Fan et al. 2006). Targeting the functional polymorphism, a CAPS marker (SF28) was reported (Fan et al. 2009), which was also converted into a PCR-based marker (Ramkumar et al. 2010a). The marker could identify all the analysed long grain Basmati rice varieties and differentiated them from short grain types (Ramkumar et al. 2010a).

Wang et al. (2012a) clearly demonstrated the use of genes and the marker-assisted selection process to modify width and length of the grain. They pyramided two genes (*GS3* – influencing grain length; *GW8* – influencing grain width) in different combinations of genes through a marker-assisted selection process in the background of Basmati385. The NIL-(*GW8+gs3*) produced longer and heavier grains, while the NIL-(*gw8+gs3*) produced narrower and more slender grains than the former one. This study explains that the traits can be altered using desired genes and respective molecular markers. In another study, the gene *gs3* was used for pyramiding along with *GW7* through marker-assisted selection, which produced evidence of significant yield increase (Wang et al. 2015a).

Another major gene for grain size is *GS5*, which also influences grain filling and weight (Li et al. 2011a). In contrast to *GS3*, this gene is a positive regulator for grain size, and higher expression of this gene leads to increased grain size. It was clearly demonstrated that the observed effect was due to the variation in the promoter and not due to the variation in the coding region of the gene (Li et al. 2011a). The nucleotide sequence of the promoter region (~2 Kb) of the *GS5* gene revealed three different haplotype groups, including narrow grain, medium-size grain and wide grain types. We may use these alleles to 'curve' the grain shape and weight, based on the consumer need. Zong et al. (2012) pyramided eight different QTLs (including *GS5* and *sd1*), responsible for thousand-grain weight, panicle size and spikelet number. The pyramided F3 and F4 plants, using a set of SSR and InDel markers and NGS data, showed better yield than the elite parents. The study revealed that 1212 bp deletion at the *GW5* allele was associated with the increased grain width and weight. The polymorphism and phenotype association was confirmed through screening 46 randomly selected different rice cultivars using InDel markers (Weng et al. 2008). The QTL, GL3.1, encodes a protein phosphatase kelch (PPKL) family – Ser/Thr phosphatase – which controls the grain length. It accelerates the cell division in the spikelet through protein phosphorylation and hence leads to increased length and thereby enhanced yield.

Grain width is another major trait that influences the grain weight and overall shape and size of the grain. *GW2* is a major gene, located on chromosome 2, which codes for a new RING-type protein with intrinsic E3 ubiquitin ligase activity. This gene negatively regulates grain width, weight and hence the yield, that is, loss of gene (or reduced expression of the gene) leads to wider grain width (Song et al. 2007). The mutated *GW2* allele, which has truncated protein with 310 amino acid deletion, increases the number of cells and accelerates grain milk filling rate and hence leads to increased grain width. A molecular marker was reported, targeting the functional single base deletion at the fourth exonic region of the gene (TD70 allele), which leads to truncated premature protein (Ya-dong et al. 2015). The premature termination of protein leads to active cell division and hence increased grain width and weight (negative regulation). This polymorphism provides opportunities for marker-assisted breeding targeting a favourable *GW2* allele. The marker clearly differentiates the large and small grains and co-segregates with the trait in the RIL population. The authors found that 82 lines with the $GW2^{TD70}$ allele had significantly larger grains than the lines with $GW2^{Kasalath}$ allele (Ya-dong et al. 2015). Ying et al. (2012) pyramided *GW2* and *GW5* using SSR and InDel markers, in which the grain width was better than the plants, with either of the individual genes (*GW2* or *GW5*). The work showed that the genes have independent pathways that affect the grain width.

GW6a is also a positive regulator of the trait, grain weight and yield, located on chromosome 6 and encodes for a new-type GNAT-like protein. Higher expression of the gene leads to enlarged spikelet hulls through increased cell number and accelerated

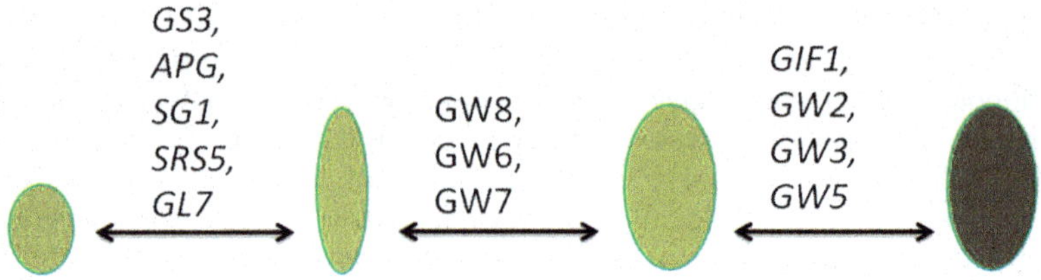

Figure 5 Role of various genes for grain length, grain width, grain filling and grain weight to increase rice yield . Many grain size-related genes have pleiotropic effect on grain width and weight.

grain milk filling (Song et al. 2015). *GW5*, located in chromosome 5, plays a significant role in controlling grain width, weight and length through involvement in the ubiquitin–proteasome pathway (Weng et al. 2008). This gene negatively regulates the trait; the absence or loss of gene function leads to increased grain width. *GW7* gene is located on chromosome 7 and encodes a TONNEAU1-recruiting motif protein; increased expression of this gene leads to slender rice grain by enhancing longitudinal cell division but decreased transverse cell division in transverse. Another gene, *GW8*, similar to OsSPL16, an SBP-domain transcription factor, promotes grain width. *GW8* binds to the promoter region of *GW7* and controls its expression and hence leads to increased grain width. By employing these genes in proper combinations, we may be able to achieve the enhanced yield with desired grain size (Fig. 5).

Thousand-grain weight 6 (*TGW6*) gene was characterized from the Indian landrace cultivar Kasalath. This gene contains only one exon that encodes a novel protein with indole-3-acetic acid (IAA)-glucose hydrolase activity. This gene expression is negatively correlated with grain weight; loss of the gene or lower expression of the gene leads to increased grain weight. One base pair deletion at nucleotide 313 causes loss of the Kasalath allele and leads to frameshift mutation and functional loss of the protein; this leads in turn to increased thousand-grain weight (Ishimaru et al. 2013). Many other genes were reported to be related to grain length, width and weight, which are listed in Supplementary Table 1.

6.2 Grain number per panicle

Grain number per panicle is a major agronomic trait, which influences the crop yield significantly. Ashikari et al. (2005) reported the *Gn1* QTL, located on chromosome 1, which influences the grain number up to 44% in segregating population derived from a cross between the indica variety Habataki and the japonica variety Koshihikari. The authors found two different QTLs within that region; *Gn1a* and *Gn1b*. *Gn1a* was cloned and characterized, and contains four exons and three introns. It encodes for cytokinin oxidase/dehydrogenase (OsCKX2). This gene regulates the grain number negatively; higher expression of the gene leads to reduced grain number while reduced expression leads to increased grain number (Ashikari et al. 2005).

The Spikelet Number (*SPIKE*)/*NAL1* gene enhances grain number per panicle by up to 36% without any negative effect on grain appearance. It has a pleiotropic effect on flag leaf width, root systems and leaf size also. The gene is located on chromosome 4 and encodes for protein containing trypsin-like serine and a cysteine protease domain.

Amino acid sequence comparison between IR64 and NIL-*SPIKE* revealed three amino acid substitutions. *Ghd7* is also a major gene that shows a pleiotropic effect on many important agronomic traits such as grain number per panicle, heading date and plant height. Enhanced expression of the gene leads to increased grain number and panicle size. The gene is located on the centromeric region of chromosome 7, encodes a CCT domain protein and co-segregates with the SSR marker RM5436. The gene also increases the plant height due to the increased cell number. However, it is reported that the performance and effect of *Ghd7* is dependent on the background genome, due to interaction with other genes in gene introgression programmes (Xue et al. 2008).

The Dense and Erect Panicle 1 (*DEP1*) gene shows a pleiotropic effect on many valuable agronomic traits such as grain number per panicle, primary and secondary branching of panicle and erect panicle. The gene is located on chromosome 9, encoding PEBP (phosphatidyl ethanolamine-binding protein) like a domain protein. The gene favours the traits of interests by gain of function through mutation. Due to the inclusion of this gene, 40.9% greater grain number/yield was achieved (Huang et al. 2009). The *DEP2* gene encodes a novel plant-specific protein, and its mutant leads to a dense and erect panicle. Unlike the *DEP1* gene, the number of spikelet and primary and secondary branches of panicle of *dep2* is the same as in wild types (Li et al. 2010). The mutant allele of *DEP3* (*dep3*) also contributes to an erect panicle; it is located on chromosome 6 and encodes the patatin-like phospholipase A2 (PLA2) super family domain-containing protein. The *dep3* gene also increases the number of grains per panicle by increasing secondary branches. It affects grain size in such a way that the *dep3* grain is rounder than the wild type, which slightly reduces thousand-grain weight. However, the overall yield was higher in the case of *dep3* than for the wild type (Qiao et al. 2011). More genes that have been reported to influence grain number per panicle are listed in Supplementary Table 1.

6.3 Panicle architecture

The aerial architecture of the rice plant is determined by the number of tillers and panicles. Panicle length and the number of productive tillers per plant are the important agronomic characteristics. The primary and secondary productive tillers of the plants determine number of panicles per plant. Large Panicle (*LP*) is a major gene that affects the panicle's architecture. The gene was located on chromosome 2, containing two exons and one intron and encodes kelch repeat-containing F-box protein. The *lp* plants showed more lodging resistance due to a stronger culm as well. The *lp* gene increases grain number per panicle and thousand-grain weight, while it decreased the number of tillers per plant. However, the overall productivity (grain yield) per plant of the *lp* gene was 11% higher than wild plants (Li et al. 2011b). Another gene that controls the number of tillers per plant is Monoculm1 (*MOC1*), located on chromosome 6, which encodes a putative GRAS family nuclear protein. The mutant *moc1* leads to a single monoculm without any secondary/further tillers (Li et al. 2003). LAX Panicle (*LAX*), *LAX2* and Small Panicle (*SPA*) genes are the major regulators of axillary meristem formation in rice (Komatsu et al. 2003a; Tabuchi et al. 2011). *LAX2* encodes for a novel nuclear protein that acts together with *LAX1* to determine the tiller architecture. Analysis of *LAX1* and *LAX2* indicated that they are involved in different pathways for controlling branching in plants (Tabuchi et al. 2011).

Panicle length and density are mainly determined by primary and secondary inflorescence branches. Many genes influence the length and position of inflorescence branches and among them, *SP1* plays a significant role, which controls panicle elongation

(Li et al. 2009). Moreover, the gene also affects grain weight significantly. The gene was located on chromosome 11, and encodes a putative transporter that belongs to the peptide transporter (PTR) (Li et al. 2009). Wealthy Farmer's Panicle (*WFP*)/*OsSPL14* is responsible for panicle branching and higher grain yield. Higher expression of this gene leads to higher panicle branching and high grain number. The gene was located on chromosome 8, and encodes the plant-specific transcription factor OsSPL14. The gene is regulated by microRNA, and has a microRNA-binding region at its third exonic region. The authors of the study reported epigenetic variations, as well as expression difference among the alleles, which may be responsible for the functional difference of the gene. Higher expression of the gene leads to a reduced number of total tillers of the plant (Miura et al. 2010). A putative functional marker for this *SPL14* gene was reported through a whole-genome sequence approach, along with SNP/InDel markers for another seven yield-related genes – *Gn1a*, *SCM2*, *Ghd7*, *DEP1*, *SPIKE*, *GS5* and *TGW6* (Kim et al. 2016). The Lonely Guy (*LOG*) gene was located on chromosome 1 and contains 7 exons and 6 introns. It encodes a cytokinin-activating enzyme containing 242 amino acids. This gene controls the growth of meristem activity, and loss of function of this gene leads to premature termination of the shoot meristem (Kurakawa et al. 2007). More genes related to panicle architecture are listed in Supplementary Table 1.

NILs play a significant role in breeding programmes, as the source for a particular gene or a set of pyramided genes without any dragging effect of the gene. They help to validate the functions of genes and enable varietal improvement. Molecular markers play a crucial role in producing NILs. Zhenshan97 is extensively used as one parent in a three-line hybridization system in China, due to its high combining ability as well as its wide environmental adaptability (Wang et al. 2012a). In the background of Zhenshan97, NILs were created for *qHD8*, *qHD7*, and *qHD6.1* and *GS3* genes, either individually or in combination with these genes by using SSR and InDel markers. The NIL (*qHD8*) not only showed an improved heading date but also increased yield per plant by 50% and leaf size by nearly 30%. The pyramided line with *qHD8* and *gs3* showed longer grains with a higher yield and a more suitable heading date (Wang et al. 2012a). However, all the gene combinations may not increase the yield; in a worst case, they may even reduce it. For example, NIL – *Gn1a* + *sd1* was produced in the background of Koshihikari, and the pyramided plants (Nil-*Gn1a* + *sd1*) showed a lower yield than a Nil-*Gn1a* plant (Ashikari et al. 2005).

7 Deploying novel alleles of known genes

Rice was probably first domesticated in north-eastern India and southern China nearly 8000 to 10000 years ago (Gowda et al. 2003; Kovach et al. 2007). Due to domestication and manmade artificial selection, many beneficial alleles have been left behind in the evolutionary process (Ramkumar et al. 2010b). Utilizing these untapped genomic components in breeding programmes will enhance the rice yield significantly. Hence, it is necessary to exploit the diversity in rice cultivars and wild *Oryza* species to discover novel and superior alleles of yield and yield-related genes. Recently, the IRRI in the Philippines revealed genome information for 3000 rice cultivars (3K RGP), and many other genome sequences are available in the public database already (http://gigadb.org/dataset/200001). However, this huge quantity of genetic information is underutilized, especially in the case of yield-related traits and genes. Better alleles should be identified and analysed for each important, significant gene and the superior alleles should be used in gene incorporation or pyramiding breeding programmes, which will lead to the desired quantity and quality of crop yield. Allele mining is a strategy to analyse and utilize the natural variations of alleles and to improve the agronomic traits using those novel alleles (Ramkumar et al. 2010b). For example, rice accessions can be classified into four different groups based on amylose content, that is, very low, low, intermediate and high content (Kumar and Khush 1987). Mikami et al. (2008) demonstrated that five different allelic forms (Wx^{a}, Wx^{in}, Wx^{b}, Wx^{op} and *wx*) of the gene showed a positive correlation with different quantities of amylose content in the grain. Hence, by deploying a particular allele, we may achieve the desired amylose content in the rice grain.

Higher expression of the *OsSPL14* (Squamosa promoter-binding Protein Like 14; also known as IPA1) gene reduces unproductive tillers and increases primary and secondary branches in the panicle, thereby increasing the number of grains and the overall yield (Jiao et al. 2010). Miura et al. (2010) reported three different *SPL14* alleles from Nipponbare, ST-12 and the japonica variety, Aikawa1. DNA sequence analysis revealed that there was no difference between the exonic region of the *SPL14* genes of ST-12 and those of Nipponbare. MicroRNA (miRNA) targets the third exonic region of the gene and cleaves the mRNA as a post-transcriptional relation process. However, Aikawa1 has a point (SNP) mutation at the third exon of the gene; consequently, miRNA cannot cleave the $SPL14^{Aikawa1}$ allele, and hence the gene has high expression, which leads to low and unproductive tiller production and high primary and secondary branches in panicles. However, the effect of regulation of the $SPL14^{ST-12}$ allele was not clear (Miura et al. 2010), and although it does not possess the SNP at the third exon, its expression was higher in comparison with $SPL14^{Aikawa1}$ and $SPL14^{Nipponbare}$ alleles at 2–5 mm panicle stage, and hence its expression leads to higher primary branches in panicles as well as higher yield (Miura et al. 2010).

The mutant allele of the *DEP1* gene increases the grain yield by increasing grain number and primary and secondary branches, and it is pleiotropically responsible for three different traits: a dense panicle, an erect panicle and a high grain number per panicle. Huang et al. (2009) analysed *DEP1* alleles from 14 different *Oryza* species, including wild species, domesticated *Oryza* species and other small grain cereal crops such as wheat. The study revealed different forms of *DEP1* alleles with different degrees of variation at nucleotide and at amino acid level that influence grain yield. There are many other allele-mining studies, especially studies of stress tolerance (Ramkumar et al. 2010b; Fukao et al. 2009). The authors of this study strongly believe that identification of superior alleles of valuable genes is as important as identification of novel genes. Exploring rice diversity may lead to the identification of appropriate alleles to achieve desirable phenotype/yield. In addition to identification of superior alleles, those alleles should be incorporated into elite cultivars to further enhance the yield.

8 Future trends and conclusion

World population is growing at the rate of 1.5 to 2% every year (Gowda et al. 2003), while the rate of rice production is increasing by a mere ~1.0% per year (Khush 2001; Jeon et al. 2011). However, recent developments in genetics and molecular breeding, as well as the availability of a wide range of diverse rice cultivars, are supporting the enhancement of rice yields. Additionally, developments in bioinformatics and next-generation sequencing technology have opened up a path for the identification of novel and superior alleles for yield and yield component traits/genes from a wide range of rice cultivars. The genetic variability of alleles to tackle some of the biotic and abiotic stresses is limited in the cultivated rice germplasm (Brar and Khush 1997) and they have only 10 to 20% of the genetic diversity present in *O. rufipogon*. Many beneficial alleles may remain untapped in wild *Oryza* species, owing to their inferior phenotypic expression and linked drag effect of the alleles. As most of the rice cultivars and many landraces have been exploited to increase yield potential, wild *Oryza* species are alternative options to utilize and enhance agronomically important traits. The authors of this study believe that wild *Oryza* species have huge potential to enhance yield as well as to protect rice yield gain from biotic and abiotic stresses. This chapter has presented a number of examples to prove that wild *Oryza* species are a reservoir for many useful alleles. Many genes for yield and biotic stress resistance have been identified from different wild *Oryza* species and are being used in many breeding programmes. Hence, the underutilized genes from wild *Oryza* should be identified and utilized to enhance yield traits. The major problem in using wild *Oryza* species is F_1 seed sterility; however, modern techniques such as embryo rescue could help scientists obtain proper germination of the embryo followed by backcrossing to *O. sativa* in order to restore fertility (Jena and Khush 1984). Moreover, advances in cytology and molecular biology will assist scientists in crossing plants with different ploidy levels. Developments in molecular markers specific to yield-enhancing genes will also facilitate use of the untapped alleles from wild *Oryza* species. These molecular markers enhance the efficiency of breeding programmes, reduce the time needed to incorporate a targeted genes in particular genetic background and are helpful in reducing the dragging effect. Hence, only beneficial alleles can be transferred without any negative effect of the allele.

Increasing yield through improving agronomic traits using genomic approaches is one potential strategy. Trait improvement can be achieved by conventional breeding combined with molecular markers to transfer desired genes from distant donors to elite rice cultivars. A number of yield-related genes that influence grain number, panicle architecture and overall yield potential have been discussed in this chapter. These genes can be used to modify traits according to the breeding need in order to enhance the crop yield. Incorporation of genes can be targeted for breeding the proposed novel ideotype in order to enhance the rice yield. Gene pyramiding also results in enhanced yield, although only a few studies have reported a combination of genes that enhances yield potential significantly.

Although many genes have been identified, studies on incorporating or pyramiding complementary yield genes are far more scarce. These must be increased if the yield of elite rice varieties is to be enhanced. However, all the genes may not work well in all the genetic background and therefore further studies are needed to validate the genes' functionality in different genomes. Moreover, the genes' performance in different combinations also needs to be validated. Using advancements in genetics and genomics technology, novel strategies must be applied to increase the rice yield to feed the growing world population.

9 Where to look for further information

Genetic improvement of rice, particularly increasing the yield potential of new varieties, is one of the main strategies in the International Rice Research Institute (IRRI). Increasing yield potential depends on increasing genetic gain in yield component traits as well as resistance or tolerances to various biotic and abiotic factors such as diseases and pests, and drought, salinity, flooding and adverse temperature. Improved rice germplasm will be developed with increased genetic gain through correct gene discovery research by developing pre-breeding products having genes for tolerance to pests and diseases of rice. At IRRI, several thousand rice lines have been sequenced and the ability of the lines for different traits of interest have been well understood (3K RGP). Thus, a representative set of rice germplasm with different value-added traits will be developed. In future, a plant system biology approach will be used to identify key genes for adaptation, yield and other traits of interest. A gene discovery approach will include multi-scale physiological description and modelling of developmental process. A new quick method will be followed to utilize new genetic stocks (MAGIC and CSSL) developed at IRRI to look for yield-contributing traits/genes to improve rice. These new populations will provide abundant genetic diversity harbouring new genes and their information for rice improvement. Emerging tools such as precision breeding, genome editing for targeting yield traits as well as development and use of diagnostic gene-based DNA markers will help towards improving traits and stacking favourable genes to improve yield potential of rice. Development through manipulation of C3 to C4 photosynthesis in rice may lead to increase in rice production.

9.1 A few useful website links are added here for any additional information:

International Rice Research Institute (IRRI) – http://irri.org/. IRRI aims to reduce poverty and hunger, improve the health of rice farmers and consumers, and ensure environmental sustainability of rice farming. It does these through collaborative research, partnerships and the strengthening of the national agricultural research and extension systems, or NARES, of the countries they work in. Many research activities can be seen in this website.

Consultative Group for International Agricultural Research (CGIAR) – http://www.cgiar.org/. CGIAR is a global research partnership for a food-secure future. CGIAR science is dedicated to reducing poverty, enhancing food and nutrition security, and improving natural resources and ecosystem services. Its research is carried out by 15 CGIAR centres in close collaboration with hundreds of partners, including national and regional research institutes, civil society organizations, academia, development organizations and the private sector.

Global Rice Science Partnership – http://www.grisp.net/main/summary. The Global Rice Science Partnership (GRiSP) provides a single strategic plan and unique new partnership platform for impact-oriented rice research for development. It is designed to more effectively solve development challenges.

10 Competing interests

The authors declare that they have no competing interest.

11 Authors' contributions

KKJ and GR have written and finalized the manuscript. KKJ has visualized the overall concepts of the manuscript.

12 Acknowledgements

We are grateful to the Global Rice Science Partnership programme of CGIAR-IRRI for financially supporting this investigation and the manuscript preparation. We thank Dr. George Kotch, Head of Plant Breeding Division, IRRI, for critically reviewing the manuscript. We thank the editorial team of IRRI Communications for editing the manuscript.

13 References

3K RGP: The 3,000 rice genomes project. *GigaScience* 2014 3:7

Ashikari, M., Sakakibara, H., Lin, S., Yamamoto, T., Takashi, T., Nishimura, A., Angeles, E. R., Qian, Q., Kitano, H. and Matsuoka, M. (2005). Cytokinin oxidase regulates rice grain production. *Science* 309:741–5.

Bandillo, N., Raghavan, C., Muyco, P. A., Sevilla, M. A. L., Lobina, I. T., Dilla-Ermita, C. J., Tung, C. W., McCouch, S., Thomson, M., Mauleon, R., Singh, R. K., Gregorio, G., Redoña, E. and Leung, H. (2013). Multi-parent advanced generation inter-cross (MAGIC) populations in rice: progress and potential for genetics research and breeding. *Rice* 6:11. doi:10.1186/1939-8433-6-11.

Boualaphanh, C., Daygon, V. D., Calingacion, M. N., Sanitchon, J., Jothityangkoon, D., Mumm, R., Hall, R. D. and Fitzgerald, M. A. (2011). Use of new generation single nucleotide polymorphism genotyping for rapid development of near-isogenic lines in rice. *Crop Sci.* 51:2067–73.

Brar, D. S. and Khush, G. S. (1997). Alien introgression in rice. *Plant Mol. Biol.* 35:35–47.

Breseghello, F. and Coelho, A. S. (2013). Traditional and modern plant breeding methods with examples in rice (*Oryza sativa* L.). *J. Agric. Food Chem.* 61:8277–86. doi:10.1021/jf305531j.

Cai, G., Yang, Q., Yang, Q., Zhao, Z., Chen, H., Wu, J., Fan, C. and Zhou, Y. (2012). Identification of candidate genes of QTLs for seed weight in *Brassica napus* through comparative mapping among *Arabidopsis* and *Brassica* species. *BMC Genet.* 13:105. doi:10.1186/1471-2156-13-105.

Chen, J., Ding, J., Ouyang, Y., Du, H., Yang, J., Cheng, K., Zhao, J., Qiu, S., Zhang, X., Yao, J., Liu, K., Wang, L., Xu, C., Li, X., Xue, Y., Xia, M., Ji, Q., Lu, J., Xu, M. and Zhang, Q. (2008). A triallelic system of *S5* is a major regulator of the reproductive barrier and compatibility of indica – japonica hybrids in rice. *Proc. Natl. Acad. Sci. USA* 105:11436–41. doi:10.1073/pnas.0804761105.

Chen, Y., Fan, X., Song, W., Zhang, Y. and Xu, G. (2012). Over-expression of *OsPIN2* leads to increased tiller numbers, angle and shorter plant height through suppression of *OsLAZY1*. *Plant Biotechnol. J.* 10:139–49.

Deng, X., Zhou, J., Xu, P., Li, J., Hu, F. and Tao, D. (2010). The role of *S1-g* allele from *Oryza glaberrima* in improving interspecific hybrid sterility between *O. sativa* and *O. glaberrima*. *Breed. Sci.* 60:342–6. doi:10.1270/jsbbs.60.342.

Doi, K., Izawa, T., Fuse, T., Yamanouchi, U., Kubo, T., Shimatani, Z., Yano, M. and Yoshimura, A. (2004). *Ehd1*, a B-type response regulator in rice, confers short-day promotion of flowering and controls FT-like gene expression independently of *Hd1*. *Genes Dev.* 18:926–36.

Fan, C., Xing, Y. Z., Mao, H. L., Lu, T. T., Han, B., Xu, C. G., Li, X. H. and Zhang, Q. F. (2006). *GS3*, a major QTL for grain length and weight and minor QTL for grain width and thickness in rice, encodes a putative transmembrane protein. *Theor. Appl. Genet.* 112:1164–71.

Fan, C., Yu, S., Wand, C. and Xing, Y. (2009). A causal C–A mutation in the second exon of *GS3* highly associated with rice grain length and validated as a functional marker. *Theor. Appl. Genet.* 118:465–72.

Fujita, D., Ebron, L. A., Araki, E., Kato, H., Khush, G. S., Sheehy, J. E., Lafarge, T., Fukuta, Y. and Kobayashi, N. (2010). Fine mapping of a gene for low-tiller number, Ltn, in japonica rice (*Oryza sativa* L.) variety Aikawa 1. *Theor. Appl. Genet.* 120:1233–40.

Fujita, D., Trijatmiko, K. R., Tagle, A. G., Sapasap, M. V., Koide, Y., Sasaki, K., Tsakirpaloglou, N., Gannaban, R. B., Nishimura, T., Yanagihara, S., Fukuta, Y., Koshiba, T., Loedin, I. H. S., Ishimaru,

T. and Kobayashi, N. (2013). *NAL1* allele from a rice landrace greatly increases yield in modern indica cultivars. *Proc. Natl. Acad. Sci. USA.* 110:20431–6.

Fukao, T., Harris, T. and Bailey-Serres, J. (2009). Evolutionary analysis of the *Sub1* gene cluster that confers submergence tolerance to domesticated rice. *Ann. Bot.* 103:143–50.

Gao, L. Z., Zhang, C. H., Chang, L. P., Jia, J. Z., Qiu, Z. E. and Dong, Y. S. (2005). Microsatellite diversity within *Oryza sativa* with emphasis on indica japonica divergence. *Genet. Res.* 85:1–14.

Gowda, M., Venu, R. C., Roopalakshmi, K., Sreerekha, M. V. and Kulkarni, R. S. (2003). Advances in rice breeding, genetics and genomics. *Mol. Breed.* 11:337–52.

Guo, L., Ma, L., Jiang, H., Zeng, D., Hu, J., Wu, L., Gao, Z., Zhang, G. and Qian, Q. (2009). Genetic analysis and fine mapping of two genes for grain shape and weight in rice. *J. Integr. Plant Biol.* 51:45–51.

Heang, D. and Sassa, H. (2012a). An atypical bHLH protein encoded by *POSITIVE REGULATOR OF GRAIN LENGTH 2* is involved in controlling grain length and weight of rice through interaction with a typical bHLH protein APG. *Breed. Sci.* 62:133–41.

Heang, D. and Sassa, H. (2012b). Antagonistic actions of HLH/bHLH proteins are involved in grain length and weight in rice. *PLoS ONE* 7:e31325.

Hochholdinger, F. and Hoecker, N. (2007). Towards the molecular basis of heterosis. *Trends Plant Sci.* 12:427–32.

Huang, B. E., Verbyla, K. L., Verbyla, A. P., Raghavan, C., Singh, V. K., Gaur, P., Leung, H., Varshney, R. K. and Cavanagh, C. R. (2015). MAGIC populations in crops: current status and future prospects. *Theor. Appl. Genet.* 128:999–1017. doi:10.1007/s00122-015-2506-0.

Huang, X., Qian, Q., Liu, Z., Sun, H., He, S., Luo, D., Xia, G., Chu, C., Li, J. and Fu, X. (2009). Natural variation at the *DEP1* locus enhances grain yield in rice. *Nat. Genet.* 41:494–7.

Ikeda, A., Ueguchi-Tanaka, M., Sonoda, Y., Kitano, H., Koshioka, M., Futsuhara, Y., Matsuoka, M. and Yamaguchi, J. (2001). Slender rice, a constitutive gibberellin response mutant, is caused by a null mutation of the *SLR1* gene, an ortholog of the height-regulating gene *GAI/RGA/RHT/D8*. *Plant Cell* 13:999–1010.

Ikeda, K., Ito, M., Nagasawa, N., Kyozuka, J. and Nagato, Y. (2007). Rice ABERRANT PANICLE ORGANIZATION 1, encoding an F-box protein, regulates meristem fate. *Plant J.* 51:1030–40.

Ikeda-Kawakatsu, K., Maekawa, M., Izawa, T., Itoh, J. I. and Nagato Y (2012). ABERRANT PANICLE ORGANIZATION 2/RFL, the rice ortholog of *Arabidopsis* LEAFY, suppresses the transition from inflorescence meristem to floral meristem through interaction with *APO1*. *Plant J.* 69:168–80. doi:10.1111/j.1365-313X.2011.04781.x.

Ikehashi, H. and Araki, H. (1984). Variety screening of compatibility types revealed in F_1 fertility of distant cross in rice. *Jpn. J. Breed.* 34:304–13.

Ishii, T., Brar, D. S., Multani, D. S. and Khush, G. S. (1994). Molecular tagging of genes for brown planthopper resistance and earliness introgressed from *Oryza australiensis* into cultivated rice, *Oryza sativa*. *Genome* 37:217–21.

Ishimaru, K., Hirotsu, N., Madoka, Y., Murakami, N., Hara, N., Onodera, H., Kashiwagi, T., Ujiie, K., Shimizu, B., Onishi, A., Miyagawa, H. and Katoh, E. (2013). Loss of function of the IAA-glucose hydrolase gene TGW6 enhances rice grain weight and increases yield. *Nat. Genet.* 45:707–11.

Jena, K. K. and Khush, G. S. (1984). Embryo rescue of interspecific hybrids and its scope in rice improvement. *Rice Genet. News Lett.* 1:133–4.

Jena, K. K., Jeung, J. U., Lee, J. H., Choi, H. C. and Brar, D. S. (2006). High-resolution mapping of a new brown planthopper (BPH) resistance gene, *Bph18(t)*, and marker-assisted selection for BPH resistance in rice (*Oryza sativa* L.). *Theor. Appl. Genet.* 112:288–97.

Jena, K. K. (2010). The species of the genus *Oryza* and transfer of useful genes from wild species into cultivated rice, *O. sativa*. *Breed. Sci.* 60:518–23. doi:10.1270/jsbbs.60.518.

Jeon, J. S., Jung, K. H., Kim, H. B., Suh, J. P. and Khush, G. S. (2011). Genetic and molecular insights into the enhancement of rice yield potential. *J. Plant Biol.* 54:1–9. doi:10.1007/s12374-011-9144-0.

Jeung, J. U., Kim, B. R., Cho, Y. C., Han, S. S., Moon, H. P., Lee, Y. T. and Jena, K. K. (2007). A novel gene, *Pi40(t)*, linked to the DNA markers derived from NBS-LRR motifs confers broad spectrum of blast resistance in rice. *Theor. Appl. Genet.* 115:1163–77.

Ji, H. S., Kim, S. R., Kim, Y. H., Suh, J. P., Park, H. M., Sreenivasulu, N., Misra, G., Hechanova, S. L., Kim, H., Lee, G. S., Yoon, Y. H., Kim, T. H., Lim, H., Suh, S. C., Yang, J., An, G. and Jena, K. K. (2016). *Sci. Reports* 6:34376.

Jiao, Y., Wang, Y., Xue, D., Wang, J., Yan, M., Liu, G., Dong, G., Zeng, D., Lu, Z., Zhu, X., Qian, Q. and Li, J. (2010). Regulation of *OsSPL14* by OsmiR156 defines ideal plant architecture in rice. *Nat. Genet.* 42:541–4. doi:10.1038/ng.591.

Kalia, R. K., Rai, M. K., Kalia, S., Singh, R. and Dhawan, A. K. (2011). Microsatellite markers : an overview of the recent progress in plants. *Euphytica* 177:309–34. doi:10.1007/s10681-010-0286-9.

Kato, S., Kosaka, H. and Hara, S. (1928). On the affinity of rice varieties as shown by the fertility of hybrid plants (in Japanese with English summary) *Sci. Bull. Fac. Agric. Kyushu Univ.* 3:132–47.

Khush, G. S. (1997). Origin, dispersal, cultivation and variation of rice. *Plant Mol. Biol.* 35:25–34.

Khush, G. S. (1999). Green revolution: preparing for the 21st century. *Genome* 42:646–55.

Khush, G. S. (2001). Challenges for meeting the global food and nutrient needs in the new millennium. *Proc. Nutr. Soc.* 60:15–26. doi:10.1079/PNS200075.

Khush, G. S. (2005). What it will take to feed 5.0 billion rice consumers in 2030. *Plant Mol. Biol.* 59:1–6. doi:10.1007/s11103-005-2159-5.

Khush, G. S. and Jena, K. K. (2009). Current status and future prospects for research on blast resistance in rice (*Oryza sativa* L.). In: Wang, G. L., Valent B (eds) *Advances in Genetics, Genomics and Control of Rice Blast Disease.* New York: Springer, pp. 1–10.

Kim, D., Lee, H., Kwon, S., Fabreag, M. E., Kang, J., Yun, Y., Chung, C. and Ahn, S. (2014). High-density mapping of quantitative trait loci for grain-weight and spikelet number in rice. *Rice* 7:14. doi:10.1186/s12284-014-0014-5.

Kim, S. R., Ramos, J., Ashikari, M., Virk, P. S., Torres, E. A., Nissila, E., Hechanova, S. L., Mauleon R and Jena, K. K. (2016). Development and validation of allele-specific SNP/indel markers for eight yield-enhancing genes using whole-genome sequencing strategy to increase yield potential of rice, *Oryza sativa* L. *Rice* 9:12.

Kitagawa, K., Kurinami, S., Oki, K., Abe, Y., Ando, T., Kono, I., Yano, M., Kitano, H. and Iwasaki, Y. (2010). A novel kinesin 13 protein regulating rice seed length. *Plant Cell Physiol.* 51:1315–29.

Kobayashi, K., Maekawa, M., Miyao, A., Hirochika, H. and Kyozuka, J. (2010). PANICLE PHYTOMER2 (*PAP2*), encoding a SEPALLATA subfamily MADS-box protein, positively controls spikelet meristem identity in rice. *Plant Cell Physiol.* 51:47–57.

Komatsu, K., Maekawa, M., Ujiie, S., Satake, Y., Furutani, I., Okamoto, H., Shimamoto, K. and Kyozuka, J. (2003a). *LAX* and *SPA*: major regulators of shoot branching in rice. *Proc. Natl. Acad. Sci. USA* 100:11765–70.

Komatsu, M., Chujo, A., Nagato, Y., Shimamoto, K. and Kyozuka, J. (2003b). *FRIZZY PANICLE* is required to prevent the formation of axillary meristems and to establish floral meristem identity in rice spikelets. *Development* 130:3841–50.

Kovach, M. J., Sweeney, M. T. and Mccouch, S. R. (2007). New insights into the history of rice domestication. *Trends Genet.* 23:578–87. doi:10.1016/j.tig.2007.08.012.

Kovach, M. J. and McCouch, S. R. (2008). Leveraging natural diversity: back through the bottleneck. *Curr. Opin. Plant Biol.* 11:193–200.

Kumar, I. and Khush, G. S. (1987). Genetic-analysis of different amylose levels in rice. *Crop Sci.* 27:1167–72.

Kumar, P. N., Sujatha, K., Laha, G. S., Rao, K. S., Mishra, B., Viraktamath, B. C., Hari, Y., Reddy, C. S., Balachandran, S. M., Ram, T., Madhav, M. S., Rani, N. S., Neeraja, C. N., Reddy, G. A., Shaik, H. and Sundaram, R. M. (2012). Identification and fine-mapping of *Xa33*, a novel gene for resistance to *Xanthomonas oryzae pv oryzae*. *Phytopathology* 102:222–8.

Kurakawa, T., Ueda, N., Maekawa, M., Kobayashi, K., Kojima, M., Nagato, Y., Sakakibara, H. and Kyozuka, J. (2007). Direct control of shoot meristem activity by a cytokinin-activating enzyme. *Nature* 445:652–5.

Li, F., Liu, W., Tang, J., Chen, J., Tong, H., Hu, B., Li, C., Fang, J., Chen, M. and Chu, C. (2010). Rice *DENSE AND ERECT PANICLE 2* is essential for determining panicle outgrowth and elongation. *Cell Res.* 20:838–49.

Li, J., Chu, H., Zhang, Y., Mou, T., Wu, C., Zhang, Q. and Xu, J. (2012). The rice *HGW* gene encodes a ubiquitin-associated (UBA) domain protein that regulates heading date and grain weight. *PLoS ONE* 7:e34231.

Li, M., Tang, D., Wang, K., Wu, X., Lu, L., Yu, H., Gu, M., Yan, C. and Cheng, Z. (2011b). Mutations in the F-box gene *LARGER PANICLE* improve the panicle architecture and enhance the grain yield in rice. *Plant Biotechnol. J.* 9:1002–13.

Li, S., Qian, Q., Fu, Z., Zeng, D., Meng, X., Kyozuka, J., Maekawa, M., Zhu, X., Zhang, J., Li, J. and Wang, Y. (2009). *Short panicle1* encodes a putative PTR family transporter and determines rice panicle size. *Plant J.* 58:592–605.

Li, X., Qian, Q., Fu, Z., Wang, Y., Xiong, G., Zeng, D., Wang, X., Liu, X., Teng, S., Hiroshi, F., Yuan, M., Luo, D., Han, B. and Li, J. (2003). Control of tillering in rice. *Nature* 422:618–21.

Li, Y., Fan, C., Xing, Y., Jiang, Y., Luo, L., Sun, L., Shao, D., Xu, C., Li, X., Xiao, J., He, Y. and Zhang, Q. (2011a). Natural variation in *GS5* plays an important role in regulating grain size and yield in rice. *Nat. Genet.* 43:1266–9.

Lin, C. H., Yeakley, J. M., McDaniel, T. K. and Shen, R. (2009). Medium-to high-throughput SNP genotyping using Vera Code microbeads. *Methods Mol. Biol.* 496:129–42. doi:10.1007/978-1-59745-553-4_10.

Liu, L., Tong, H., Xiao, Y., Che, R., Xu, F., Hu, B., Liang, C., Chu, J., Li, J. and Chu, C. (2015). Activation of BIG GRAIN1 significantly improves grain size by regulating auxin transport in rice. *PNAS* 112:11102–7.

Liu, T., Mao, D., Zhang, S., Xu, C. and Xing, Y. (2009). Fine mapping *SPP1*, a QTL controlling the number of spikelets per panicle, to a BAC clone in rice (*Oryza sativa*). *Theor. Appl. Genet.* 118:1509–17.

Long, Y., Zhao, L., Niu, B., Su, J., Wu, H., Chen, Y., Zhang, Q., Guo, J., Zhuang, C., Mei, M., Xia, J., Wang, L., Wu, H. and Liu, Y. G. (2008). Hybrid male sterility in rice controlled by interaction between divergent alleles of two adjacent genes. *Proc. Natl. Acad. Sci. USA* 105:18871–6.

Mao, H., Sun, S., Yao, J., Wang, C., Yu, S., Xu, C., Li, X. and Zhang, Q. (2010). Linking differential domain functions of the *GS3* protein to natural variation of grain size in rice. *Proc. Natl. Acad. Sci. USA.* 107:19579–84.

Matsuzaki, H., Dong, S., Loi, H., Di, X., Liu, G., Hubbell, E., Law, J., Berntsen, T., Chadha, M., Hui, H., Yang, G., Kennedy, G. C., Webster, T. A., Cawley, S., Walsh, P. S., Jones, K. W., Fodor, S. P. and Mei, R. (2004). Genotyping over 100,000 SNPs on a pair of oligonucleotide arrays. *Nat. Methods* 1:109–11.

McCouch, S. R., Teytelman, L., Xu, Y., Lobos, K. B., Clare, K., Walton, M., Fu, B., Maghirang, R., Li, Z., Xing, Y., Zhang, Q., Kono, I., Yano, M., Fjellstrom, R., DeClerck, G., Schneider, D., Cartinhour, S., Ware, D. and Stein, L. (2002). Development and mapping of 2240 new SSR markers for rice (*Oryza sativa* L.). *DNA Res.* 9:199–207.

McCouch, S. R., Sweeney, M., Li, J., Jiang, H., Thomson, M., Septiningsih, E. M., Edwards, J., Moncada, P., Xiao, J., Garris, A. J., Tai, T., Martinez, C., Tohme, J., Sugiono, M., McClung, A., Yuan, L. P. and Ahn, S. N. (2007). Through the genetic bottleneck : *O . rufipogon* as a source of trait-enhancing alleles for *O. sativa*. *Euphytica* 154:317–39. doi:10.1007/s10681-006-9210-8.

Mikami, I., Uwatoko, N., Ikeda, Y., Yamaguchi, J., Hirano, H. Y., Suzuki, Y. and Sano, Y. (2008). Allelic diversification at the *wx* locus in landraces of Asian rice. *Theor. Appl. Genet.* 116:979–89.

Minakuchi, K., Kameoka, H., Yasuno, N., Umehara, M., Luo, L., Kobayashi, K., Hanada, A., Ueno, K., Asami, T., Yamaguchi, S. and Kyozuka, J. (2010). FINE CULM1 (*FC1*) works downstream of strigolactones to inhibit the outgrowth of axillary buds in rice. *Plant Cell Physiol.* 51:1127–35.

Miura, K., Ikeda, M., Matsubara, A., Song, X. J., Ito, M., Asano, K., Matsuoka, M., Kitano, H. and Ashikari, M. (2010). *OsSPL14* promotes panicle branching and higher grain productivity in rice. *Nat. Genet.* 42:545–9.

Miura, K., Ashikari, M. and Matsuoka, M. (2011). The role of QTLs in the breeding of high-yielding rice. *Trends Plant Sci.* 16:319–26.

Monna, L., Kitazawa, N., Yoshino, R., Suzuki, J., Masuda, H., Maehara, Y., Tanji, M., Sato, M., Nasu, S. and Minobe, Y. (2002a). Positional cloning of rice semidwarf gene, sd-1: Rice 'Green Revolution Gene' encodes a mutant enzyme involved in gibberellin synthesis. *DNA Res.* 9:11–17.

Monna, L., Lin, X., Kojima, S., Sasaki, T. and Yano, M. (2002b). Genetic dissection of a genomic region for a quantitative trait locus, *Hd3*, into two loci, *Hd3a* and *Hd3b*, controlling heading date in rice. *Theor. Appl. Genet.* 104:772 – 8.

Nakagawa, H., Tanaka, A., Tanabata, T., Ohtake, M., Fujioka, S., Nakamura, H., Ichikawa, H. and Mori, M. (2012). *Short grain1* decreases organ elongation and brassinosteroid response in rice. *Plant Physiol.* 158:1208–19.

Ookawa, T., Hobo, T., Yano m, Murata, K., Ando, T., Miura, H., Asano, K., Ochiai, Y., Ikeda, M., Nishitani, R., Ebitani, T., Ozaki, H., Angeles, E. R., Hirasawa, T. and Matsuoka, M. (2010). New approach for rice improvement using a pleiotropic QTL gene for lodging resistance and yield. *Nat. Commun.* 1:132. doi:10.1038/ncomms1132.

Ouyang, Y., Liu, Y. G. and Zhang, Q. (2010). Hybrid sterility in plant: stories from rice. *Curr. Opin. Plant. Biol.* 13:186–92.

Piao, R., Jiang, W., Ham, T. H., Choi, M. S., Qiao, Y., Chu, S. H., Park, J. H., Woo, M. O., Jin, Z., An, G., Lee, J. and Koh, H. J. (2009). Map-based cloning of the *ERECT PANICLE 3* gene in rice. *Theor. Appl. Genet.* 119:1497–1506.

Qi, P., Lin, Y. S., Song, X. J., Shen, J. B., Huang, W., Shan, J. X., Zhu, M. Z., Jiang, L., Gao, J. P. and Lin, H. X. (2012). The novel quantitative trait locus GL3.1 controls rice grain size and yield by regulating Cyclin-T1;3. *Cell Res.* 22:1666–80.

Qiao, Y., Piao, R., Shi, J., Lee, S. I., Jiang, W., Kim, B. K., Lee, J., Han, L., Ma, W. and Koh, H. J. (2011). Fine mapping and candidate gene analysis of dense and erect panicle 3, *DEP3*, which confers high grain yield in rice (*Oryza sativa* L.). *Theor. Appl. Genet.* 122:1439–49.

Qu, S., Liu, G., Zhou, B., Bellizz, M., Zeng, L., Dai, L., Han, B. and Wang, G. L. (2006). The broad-spectrum blast resistance gene *Pi9* encodes a nucleotide-binding site–leucine-rich repeat protein and is a member of a multigene family in rice. *Genetics* 172:1901–14.

Rahman, M. L., Jiang, W., Chu, S. H., Qiao, Y., Ham, T. H., Woo, M. O., Lee, J., Khanam, M. S., Chin, J. H., Jeung, J. U., Brar, D. S., Jena, K. K. and Koh, H. J. (2009). High-resolution mapping of two rice brown planthopper resistance genes, *Bph20(t)* and *Bph21(t)*, originating from *Oryza minuta*. *Theor. Appl. Genet.* 119:1237–46. doi:10.1007/s00122-009-1125-z.

Rajendrakumar, P., Biswal, A. K., Balachandran, S. M., Ramesha, M. S., Viraktamath, B. C. and Sundaram, R. M. (2007). A mitochondrial repeat specific marker for distinguishing wild abortive type cytoplasmic male sterile rice lines from their cognate isogenic maintainer lines. *Crop Sci.* 47:207–11. doi:10.2135/cropsci2006.06.0365.

Rajendran, N., Gandhimani, R., Singh, S. and Palchamy, K. (2007). Development of a DNA marker for distinguishing CMS lines from fertile lines in rice (*Oryza sativa* L.). *Euphytica* 156:129–39. doi:10.1007/s10681-007-9360-3.

Ramkumar, G., Sivaranjani, A. K. P., Pandey, M. K., Sakthivel, K., Rani, N. S., Sudarshan, I., Prasad, G. S. V., Neeraja, C. N., Sundaram, R. M., Viraktamath, B. C. and Madhav, M. S. (2010a). Development of a PCR-based SNP marker system for effective selection of kernel length and kernel elongation in rice. *Mol. Breed.* 26:735–40.

Ramkumar, G., Sakthivel, K., Sundaram, R. M., Neeraja, C. N., Balachandran, S. M., Rani, N. S., Viraktamath, B. C. and Madhav, M. S. (2010b). Allele mining in crops: Prospects and potentials. *Biotechnol. Adv.* 28:451–61. doi:10.1016/j.biotechadv.2010.02.007.

Ramkumar, G., Prahalada, G. D., Hechanova, S., Vinarao, R. and Jena, K. K. (2015). Development and validation of SNP-based functional codominant markers for two major disease resistance genes in rice (*O. sativa* L.). *Mol. Breed.* 35:129. doi:10.1007/s11032-015-0323-4.

Ren, X., Wang, X., Yuan, H., Weng, Q., Zhu, L. and He, G. (2004). Mapping quantitative trait loci and expressed sequence tags related to brown planthopper resistance in rice. *Plant Breed.* 123:342–8.

Renganayaki, K., Friz, A. K., Sadasivam, S., Pammi, S., Harrington, S. E., McCouch, S. R., Kumar, S. M. and Reddy, A. S. (2002). Mapping and progress toward map-based cloning of brown planthopper biotype-4 resistance gene introgressed from *Oryza officinalis* into cultivated rice, *O. sativa. Crop Sci.* 42:2112–17.

Revathi, P. (2015). Marker assisted selection for S5 neutral allele in inter-subspecific hybridization of rice. *Mol. Plant Breed.* 6:1–7. doi:10.5376/mpb.2015.06.0001.

Rutkoski, J., Singh, R. P., Huerta-Espino, J., Bhavani, S., Poland, J., Jannink, J. L. and Sorrells, M. E. (2015). Genetic gain from phenotypic and genomic selection for quantitative resistance to stem rust of wheat. *Plant Genome* 8:1–10.

Segami, S., Kono, I., Ando, T., Yano, M., Kitano, H., Miura, K. and Iwasaki, Y. (2012). *Small and round seed 5* gene encodes alpha-tubulin regulating seed cell elongation in rice. *Rice* 5:4. doi:10.1186/1939-8433-5-4.

Septiningsih, E. M., Prasetiyono, J., Lubis, E., Tai, T. H., Tjubaryat, T., Moeljopawiro, S. and McCouch, S. R. (2003). Identification of quantitative trait loci for yield and yield components in an advanced backcross population derived from the *Oryza sativa* variety IR64 and the wild relative *O. rufipogon. Theor. Appl. Genet.* 107:1419–32.

Shakiba, E., Eizenga, G. C. (2014). Unraveling the secrets of rice wild species. In: Yan, W., Bao, J., editors. *Rice - Germplasm, Genetics and Improvement.* InTech. 1-58. DOI. 10.5772/58393.

Sharma, D. R., Kaur, R. and Kumar, K. (1996). Embryo rescue in plants – a review. *Euphytica* 89:325–37.

She, K. C., Kusano, K., Koizumi, K., Yamakawa, H., Hakata, M., Imamura, T., Fukuda, M., Naito, N., Tsurumaki, Y., Yaeshima, Y., Tsuge, T., Matsumoto, K., Kudoh, M., Itoh, E., Kikuchi, S., Kishimoto, N., Yazaki, J., Ando, T., Yano, M., Aoyama, T., Sasaki, T., Satoh, H. and Shimada, H. (2010). A novel factor *FLOURY ENDOSPERM2* is involved in regulation of rice grain size and starch quality. *Plant Cell* 22:3280–94.

Shomura, A., Izawa, T., Ebana, K., Ebitani, T., Kanegae, H., Konishi, S. and Yano, M. (2008) Deletion in a gene associated with grain size increased yields during rice domestication. *Nat. Genet.* 40:1023–8.

Singh, A. K., Singh, P. K., Arya, M., Singh, N. K. and Singh, U. S. (2015). Molecular screening of blast resistance genes in rice using SSR markers. *Plant Pathol.* J 31:12–24.

Singh, S. P., Sundaram, R. M., Biradar, S. K., Ahmed, M. I., Viraktamath, B. C. and Siddiq, E. A. (2006). Identification of simple sequence repeat markers for utilizing wide-compatibility genes in inter-subspecific hybrids in rice (*Oryza sativa* L.). *Theor. Appl. Genet.* 113:509–17.

Song, X. J., Huang, W., Shi, M., Zhu, M. Z. and Lin, H. X. (2007). A QTL for rice grain width and weight encodes a previously unknown RING-type E3 ubiquitin ligase. *Nat. Genet.* 39:623–30.

Song, X. J., Kuroha, T., Ayano, M., Furuta, T., Nagai, K., Komeda, N., Segami, S., Miura, K., Ogawa, D., Kamura, T., Suzuki, T., Higashiyama, T., Yamasaki, M., Mori, H., Inukai, Y., Wu, J., Kitano, H., Sakakibara, H., Jacobsen, S. E. and Ashikari, M. (2015). Rare allele of a previously unidentified histone H4 acetyltransferase enhances grain weight, yield, and plant biomass in rice. *Proc. Natl. Acad. Sci. USA* 112:76–81.

Spielmeyer, W., Ellis, M. H. and Chandler, P. M. (2002). Semidwarf (*sd-1*), 'green revolution' rice, contains a defective gibberellin 20-oxidase gene. *Proc. Natl. Acad. Sci. USA* 99:9043–8.

Steemers, F. J. and Gunderson, K. L. (2007). Whole genome genotyping technologies on the BeadArray platform. *Biotechnol. J.* 2:41–9.

Sundaram, R. M., Sakthivel, K., Hariprasad, A. S., Ramesha, M. S., Viraktamath, B. C., Neeraja, C. N., Balachandran, S. M., Rani, N. S., Revathi, P., Sandhya, P. and Hari, Y. (2010). Development and validation of a PCR-based functional marker system for the major wide-compatible gene locus *S5* in rice. *Mol. Breed.* 26:719–27. doi:10.1007/s11032-010-9482-5.

Sweeney, M. and McCouch, S. (2007). The complex history of the domestication of rice. *Ann. Bot.* 100:951–7.

Tabuchi, H., Zhang, Y., Hattori, S., Omae, M., Shimizu-Sato, S., Oikawa, T., Qian, Q., Nishimura, M., Kitano, H., Xie, H., Fang, X., Yoshida, H., Kyozuka, J., Chen, F. and Sato, Y. (2011). *LAX PANICLE2* of rice encodes a novel nuclear protein and regulates the formation of axillary meristems. *Plant Cell* 23:3276–87.

Tan, L., Li, X., Liu, F., Sun, X., Li, C., Zhu, Z., Fu, Y., Cai, H., Wang, X., Xie, D. and Sun, C. (2008). Control of a key transition from prostrate to erect growth in rice domestication. *Nat. Genet.* 40:1360–4.

Thalapati, S., Batchu, A. K., Neelamraju, S. and Ramanan, R. (2012). *Os11Gsk* gene from a wild rice, *Oryza rufipogon* improves yield in rice. *Funct. Integr. Genomics* 12:277–89.

Thomson, M. J., Zhao, K., Wright, M., McNally, K. L., Rey, J., Tung, C. W., Reynolds, A., Scheffler, B., Eizenga, G., McClung, A., Kim, H., Ismail, A. M., Ocampo, M. D., Mojica, C., Reveche, M. Y., Ermita, C. J. D., Mauleon, R., Leung, H., Bustamante, C. and McCouch, S. R. (2012). High-throughput single nucleotide polymorphism genotyping for breeding applications in rice using the BeadXpress platform. *Mol. Breed.* 29:875–86. doi:10.1007/s11032-011-9663-x.

Thomson, M. J. (2014). High-throughput SNP genotyping to accelerate crop improvement. *Plant Breed. Biotechnol.* 2:195–212.

Tong, H., Liu, L., Jin, Y., Du, L., Yin, Y., Qian, Q., Zhu, L. and Chu, C. (2012). DWARF AND LOW-TILLERING acts as a direct downstream target of a GSK3/SHAGGY-like kinase to mediate brassinosteroid responses in rice. *Plant Cell* 24:2562–77.

Wang, C., Fan, Y., Zheng, C., Qin, T., Zhang, X. and Zhao, K. (2014). High-resolution genetic mapping of rice bacterial blight resistance gene *Xa23*. *Mol. Genet. Genomics* 289:745-53.

Wang, E., Wang, J., Zhu, X., Hao, W., Wang, L., Li, Q., Zhang, L., He, W., Lu, B., Lin, H., Ma, H., Zhang, G. and He, Z. (2008b). Control of rice grain-filling and yield by a gene with a potential signature of domestication. *Nat. Genet.* 40:1370–4.

Wang, J., Nakazaki, T., Chen, S., Chen, W., Saito, H., Tsukiyama, T., Okumoto, Y., Xu, Z. and Tanisaka, T. (2009). Identification and characterization of the erect-pose panicle gene *EP* conferring high grain yield in rice (*Oryza sativa* L.). *Theor. Appl. Genet.* 119:85–91.

Wang, P., Xing, Y., Li, Z. and Yu, S. (2012a). Improving rice yield and quality by QTL pyramiding. *Mol. Breed.* 29:903–13.

Wang, S., Wu, K., Yuan, Q., Liu, X., Liu, Z., Lin, X., Zeng, R., Zhu, H., Dong, G., Qian, Q., Zhang, G. and Fu, X. (2012b). Control of grain size, shape and quality by *OsSPL16* in rice. *Nat. Genet.* 44:950–4.

Wang, S., Li, S., Liu, Q., Wu, K., J Zhang, Wang, S., Wang, Y., Chen, X., Zhang, Y., C Gao, Wang, F., Huang, H. and Fu, X. (2015a). The *OsSPL16-GW7* regulatory module determines grain shape and simultaneously improves rice yield and grain quality. *Nat. Genet.* 47:949–54.

Wang, X., Jia, Y., Shu, Q. Y. and Wu, D. (2008a). Haplotype diversity at the *Pi-ta* locus in cultivated rice and its wild relatives. *Phytopathology* 98:1305–11.

Wang, Y., Xiong, G., Hu j, Jiang, L., Yu, H., Xu, J., Fang, Y., Zeng, L., Xu, E., Xu, J., Ye, W., Meng, X., Liu, R., Chen, H., Jing, Y., Wang, Y., Zhu, X., Li, J. and Qian, Q. (2015b). Copy number variation at the *GL7* locus contributes to grain size diversity in rice. *Nat. Genet.* 47:944–8. doi:10.1038/ng.3346.

Weng, J., Gu, S., Wan, X., Gao, H., Guo, T., Su, N., Lei, C., Zhang, X., Cheng, Z., Guo, X., Wang, J., Jiang, L., Zhai, H. and Wan, J. (2008). Isolation and initial characterization of *GW5*, a major QTL associated with rice grain width and weight. *Cell Res.* 18:1199–1209. doi:10.1038/cr.2008.307.

Xie, X., Song, M. H., Jin, F., Ahn, S. N., Suh, J. P., Hwang, H. G. and McCouch, S. R. (2006). Fine mapping of a grain weight quantitative trait locus on rice chromosome 8 using near-isogenic lines de-rived from a cross between *Oryza sativa* and *Oryza rufipogon*. *Theor. Appl. Genet.* 113:885–94.

Xing, Y. Z., Tang, W. J., Xue, W. Y., Xu, C. G. and Zhang, Q. (2008). Fine mapping of a major quantitative trait loci, *qSSP7*, controlling the number of spikelets per panicle as a single Mendelian factor in rice. *Theor. Appl. Genet.* 116:789–96.

Xu, C., Wang, Y., Yu, Y., Duan, J., Liao, Z., Xiong, G., Meng, X., Liu, G., Qian, Q. and Li, J. (2012). Degradation of *MONOCULM 1* by APC/C^{TAD1} regulates rice tillering. *Nat. Commun.* 3:750.

Xu, Y., Mccouch, S. R. and Zhang, Q. (2005). How can we use genomics to improve cereals with rice as a reference genome? *Plant Mol. Biol.* 59:7–26. doi:10.1007/s11103-004-4681-2.

Xue, W., Xing, Y., Weng, X., Zhao, Y., Tang, W., Wang, L., Zhou, H., Yu, S., Xu, C., Li, X. and Zhang, Q. (2008). Natural variation in *Ghd7* is an important regulator of heading date and yield potential in rice. *Nat. Genet.* 40:761–7.

Ya-dong, Z., Jia, Z., Yan-li, L., Chun-fang, Z., Tao, C., Qing-yong, Z., Zhen, Z., Li-hui, Z., Shu, Y., Ling, Z., Xing, Y. and Cai-lin, W. (2015). Functional marker development and effect analysis of grain size gene *GW2* in extreme grain size germplasm in rice. *Rice Sci.* 22:65–70. doi:10.1016/S1672-6308(14)60280-8.

Yan, W. H., Wang, P., Chen, H. X., Zhou, H. J., Li, Q. P., Wang, C. R., Ding, Z. H., Zhang, Y. S., Yu, S. B., Xing, Y. Z. and Zhang, Q. F. (2011). A major QTL, *Ghd8*, plays pleiotropic roles in regulating grain productivity, plant height, and heading date in rice. *Mol. Plant* 4:319–30.

Yang, H. Y., Ren, X., Weng, Q. M., Zhu, L. L. and He, G. C. (2002). Molecular mapping and genetic analysis of a rice brown planthopper (*Nilarparvata lugens* Stål) resistance gene. *Hereditas* 136:39–43.

Ying, J. Z., Gao, J. P., Shan, J. X., Zhu, M. Z., Shi, M. and Lin, H. X. (2012). Dissecting the genetic basis of extremely large grain shape in rice cultivar 'JZ1560'. *J. Genet. Genomics* 39:325–33.

Yoon, D. B., Kang, K. H., Kim, H. J., Ju, H. G., Kwon, S. J., Suh, J. P., Jeong, O. Y. and Ahn, S. N. (2006). Mapping quantitative trait loci for yield components and morphological traits in an advanced backcross population between *Oryza grandiglumis* and the *O. sativa* japonica cultivar Hwaseongbyeo. *Theor. Appl. Genet.* 112:1052–62.

Yoshida, A., Ohmori, Y., Kitano, H., Taguchi-Shiobara, F. and Hirano, H. Y. (2012). *ABERRNT SPIKELET AND PANICLE1*, encoding a TOPLESS related transcriptional co-repressor, is involved in the regulation of meristem fate in rice. *Plant J.* 70:327–39.

Yu, B., Lin, Z., Li, H., Li, X., Li, J., Wang, Y., Zhang, X., Zhu, Z., Zhai, W., Wang, X., Xie, D. and Sun, C. (2007). *TAC1*, a major quantitative trait locus controlling tiller angle in rice. *Plant J.* 52:891–8.

Yuan, P. R., Kim, H. J., Chen, Q. H., Ju, H. G., Lee, S. J., Ji, S. D. and Ahn, S. N. (2009). QTL dissection of agronomic and domestication traits using introgression lines carrying wild rice (*Oryza rufipogon* Griff.) segments in cultivated rice (*O. sativa* L.) background. *J. Crop Sci. Biotechnol.* 12:245–52.

Zha, X., Luo, X., Qian, X., He, G., Yang, M., Li, Y. and Yang, J. (2009). Over-expression of the rice *LRK1* gene improves quantitative yield components. *Plant Biotechnol. J.* 7:611–20.

Zhang, L., Wang, J., Wang, J., Wang, L., Ma, B., Zeng, L., Qi, Y., Li, Q. and He, Z. (2015). Quantitative trait locus analysis and fine mapping of the *qPL6* locus for panicle length in rice. *Theor. Appl. Genet.* 128:1151–61.

Zhang, Q. (2007). Strategies for developing green super rice. *Proc. Natl. Acad. Sci. USA* 104:16402–9. doi:10.1073/pnas.0708013104.

Zhang, S. W., Li, C. H., Cao, J., Zhang, Y. C., Zhang, S. Q., Xia, Y. F., Sun, D. Y. and Sun, Y. (2009). Altered architecture and enhanced drought tolerance in rice via the down-regulation of indole-3-acetic acid by *TLD1/OsGH3*.13 activation. *Plant Physiol.* 151:1889–1901.

Zhang, Z. H., Wang, K., Guo, L., Zhu, Y. J., Fan, Y. Y., Cheng, S. H. and Zhuang, J. Y. (2012). Pleiotropism of the photoperiod-insensitive allele of *Hd1* on heading date, plant height and yield traits in rice. *PLoS ONE* 7:e52538.

Zhou, G., Chen, Y., Yao, W., Zhang, C., Xie, W., Hua, J., Xing, Y., Xiao, J. and Zhang, Q. (2012). Genetic composition of yield heterosis in an elite rice hybrid. *Proc. Natl. Acad. Sci. USA* 109:15847–52. doi:10.1073/pnas.1214141109.

Zhou, Y., Zhu, J., Li, Z., Yi, C., Liu, J., Zhang, H., Tang, S., Gu, M. and Liang, G. (2009). Deletion in a quantitative trait gene q*PE9-1* associated with panicle erectness improves plant architecture during rice domestication. *Genetics* 183:315–24.

Zhu, K., Tang, D., Yan, C., Chi, Z., Yu, H., Chen, J., Liang, J., Gu, M. and Cheng, Z. (2010). *ERECT PANICLE2* encodes a novel protein that regulates panicle erectness in indica rice. *Genetics* 184: 343–50.

Zong, G., Wang, A., Wang, L., Liang, G., Gu, M., Sang, T. and Han, B. (2012). A pyramid breeding of eight grain-yield related quantitative trait loci based on marker-assistant and phenotype selection in rice (*Oryza sativa* L.). *J. Genet. Genomics* 39:335–50.

Chapter 4

Improving photosynthesis in rice: from small steps to giant leaps

R. F. Sage, University of Toronto, Canada; and Shunsuke Adachi and Tadashi Hirasawa, Tokyo University of Agriculture and Technology, Japan

1 Introduction

There is currently much interest in improving crop yields to meet projected demands for agricultural commodities, which are expected to roughly double by 2050 (Tilman et al., 2011; Ray et al., 2013). The growth in projected demand results from a combination of increase in food requirements due to population growth, changing economic and social conditions that favour meat consumption, and the cultivation of new biofuel and fibre crops (Ray et al., 2012; Long et al., 2015). In addition, productivity enhancements will need to compensate for yield stagnation due to soil erosion, salinization, higher fertilizer costs, urbanization and climate change-induced stress (Sheehy and Mitchell, 2011; Ray et al., 2012). As a crop of warm climates with high water requirements, rice is particularly vulnerable to yield losses in future climates; for example, for each 1°C rise in mean minimum temperature during the growing season, rice yields are predicted to decline by 10% (Peng et al., 2004). Much of the increase in production can be realized through improved crop management, introduction of new varieties with superior disease resistance and elevated CO_2 concentrations. These options are not expected to fully meet future demand because they are approaching their respective limits of effectiveness, at least in developed nations (Sheehy and Mitchell, 2011). The success of new breeds, modern fertilization regimes, pesticide use and scientific management has enabled crop production to approach yield ceilings determined by available sunlight, local climate and the ability of rice plants to convert solar radiation into usable biomass (Sheehy et al., 2007;

http://dx.doi.org/10.19103/AS.2016.0003.04

Sheehy and Mitchell, 2011). As a crop approaches its yield ceiling, traditional improvement strategies will encounter diminishing returns, and it will become necessary to raise the ceiling to significantly enhance yield. To increase yield ceilings, improvements are required in the efficiency at which photosynthesis converts solar energy into biomass (Sheehy and Mitchell, 2011).

A significant constraint on the utilization of solar energy is the crop conversion efficiency (ε_c), which reflects the efficiency by which photosynthesis converts photons into new biomass (Zhu et al., 2010). As described by Monteith (1977), crop yield (Y) equals

$$Y = 0.487 S_t \varepsilon_i \varepsilon_c \varepsilon_p \qquad \text{eq. 1}$$

Where 0.487 corrects for the usable fraction of incoming solar radiation, S_t is total solar input onto a crop field, ε_i is the efficiency of light interception, ε_c is the conversion efficiency represented by gross photosynthesis minus all respiratory processes, and ε_p is partitioning efficiency of carbohydrate to the yield product. In a well-managed field with modern varieties, ε_i and ε_p are close to their theoretical maxima (Zhu et al., 2010). By contrast, ε_c is well below the theoretical maximum, indicating this term represents the greatest potential for enhancing the yield ceiling of a crop. The most important limit on ε_c is photorespiration, which for rice and other C_3 crops of warm climates can reduce ε_c by close to 50% above 30°C (Zhu et al., 2008).

With recent advances in the understanding of the photosynthetic biochemistry and physics of light harvesting, new strategies for improving photosynthesis are being evaluated in numerous C_3 crops, including rice (Makino et al., 1997; Makino, 2003; Sudo et al., 2014; Zhu et al., 2010; Ort et al., 2015; RIPE, 2016). Advances in molecular engineering have made possible the rapid engineering of small suites of genes to overcome limitations and improve photosynthetic capacity (Ort et al., 2015; Simkin et al., 2015). Genomic technologies, coupled with quantitative genetics, are identifying numerous suites of genes that control photosynthetic performance and growth in rice and other crops (Flood et al., 2011; Raines, 2011; Adachi et al., 2014a; Reynolds and Langridge, 2016). With the rapid development of genomic screening and molecular engineering capabilities, it is now possible to envision sophisticated engineering efforts which could introduce the C_4 pathway and other carbon concentrating systems into C_3 plants of warm climates, which would solve the photorespiration problem and enhance yields by close to 50% (Sheehy et al., 2007; Covshoff and Hibberd, 2012). Because human activities are changing atmospheric CO_2 concentration and warming the global climate, researchers also need to be aware that strategies to improve photosynthesis in the past may no longer apply (Zhu et al., 2004, 2007). Therefore, it is important to consider the environments in which crops will be grown when designing strategies to improve photosynthesis for future conditions.

In this chapter, we review various strategies for improving rice photosynthesis, from the small steps to dramatic leaps. Much of the discussion will address the issue of engineering the C_3 photosynthetic mechanism of the rice plant to optimally exploit elevated CO_2 conditions expected in the coming century; however, because engineering C_4 photosynthesis into rice has the potential for a large payoff, we will also consider this strategy for improving rice photosynthesis. To begin, we present a theoretical discussion of photosynthetic limitations that will allow readers to understand the rationale behind efforts to improve photosynthesis in rice and other C_3 crops.

2 Demand functions in C_3 plants: biochemical limitations

Rice uses the C_3 photosynthetic pathway, and given its widespread cultivation in warm climates of low latitude, it stands out as an exception to the common pattern that C_4 plants predominate in warm, high light environments of the tropics and subtropics. The success of rice in warm environments is surprising given that atmospheric CO_2 levels have been relatively low, averaging 275 ppm, between the dawn of rice cultivation 8000 years ago and the beginning of the nineteenth century (Sage, 1995). During the era of rice cultivation, atmospheric CO_2 concentrations exceeded 300 ppm only in the past century, and at the time of this writing, they just rose above 400 ppm (CO2now.org). At these low CO_2 levels of recent time, net CO_2 assimilation rate (*A*) is strongly limited by the carboxylation capacity of photosynthesis, and the inhibitory reactions of photorespiration are high in the warm environments where rice grows (Sage, 2013). With CO_2 levels on the rise, the profile of limitation is shifting, and thus the strategies for photosynthetic enhancement will change (Woodrow, 1994; Zhu et al., 2007). It is therefore useful to discuss patterns of photosynthetic limitation across a range of CO_2 availabilities, so the reader will be able to appreciate why past strategies no longer apply, while understanding the logic behind future possibilities for improving rice photosynthesis.

The limitations on C_3 photosynthesis can be subdivided into two general categories: biochemical and diffusion limitations (Table 1). Biochemical limitations have been further subdivided into the capacity of Rubisco to consume RuBP, or what is often termed a 'Rubisco limitation', and the capacity to regenerate RuBP (Fig. 1; Farquhar et al., 1980). Together, biochemical limitations establish the demand for CO_2 within C_3 leaves in what has been termed the 'demand function' of photosynthesis (Farquhar and Sharkey, 1982). Demand functions can be measured using whole-leaf gas exchange techniques to factor out the stomatal limitations on photosynthesis, enabling researchers to plot *A* as a function of intercellular CO_2 concentration (C_i) within leaves. Because Rubisco and RuBP regeneration limitations show distinct gas exchange responses to C_i, the *A* versus C_i response can be analysed to quantify Rubisco versus RuBP regeneration capacities (Fig. 2; von Caemmerer and Farquhar, 1981; Sharkey et al., 2007; von Caemmerer, 2013).

In addition to the biochemical limitations, the rate at which the supply of CO_2 within chloroplasts is replenished depends upon a series of diffusion conductances that are present in the transport path from the atmosphere to the site of RuBP carboxylation in the chloroplast stroma (Evans and Loreto, 2000; Terashima et al., 2011). The best known of these is the highly regulated stomatal conductance (g_s) that controls the diffusion of CO_2 into the intercellular spaces; however, an equally significant series of diffusion steps are collectively termed 'mesophyll conductance' (g_m), which controls flux from the intercellular spaces to the site of carboxylation in the chloroplast stroma (Terashima et al., 2011). Because of these diffusion limitations, chloroplast CO_2 concentrations are about half of atmospheric CO_2 concentrations (Scafaro et al., 2011).

We now address the major subcategories of photosynthetic limitation in order, beginning with the Rubisco limitation.

2.1 Rubisco capacity limitations

Rubisco is the most abundance enzyme in the photosynthetic apparatus, accounting for about a quarter of all leaf protein in rice and other C_3 plants (Sage et al., 1987; Evans, 1989;

Table 1 A hierarchical outline of the processes and sub-processes that can control photosynthetic rate in C_3 plants

A) Biochemical limitations (Demand function limitations)		B) Diffusion limitations (Supply function limitations)	
A1) Rubisco capacity to consume RuBP	A2) RuBP regeneration capacity	B1) Stomatal and boundary layer conductance	B2) Mesophyll conductance
a) Rubisco content	a) Electron transport capacity (can include any step in the electron transport chain)	a) Stomatal number	a) Length, tortuosity of the intercellular air spaces
b) Rubisco activation state (includes limitations in activity of Rubisco activase and binding of Rubisco inhibitors)	b) Light harvesting (reflects antennae size, chlorophyll content, photoinhibition. Light harvesting limitations are common in low light)	b) Stomatal aperture	b) Cell wall thickness
	c) Calvin cycle (can include any step in the Calvin cycle, but SBPase has greatest control)	c) Leaf boundary layer conductance (reflects leaf size, shape and pubescence)	c) CO_2 porin concentration and activity in plasmalemma and chloroplast membranes
	d) P_i regeneration by starch and sucrose synthesis	d) Crop canopy boundary layer	d) Chloroplast coverage of the plasmalemma
c) Photorespiration i) competition for Rubisco active sites ii) loss of previously fixed CO_2	e) Photorespiration i) RuBP consumption ii) loss of previously fixed CO_2 iii) energy for reassimilating NH_4^+ and PGA formation		e) Carbonic anhydrase concentration

P_i, orthophosphate; SBP, sedoheptulose bisphosphate.

Makino et al., 2000). Despite this, it is often limiting for carbon gain in recent, low CO_2 atmospheres (Fig. 2; von Caemmerer and Evans, 2010). In warm environments favouring rice, and at the low CO_2 concentration of the pre-industrial era (275 ppm), Rubisco is limited by the supply of its CO_2 substrate, and it functions below its K_m for CO_2 (Woodrow, 1994; Sage, 2002). To overcome this limitation, plants either make an abundance of Rubisco, or enhance the supply functions to increase the concentration of CO_2 around Rubisco, for example, by opening stomata (Makino and Mae, 1999; Sage and Coleman, 2001; Zhu et al., 2004). Modern crop practices compensate for the Rubisco limitation through heavy use of fertilizers, which increase leaf protein content and photosynthetic capacity (Evans, 1989). Selectively overexpressing Rubisco using super-sense transgenic approaches represent an

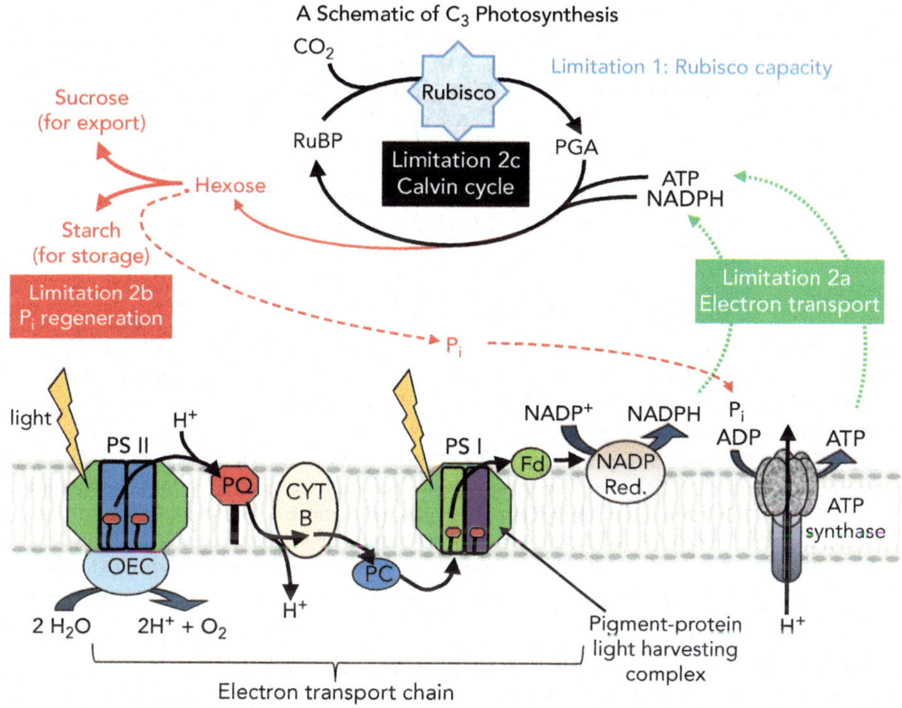

Figure 1 A schematic of the photosynthetic apparatus in C_3 plants, showing the main steps in the electron transport chain, the limitation associated with Rubisco capacity to consume RuBP (shown in light blue), and the main components contributing to the RuBP regeneration capacity. Electron transport limitations are delineated in black, Calvin cycle limitations delineated in green, and the P_i regeneration limitations are delineated in red.

alternative strategy that has already been developed to overcome Rubisco limitations in rice at lower CO_2 concentrations (Suzuki et al., 2007; Sudo et al., 2014). With the ongoing rise in atmospheric CO_2, Rubisco content limitations are becoming less pronounced, given the greater atmospheric availability of CO_2 (Zhu et al., 2004). As a result, overexpression of Rubisco may no longer be a useful strategy, because it produces small returns yet ties up substantial resources in the Rubisco enzyme (Zhu et al., 2007; Sudo et al., 2014). Instead, it may be more suitable to alter Rubisco kinetics to accelerate its turnover rate in a higher CO_2 environment, as will be further discussed subsequently (Sharwood et al., 2016).

Rubisco is a heavily regulated enzyme that must be activated in order to be catalytically functional (Spreitzer and Salvucci, 2002). This occurs when the active site is carbamylated by the binding of a non-substrate CO_2 molecule and a magnesium atom to a lysine residue in the active site. In the absence of carbamylation, Rubisco can become locked in a catalytically functionless condition by the binding of RuBP or similar pentose phosphates, while in the carbamylated state, it can become inactive if a phosphorylated sugar that resembles its transition state intermediate gets into the active site and binds tightly, blocking RuBP entry (Spreitzer and Salvucci, 2002). To remove these sugars from both carbamylated and non-carbamylated active sites, plants employ

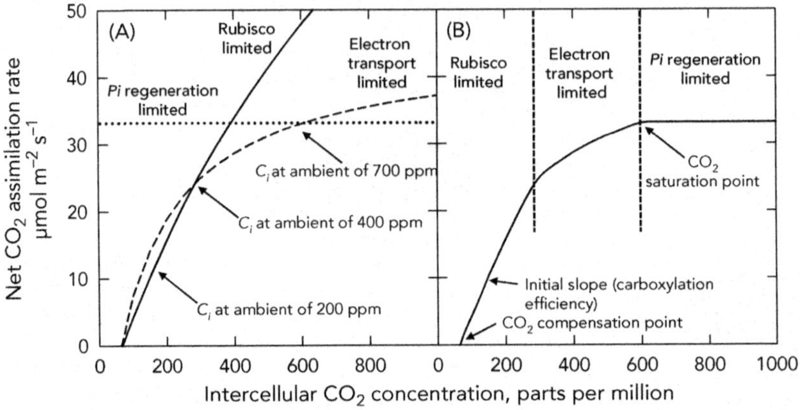

Figure 2 (Panel A) Modelled responses of net CO_2 assimilation rate (*A*) at 21% O_2 and 30°C. The responses of biochemical limitations on *A* are shown for Rubisco capacity limitations (solid line), RuBP regeneration limitations as determined by electron transport capacity (dashed line), and RuBP regeneration limitations as determined by the capacity of starch and sucrose synthesis to regenerate P_i (dotted line). (Panel B) The corresponding response of *A* in a model leaf to variation in intercellular CO_2 concentration, showing regions of the curve where Rubisco capacity, electron transport capacity and P_i regeneration capacity limit *A*. The response in panel B is comprised of the minimum value of the three curves at a given C_i in panel A. Also shown are the approximate C_i values corresponding to atmospheric CO_2 concentrations of 200 ppm during the late Pleistocene epoch 20 000 years ago, the current CO_2 of 400 ppm, and CO_2 levels of 800 ppm projected for the next century. Note, at 200 ppm, Rubisco capacity is the dominant limitation, while at 800 ppm, one of the RuBP regeneration processes is limiting. Responses here and in Fig. 3 and 7 were modelled according to Farquhar et al. (1980) as modified by Sage (1990) and Bernacchi et al. (2003). Rubisco V_{cmax} was assumed to be 100 µmol m^{-2} s^{-1}, Jmax was 160 µmol m^{-2} s^{-1}, and triose phosphate use rate 7.5 µmol m^{-2} s^{-1}.

Rubisco activase which uses ATP to drive conformational changes that evict inhibitory sugars from the active site of Rubisco and thus enable carbamylation and RuBP entry (Carmo-Silva et al., 2015). Activase, however, is a heat (and possibly cold)-labile enzyme that loses integrity above approximately 35°C, allowing for Rubisco deactivation and a reduction in the capacity of Rubisco to consume RuBP (Salvucci and Crafts-Brandner, 2004). This realization has led to efforts to improve photosynthesis by engineering a heat-stable form of Rubisco activase into the genome of crops destined for warm to hot environments (Yamori et al., 2012; Carmo-Silva et al., 2015). In rice, over-expression of the activase gene enhanced Rubisco activation state at warm temperatures, although Rubisco content was reduced (Fukayama et al., 2012; Yamori et al., 2012). Activation limitations are also apparent in variable light environments, because slow activation of Rubisco following increases in light intensity can limit *A* (Pearcy, 1990). In rice at 40°C, over-expressing Rubisco activase accelerates the photosynthetic response to light fluctuation and may thus be a means to improve carbon gain within rice canopies (Yamori et al., 2012). In most rice-growing regions, light intensity is highly variable because of passing clouds and leaf movement within the canopy, indicating faster Rubisco activation could provide important benefits.

2.2 RuBP regeneration capacity

Processes limiting the capacity to regenerate RuBP control *A* at sub-saturating light intensities, or if light saturated, at atmospheric CO_2 concentrations typically greater than current values (Fig. 2; von Caemmerer and Farquhar, 1981). Limitations in RuBP regeneration capacity will also occur when leaves become photoinhibited, which can result from either light-induced damage or mechanisms to quench excess light energy within the light harvesting apparatus. RuBP regeneration capacity is often equated with an electron transport limitation, since the flux of electrons in support of ATP and NADPH synthesis is limiting at both sub-saturating light intensity and in saturating light at elevated CO_2 levels (Farquhar et al., 1980). Electron transport limitation is a generalized phrase that encompasses many potential points of limitation, including photosystem I and II function, activity of inter-chain electron carriers, and light absorbance by the chlorophyll-containing antennae complexes (Table 1; Fig. 1). Multiple Calvin cycle enzymes – sedoheptulose 1-6-bisphosphate (SBPase), fructose-1,6-bisphophatase (FBPase) and FBP aldolase – also exert control over photosynthesis, and over-expressing these in tobacco enhances photosynthesis and growth by up to 30% at elevated CO_2, when RuBP regeneration capacity is limiting (Tamoi et al., 2006; Raines, 2011; Rosenthal et al., 2011; Uematsu et al., 2012; Simkin et al., 2015). SBPase exhibits the greatest control of the Calvin cycle enzymes over *A*, leading to conclusions that increasing its content could be one of the easiest ways to boost *A* in crops destined for future conditions (Tamoi et al., 2006; Raines, 2011).

In order to sustain ATP synthesis and in turn RuBP regeneration, inorganic phosphate (P_i) has to be released from triose phosphates by starch and sucrose synthesis (Fig. 1). Should the production of triose phosphates exceed their metabolism, then the regeneration of P_i becomes limiting for photosynthesis (Sharkey, 1985). This situation generally occurs in cooler temperatures (<15°C) at current levels of CO_2, or at a wider range of temperatures at elevated CO_2, when the production of triose phosphate is stimulated by accelerated carboxylation rates enabled by high CO_2 (Sage and Kubien, 2007). P_i regeneration limitations (also known as triose phosphate use limitations, or TPU limitations; Sharkey, 1985) impose a CO_2-saturated ceiling on *A* that stands out in *A* versus C_i curves (Fig. 2). These responses represent unique O_2- and CO_2-insensitive signatures that enable one to identify a P_i regeneration limitation, and hence it is often recognized as a biochemical limitation distinct from RuBP regeneration (Sharkey, 1985; Sage, 1990). Because P_i regeneration limitation limits *A* by slowing the regeneration of RuBP; however, it is important to remember it is technically a sub-category of the RuBP regeneration mechanism (Table 1).

Once a Rubisco or RuBP regeneration limitation is indicated by gas exchange and other physiological assessments such as pulse-modulated fluorescence (Long and Bernacchi, 2003; Baker et al., 2007), the specific limiting step can be identified by either genetically manipulating levels of individual photosynthetic enzymes or by extracting enzymes and metabolites from leaves and examining their respective activities and concentrations (Sharkey et al., 1986; Sage et al., 1988; Ruuska et al., 2000; Yamori, 2013; Yamori et al., 2016). For example, an A/C_i response that shows *A* to be CO_2 saturated at 400 ppm would be indicative of a P_i regeneration limitation, thus implicating one of the steps in starch and sucrose synthesis as the critical limitation (Sharkey, 1985). Biochemical assays could follow to further pinpoint the limiting step. For example, limiting enzyme activities would be well correlated with *A* values and metabolite concentrations would increase upstream from limiting enzymatic steps (Sharkey et al., 1986; Sage et al., 1988). Once the limiting step is identified, over-expression can enhance the content of the limiting enzyme, as has been demonstrated with SBPase over-expression (Raines, 2011). If the limitation is a thermal lability, an isozyme with a more stable structure could be introduced via breeding or molecular transformation (Yamori, 2013).

It is also possible that biochemical limitations reflect regulatory feedback from limitations in carbohydrate loading into the phloem, or limitations in the ability of sink tissues to metabolize photosynthate (Ainsworth and Busch, 2011). Sink tissues are regions of the plant where carbohydrates are utilized or stored, for example, in roots, meristems, tubers and reproductive tissues. In this case, it will not be possible to enhance *A* unless the limitations in the phloem or sinks are addressed. Often, sink limitations are not immediately apparent because high carbohydrate levels in leaves trigger reductions in photosynthetic gene expression, allowing Rubisco or RuBP regeneration capacities to apparently limit carbon gain (Moore et al., 1999). In such cases, over-expressing the proximal limiting steps within the photosynthetic apparatus will not have much effect, because it will aggravate the sink limitation and strengthen feedback regulation. Instead, the sink capacity will need to be manipulated to remove the feedback limitations (Makino and Mae, 1999; White et al., 2016).

2.3 Photorespiration

The inhibition of photosynthesis by photorespiration has to be specifically recognized given its ability to reduce carbon gain by 20–50% above 25°C in recent atmospheres (Sage, 2013). In theoretical models, photorespiratory effects are integrated into the biochemical limitations outlined in Table 1 and are not immediately obvious (Farquhar et al., 1980; von Caemmerer, 2013). It is thus worth discussing photorespiration in terms of its impact on Rubisco and RuBP regeneration limitations.

Photorespiration refers to the oxygenation of RuBP by Rubisco and the metabolism of the P-glycolate product to PGA, which then enters the Calvin cycle to be converted back to RuBP (Peterhansel and Maurino, 2011). In the process, ammonia and previously fixed CO_2 are given off by glycine decarboxylase, a mitochondrial enzyme. ATP is required to phosphorylate glycerate to PGA, and energy is required to reassimilate ammonia into amino acids. The oxygenation reaction that initiates photorespiration increases with temperature, such that above 30°C in recent CO_2 atmospheres, it can inhibit *A* by over 30% (Sage, 2013). Rising CO_2 suppresses the oxygenase reaction because Rubisco has a much greater affinity for CO_2 relative to O_2, so small increments in CO_2 increase its ability to outcompete O_2 for Rubisco active sites (Sharkey, 1988). Hence, at double to triple the pre-industrial atmospheric CO_2 levels, photorespiration is reduced two-to-three fold. In C_4 plants, which can increase CO_2 tenfold around Rubisco, the suppression of photorespiration exceeds 90% above 30°C (Sage, 2013).

Photorespiration reduces *A* when RuBP regeneration or Rubisco capacity are limiting, with the exception that it does not reduce *A* when P_i regeneration capacity is limiting (Fig. 3; Sharkey, 1985). When Rubisco capacity is limiting, photorespiration inhibits *A* because O_2 competes with CO_2 for limiting Rubisco catalytic sites, and previously fixed CO_2 is given off (Sharkey, 1988). When RuBP regeneration capacity is limiting, photorespiration consumes limiting supplies of RuBP, previously fixed CO_2 is given off, and ATP and NADPH are consumed by the re-assimilation of the ammonia produced by glycine decarboxylase and in PGA regeneration from pyruvate. When P_i regeneration capacity is limiting, photorespiration still occurs, but its impact is not observed because the ceiling on photosynthesis is set by the

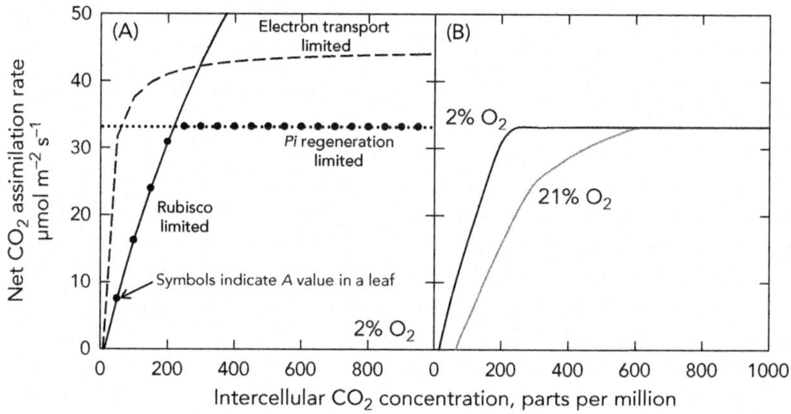

Figure 3 (Panel A) Modelled responses of net CO_2 assimilation rate (*A*) at 2% O_2 and 30°C as determined by the limiting biochemical processes to intercellular CO_2 concentration (C_i) for a C_3 plant. The responses of biochemical limitations on *A* are shown for Rubisco capacity limitations (solid line), RuBP regeneration limitations as determined by electron transport capacity (dashed line), and RuBP regeneration limitations as determined by the capacity of starch and sucrose synthesis to regenerate P_i (dotted line). In panel B, the corresponding modelled C_i response of *A* at 2% O_2 (black line) is compared with the modelled response at 21% O_2 (grey line) from Fig. 2B. Note that P_i regeneration-limited *A* is unaffected by O_2 variation, so it becomes the main limitation above 200 ppm CO_2. Electron-transport-limited *A* is relatively unaffected by CO_2 until below 200 ppm. Rubisco-limited *A* has a lower CO_2 compensation point and a much steeper initial response to CO_2 at 2% O_2 compared to 21% O_2.

rate at which triose phosphates are metabolized (Sharkey, 1985). Hence, *A* is insensitive to varying CO_2 and O_2 under a P_i regeneration limitation, as shown in Fig. 2 and 3.

Both Rubisco and RuBP regeneration limitations (minus the P_i regeneration exception) show a pronounced CO_2 response below a C_i of 500 ppm (Fig. 2). This is confusing at first, because there is a tendency to view the RuBP regeneration limitation as reflecting only the supply rate of the RuBP substrate, and thus the CO_2 dependence of RuBP regeneration is surprising. When RuBP regeneration capacity is limiting, the CO_2 dependency of *A* arises because photorespiration consumes RuBP and energy in addition to giving off previously fixed CO_2. Because increasing CO_2 inhibits photorespiration, *A* increases with rising CO_2 when RuBP regeneration capacity is limiting. When Rubisco capacity is limiting, the CO_2 dependency is due to the direct effect of CO_2 as a substrate and the inhibitory effects of rising CO_2 on photorespiration. Differences in the CO_2 response of Rubisco and RuBP regeneration limitations become apparent when photorespiration is minimal, as occurs at low O_2 or CO_2 levels above 1000 ppm. Reducing oxygen to 2% eliminates the CO_2 dependency of RuBP-regeneration limited *A* (except at very low CO_2), while at 2% O_2, Rubisco-limited *A* shows a strong CO_2 dependency that reflects the substrate response of the enzyme (Fig. 3). To avoid confusion, the term 'CO_2 limitation' and 'RuBP limitation' are not used in favour of Rubisco capacity limitation and RuBP regeneration limitation (von Caemmerer and Farquhar, 1981).

Because of the strong limitation that photorespiration imposes in warm climates of recent time, there has been much discussion of how to reduce the oxygenation reaction of Rubisco (Spreitzer and Salvucci, 2002; Parry et al., 2013). As shown by equation 2, three variables determine the rate of photorespiration relative to photosynthesis ($0.5V_o/V_c$):

$$0.5V_o/V_c = (1/S_{rel})(O/C) \quad \text{eq. 2}$$

V_o is the rate of oxygenation, V_c is the carboxylation rate, S_{rel} is the specificity of Rubisco for CO_2 relative to O_2, O is the O_2 concentration in the stroma and C is the CO_2 concentration in the stroma (von Caemmerer and Quick, 2000). $0.5V_o$ is the rate of photorespiration, reflecting that for every two oxygenation reactions, one CO_2 is given off at glycine decarboxylase. The significance of equation 2 is that it demonstrates how photorespiration can be reduced – either S_{rel} must be increased, or the O to C ratio must be reduced. Reducing O is not practical given the atmospheric concentration is very high at 210 000 ppm (Sage, 2013). Changing S_{rel} is practical given that it varies widely in nature (Galmes et al., 2005; Parry et al., 2013). There is, however, a trade-off between affinity and catalytic speed of Rubisco (Sharwood et al., 2016), and it appears this trade-off may already be optimized in plants within their natural habitat at the CO_2 conditions of recent times (Sage, 2002; Tcherkez et al., 2006). With increasing atmospheric CO_2, the optimal values for Rubisco kinetics change, which creates opportunities for introducing isoforms of Rubisco's that will perform better in future atmospheres (Zhu et al., 2004, 2007; Sharwood et al., 2006).

The third option, increasing C, is the one that Mother Nature has repeatedly evolved, through the CO_2 concentrating systems of C_2, C_4 and CAM photosynthesis, and in aquatic environments, the ability to actively accumulate bicarbonate (Sage and Stata, 2014). C_2 photosynthesis occurs when photorespiratory glycine is shuttled into bundle sheath cells for decarboxylation, causing a localized increase in CO_2 within this compartment (Sage and Stata, 2014). The C_4 option has been identified as the best way to overcome photorespiration in rice (Sheehy et al., 2007), and it will be discussed in some depth in the C_4 engineering section below. Within the context of C_3 photosynthesis there are a number of options for increasing C by modest amounts. A promising strategy that accounts for a 10% to 20% enhancement in chloroplast CO_2 levels is to introduce bacterial enzymes that completely metabolize photorespiratory glycolate within the chloroplast, and in doing so short-circuit the photorespiratory pathway (Long et al., 2015). These bacterial pathways release photorespired CO_2 directly within the chloroplast rather than in the mitochondria, which enhances chloroplast CO_2 levels and A (Kebeish et al., 2007; Peterhansel and Maurino, 2011). Releasing photorespiratory CO_2 within the chloroplast allows Rubisco to immediately refix it, while its release in mitochondria increases the chance this CO_2 will escape the cell and not immediately be refixed. A number of photorespiratory bypasses have been introduced into *Arabidopsis* and the biofuel crop *Camelina sativa*, indicating they may be widely available in a range of C_3 plants in the near future, including rice (Kebeish et al., 2007; Peterhansel and Maurino, 2011; Dalal et al., 2015).

3 Supply functions in C_3 plants: diffusion limitations

To enter the chloroplast from the atmosphere, CO_2 has to diffuse through the boundary layer, stomata, intercellular air spaces, the mesophyll cell wall and cytosol, and finally the chloroplast envelope (Fig. 4). Impediments to diffusion at each of these steps reduce the

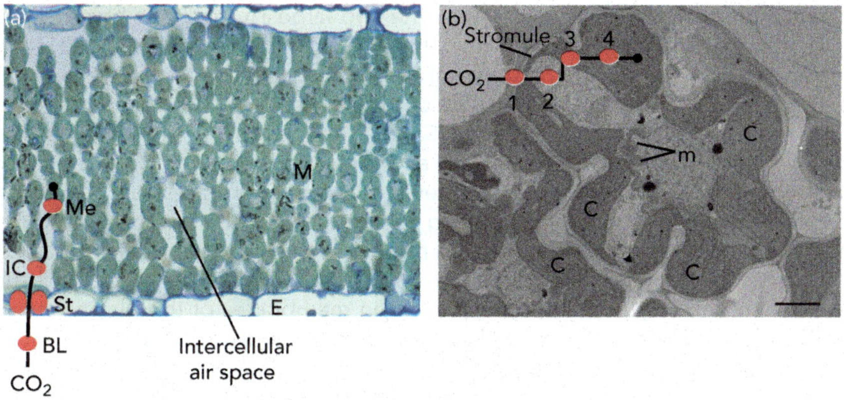

Figure 4 (a) A longitudinal section of a rice leaf showing the close arrangement of the rice mesophyll cells and the diffusion path a CO_2 molecule would take from the outside air to a chloroplast. The major diffusion conductances in the flux pathway are shown as red ovals where BL is boundary layer conductance, St is stomatal conductance, IC is conductance through the intercellular air spaces and Me is mesophyll cell conductance. (b) A cross section of a rice mesophyll cell showing the deep lobes and chloroplast arrangements in the cell, and the flux pathway for CO_2 into the cell from the intercellular air spaces. The components of the mesophyll conductance are shown as red ovals where 1 is the cell wall conductance, 2 is the cytoplasm liquid phase conductance, 3 is the chloroplast membrane conductance, and 4 is the chloroplast stromal conductance. Note the lack of significant vacuoles in the mesophyll cells. Other abbreviations, M, mesophyll cells, E, epidermis, C, chloroplast; m, mitochondria.

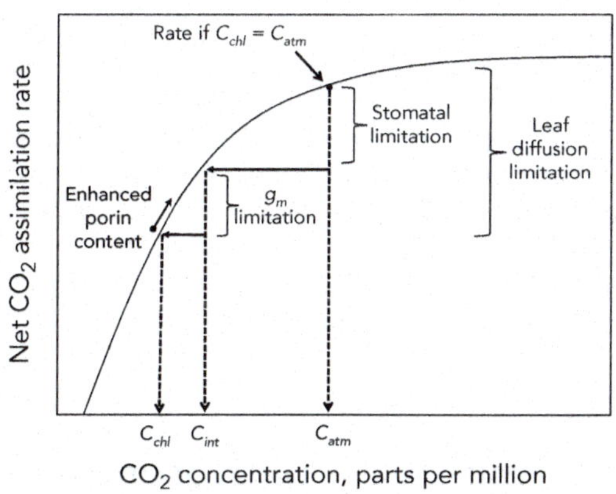

Figure 5 A modelled net CO_2 assimilation rate (A) versus CO_2 concentration response for a typical C_3 plant, showing the reduction in *A* and compartment CO_2 that would occur due to limiting stomatal and mesophyll conductance values. C_{chl} indicates the chloroplast CO_2 concentration corresponding to atmospheric CO_2 values (C_{atm}); C_{int} show the CO_2 concentration in the intercellular air spaces corresponding to C_{atm}. The putative effect of over-expressing porin content in the plasmalemma or chloroplast envelope membrane is shown as an increase in the C_{chl} value.

CO_2 concentration in the chloroplast, with the two largest reductions being associated with the stomata and mesophyll conductances (Fig. 5; Terashima et al., 2011). In rice grown in controlled conditions, the stomatal are regulated to maintain CO_2 concentrations in the intercellular spaces that are 75–85% of the atmospheric CO_2 concentration, which is on the higher side of typical C_3 plants (Sage, 2000; Scafaro et al., 2011). As a result, stomatal limitations are not severe in rice, and there may not a lot of room to enhance stomatal conductance without risking excessive transpiration rates that may injure the plant. Enhancing mesophyll conductance, g_m, potentially has the advantage of increasing stromal CO_2 without affecting transpiration rate. The most direct approach for increasing g_m is to increase chloroplast surface area exposure to intercellular air spaces and enhance the number and activity of CO_2 porins present in membranes of the plasmalemma and chloroplasts (Yamori, 2013; Flexas et al., 2016). Over-expressing CO_2 porins in rice led to a 40% increase in g_m (Hanba et al., 2004). In tobacco and *Arabidopsis*, over-expression or other transgenic enhancement of porin content by 20–45% were associated with increases in *A* of 8–21% (Flexas et al., 2016). Similarly, over-expressing a cyanobacterial inorganic carbon transporter in tobacco increased *A* over 20% at elevated CO_2, possibly by accelerating a parallel, bicarbonate flux pathway into the chloroplast (Simkin et al., 2015).

In rice, half of the g_m value is associated with the thickness of mesophyll cell walls; thicker walls have lower g_m, indicating wall design may be a means to manipulate g_m (Scafaro et al., 2011). One could also manipulate leaf anatomy to enhance chloroplast surface area contact with the cell periphery, which would reduce the diffusion distance through the liquid phase of the mesophyll cells (Terashima et al., 2011; Tholen et al., 2012; Flexas et al., 2016). Mesophyll conductance is directly proportional to chloroplast surface area exposure, so maintaining high chloroplast coverage of the mesophyll periphery would help maintain high g_m (Evans et al., 2009). However, rice already has one of the highest g_m values recorded in higher plants, which is related in part to its specialized anatomy (von Caemmerer and Evans, 1991; Sage and Sage, 2009; Scafaro et al., 2011). In rice, mesophyll cells exist as lobed disks that are arrayed close to each other in rows along the vascular tissue (Fig. 4a). The deep lobes increase surface area exposure, and chloroplasts are packed into the lobes, giving high chloroplast coverage of the cell periphery (S_c/S_m) that enhances g_m (Fig. 4b; Terashima et al., 2011). This anatomy also places greater chloroplast volume within the mesophyll cell volume, such that rice has a photosynthetic capacity close to that of wheat for a given leaf N supply, despite having leaves that are half to two-thirds as thick (Sage and Sage, 2009). Hence, it is possible that the rice leaf is already well-designed to maximize g_m in the warm, paddy environment, and manipulating its anatomy may not be the most promising way to enhance *A*.

4 Strategies for improving photosynthetic performance

Table 2 lists a number of leading strategies under discussion or in development for improving photosynthesis in C_3 plants (Zhu et al., 2010; Yamori, 2013; Ort et al., 2015; Furbank et al., 2015; RIPE, 2016). Some of the options in Table 2 have been discussed above and need not be repeated, and to cover all would exceed the space allotment

Table 2 Strategies to improve photosynthesis in rice by enhancing photosynthetic conversion efficiency (ε_c)

I. QTL analysis to exploit existing genetic variation and identify novel genetic controls over photosynthetic capacity (mechanisms for improvement are initially unknown).

II. Optimize C_3 metabolism using transgenic approaches (mainly for elevated CO_2 conditions).

- A) Optimize Rubisco performance
    - 1) Reduce content of Rubisco and photorespiratory enzymes
    - 2) Introduce a faster form of Rubisco
    - 3) Introduce Rubisco with greater S_{rel} (hot environment)
    - 4) Enhance Rubisco activase stability/content in warm environments
- B) Enhance RuBP regeneration capacity
    - 1) Overexpress the Calvin cycle enzymes SBPase, FBPase and FBP aldolase
    - 2) Enhance electron transport (e.g. by overexpressing cytochrome b_6f)
    - 3) Increase starch and/or sucrose synthesis to avoid P_i regeneration limitation
    - 4) Enhance relaxation rate of photoprotection processes
    - 5) Reallocate resources from upper to mid-canopy to improve canopy light distribution
    - 6) Enhance sink strength to overcome whole-plant feedback limitations

III. Increased CO_2 delivery to Rubisco (may become superfluous in high CO_2)

- 1) Introduce a carbon concentrating mechanism
    - a) Engineer the C_4 photosynthetic pathway
    - b) Engineer an algal-type bicarbonate pump
- 2) Introduce a photorespiratory bypass to release photorespired CO_2 in chloroplasts
- 3) Accelerate CO_2 delivery by increased mesophyll conductance
    - a) Introduce bicarbonate transporters from cyanobacteria
    - b) Overexpress membrane-bound CO_2 porins
    - c) Reduce the thickness of mesophyll cell walls

FBP, fructose bisphosphate; SBP, sedoheptulose bisphosphate.

for this article (see Ort et al., 2015 for a discussion of options IIB4 and IIB5 which are not discussed here). In this section, we restrict our discussion of improvement strategies to three specific topics that highlight work currently underway. The first is to exploit existing variation in rice using quantitative trait locus (QTL) analysis, in order to identify genetic regions that increase photosynthetic capacity. This approach already shows promise and may lead to discoveries of unknown regulatory and development genes that could enhance *A* and whole plant performance. The second is to re-engineer C_3 photosynthesis in the rice plant in order to optimally adapt it to elevated CO_2 atmospheres that will likely occur within a few decades. The third is the introduction of the C_4 photosynthetic pathway into rice.

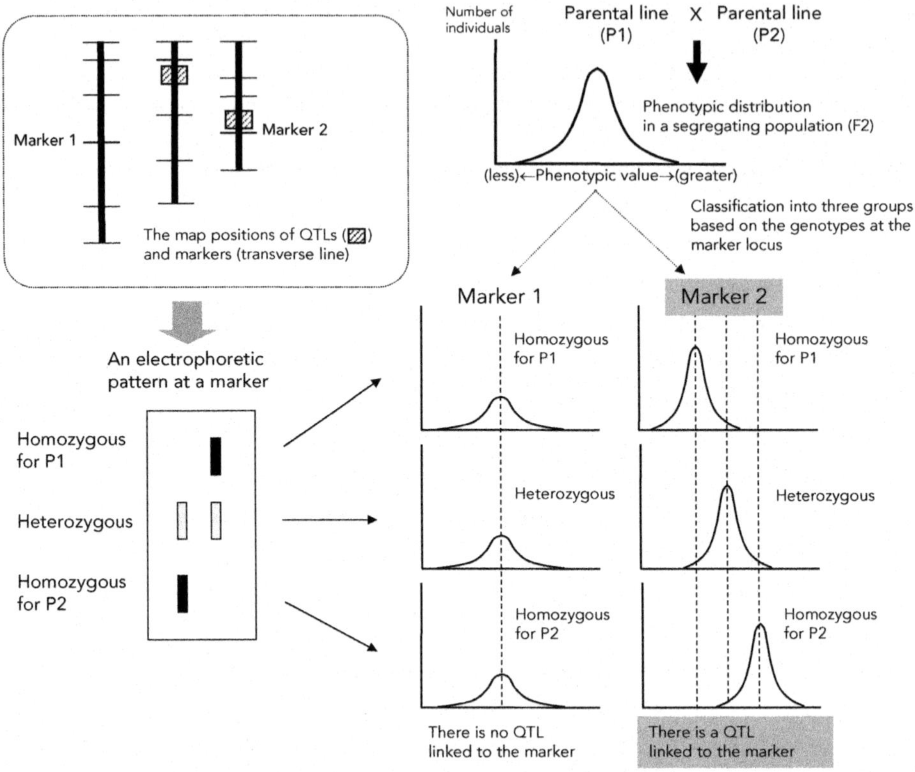

Figure 6 The methods for mapping of quantitative trait loci (QTLs). The phenotypic values for a particular trait of interest are typically evaluated using F_2 segregating populations. Subsequently, genotypes with DNA markers evenly spaced over the genome are analysed. When no QTL locates near a marker, the average phenotypic value will be similar among the three groups that are classified by genotypes at the marker locus (Marker 1). When a QTL locates near the marker, the average phenotypic value will be significantly different among the groups (Marker 2). QTL analysis applies this approach to all DNA markers in the study and evaluates the number and location of QTLs and the phenotypic effect of each QTL.

4.1 Improving rice photosynthetic capacity by QTL analysis

It is well recognized that there is wide variation in photosynthesis rate in rice that could be exploited to maximize carbon gain and resource use efficiency (Cook and Evans, 1983; Kanemura et al., 2007; Hirasawa et al., 2010). With the publication of the rice genome (International Rice Genome Sequencing Project, 2005), DNA markers for enhanced *A* can be identified, thereby facilitating rapid identification of QTL associated with improved photosynthesis (Yamamoto and Yano, 2008; Flood et al., 2011). QTL analysis identifies the number and location of desired QTLs in the genome, and the phenotypic effect of each QTL (Tanksley, 1993). QTL analysis is conducted as follows (Fig. 6): (1) develop a segregating population such as an F_2 hybrid population, (2) develop a linkage map using DNA markers evenly distributed over the genome, (3) evaluate the phenotypic values and conduct a statistical test at each marker using computer programs such as QTL Cartographer and R/qtl (Basten et al., 2005; Broman and Sen, 2009). The genes within each QTL can then be identified by high-resolution mapping, namely map-based cloning (Yano, 2001). QTL analysis requires a segregating population comprising different individuals. F_2 and BC_1F_1 populations are most frequently used because of the reduced time needed for their development. Backcrossed inbred lines (BILs) and chromosome segment substitution lines are used in studies that aim to identify QTLs for complex traits (Yamamoto et al., 2014). Genome-wide association studies, which do not rely on a controlled bi-parental segregating population, are also used to identify QTLs. This capacity reflects the improved availability of large numbers of genetic markers such as single-nucleotide polymorphisms (Myles et al., 2009).

There are two options to identify QTLs associated with increased leaf photosynthesis. The first is direct measurements of *A* in a segregating population using gas exchange measurement systems, such as the Li-Cor 6400 system (www.Licor.com). In rice, Teng et al. (2004) found two QTLs for CO_2 assimilation rate using Double Haploid lines derived from a cross between *japonica* and *indica* cultivars. Hu et al. (2009) and Gu et al. (2012) found several QTLs for rice photosynthesis using RILs and BILs lines, respectively, under both water-limited and well-watered conditions. Another option to identify QTLs is to determine phenotypic values of traits related to *A*. These include screening for SPAD values which is a non-destructive index of leaf chlorophyll content (Takai et al., 2013b). Pulse-modulated chlorophyll fluorescence is also used to identify QTLs for several photosynthesis-related traits (Guo et al., 2008; Liang et al., 2010; Czyczyło-Mysza et al., 2011; Kasajima et al., 2011). Rubisco and nitrogen content as well as carbon isotope discrimination ($\Delta^{13}C$), which reflects water use efficiency, have also been used for QTL analysis (Ishimaru et al., 2001; Laza et al., 2006; Takai et al., 2009). Using gas exchange approaches, Takai et al. (2013a) conducted high-resolution mapping for *A* and identified the *Green for Photosynthesis* (*GPS*) gene that correlates with different *A* values between *indica* and *japonica* cultivars. Their analysis showed that the *indica*-type *GPS* gene increases *A* by increasing Rubisco content, which might be attributed to enhancement of mesophyll numbers and leaf thickness. The sequencing analysis revealed that *GPS* encodes a plant-specific protein that may be involved in polar auxin transport. Another report shows that some QTLs enhance *A* via increased g_s as well as increased hydraulic conductance (Adachi et al., 2011). These results support hypotheses that genes responsible for the varietal differences in *A* commonly regulate non-photosynthetic attributes such as root growth and leaf anatomy, rather than encoding genes for proteins within the photosynthetic apparatus. As such, QTL analysis will be important for identifying genetic controls over allocation and development that would otherwise be difficult to identify. When combined with molecular strategies that target single photosynthetic limitations (e.g. by over-expressing SPBase), it should be possible to maximize the improvement of C_3 photosynthesis in rice. Already, multiple QTLs have been pyramided into rice varieties to enhance gains in *A* over that possible with one QTL alone (Adachi et al., 2013). Adachi et al. (2014a), for example, introduced two genetic regions from *indica* rice into the genetic background of *japonica* rice, which increased *A* by 7–18% beyond the enhancement associated with each line alone. Over-expressing SPBase could provide another 15–27% enhancement in *A*, if results with tobacco can be reproduced with rice (Tamoi et al., 2006).

Two developments are needed to accelerate identification of photosynthetically relevant QTLs. The first is to enhance the measurement efficiency of gas exchange and related traits because these measurements are still labour-intensive and time-consuming (Adachi et al., 2011). This reflects a general need for high-throughput phenotyping platforms that screen plants at greatly accelerated rates compared to single measurement systems such as the Li-Cor 6400 (Furbank and Tester, 2009). The second is to decrease measurement variation since photosynthesis parameters are often influenced by environmental conditions and plant-to-plant variability that prevent accurate evaluation of QTLs (Tanksley, 1993). To meet this objective, 'phenomic' platforms have been designed to minimize environmental variation over the plant population while enabling rapid measurements with minimal interference (Yang et al., 2013). In phenomics, pulse-modulated leaf fluorescence, leaf colour and temperature, and plant size and morphology are remotely measured by automated sensors that pass over the plant population, and sophisticated imaging programs analyse the results to quantify trait values, or in breeding and mutagenic studies, to identify individuals who stand out from the general pattern (Furbank and Tester, 2011; Yang et al., 2013).

4.2 Adapting rice to elevated CO_2

Atmospheric CO_2 concentrations are now 400 ppm and should exceed 600 ppm by the middle of this century; they could approach 800 ppm by the end of the century assuming 'business-as-usual' CO_2 emission scenarios (Collins et al., 2013). Photorespiration at these future CO_2 levels will be reduced by over half, while the Rubisco limitation pattern will weaken if not altogether disappear as RuBP regeneration limitations predominate (Woodrow, 1994; von Caemmerer and Evans, 2010). The rise to date in atmospheric CO_2 to 400 ppm is predicted to have already altered limitation profiles in the current flora, and further perturbation is expected with the higher CO_2 levels later in this century (Zhu et al., 2004, 2007).

The CO_2 stimulation of *A* in C_3 plants often induces excessive carbohydrate build-up that either directly impairs photosynthesis (e.g. from massive starch grains), or triggers regulatory processes that reduce expression of photosynthetic genes, in what is termed photosynthetic CO_2 acclimation (Makino and Mae, 1999; Moore et al., 2009). Because of this, *A* often declines with prolonged exposure to elevated CO_2, and the significant stimulation originally observed following a rise in CO_2 is weakened if not altogether lost (Long et al., 2004; Leakey et al., 2009). Well-fertilized crop plants tend to show less downward acclimation of *A* when grown in elevated CO_2, because the large supply of nutrients promotes high sink strength in growing tissues; however, in both fertilized- and unfertilized plants, nitrogen invested into leaves is inefficiently used in part because Rubisco levels remain excessive (Makino and Mae, 1999; Zhu et al., 2004). C_3 plants thus underperform at elevated CO_2 relative to what they could if fully optimized for high CO_2 conditions (Zhu et al., 2007; Leakey et al., 2009). This recognition has led to research to redesign the C_3 plant to optimally perform in future atmospheres of elevated CO_2 (Woodrow, 1994; Zhu et al., 2007, 2010; Long et al., 2015).

Rice has been the subject of much research on the effects of elevated CO_2, in both lab and field settings (Makino and Mae, 1999; Chen et al., 2005; Ainsworth, 2008; Hasegawa et al., 2013; 2015; Sudo et al., 2014). When grown at elevated CO_2, leaf nitrogen content declines and reduces Rubisco and electron transport capacity by 5–20% (Makino and Mae, 1999; Chen et al., 2005; Hasegawa et al., 2015). Qualitative acclimation does not appear to be strong in rice, as the ratio of Rubisco capacity to electron transport capacity declines slightly if at all as CO_2 increases and an imbalance between Rubisco and RuBP

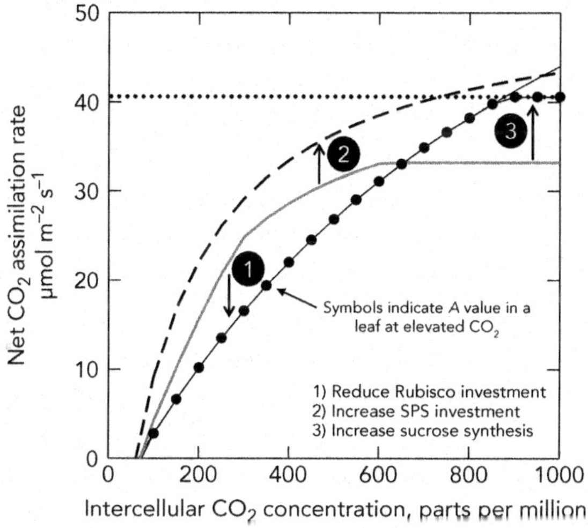

Figure 7 Modelled responses of net CO_2 assimilation rate (*A*) at 21% O_2 and 30°C for a C_3 plant engineered for optimal performance at elevated CO_2. The responses of biochemical limitations on *A* are shown for Rubisco capacity limitations (solid line), RuBP regeneration limitations as determined by electron transport capacity (dashed line), and RuBP regeneration limitations as determined by the capacity of starch and sucrose synthesis to regenerate P_i (dotted line). The responses are compared to the *A versus* C_i response for plants in current CO_2 levels from Fig. 2B (grey line). Statements 1-3 indicate specific strategies to re-optimize a C_3 leaf for elevated CO_2 In this particular simulation, the Rubisco V_{Cmax} capacity has been reduced from 100 μmol m^{-2} s^{-1} in Fig. 2 to 65 μmol m^{-2} s^{-1}, while the RuBP regeneration capacity has been increased from 160 to 180 μmol m^{-2} s^{-1} by assuming enhanced SPS content. The P_i regeneration capacity has been increased from 7.5 in Fig. 2 to 9 μmol m^{-2} s^{-1}.

regeneration limitations are maintained at elevated CO_2 (Makino et al., 1997; Makino and Mae, 1999; Adachi et al., 2014b; Hasegawa et al., 2015). Therefore, an obvious objective for improving future rice photosynthesis is to reduce investment in Rubisco, Rubisco activase and photorespiratory enzymes in order to redistribute the resources into components of electron transport, the Calvin cycle, P_i regeneration and sink metabolism (Fig. 7; Woodrow, 1994; Zhu et al., 2007). The validity of this approach is supported by transgenic studies demonstrating significant enhancements in *A* at elevated CO_2 when RuBP regeneration components are over-expressed and Rubisco expression is reduced. In tobacco, over-expression of the Calvin cycle enzyme SBPase and FPB aldolase boosts *A* by up to 30% at elevated CO_2 (Tamoi et al., 2006; Rosenthal et al., 2011; Simkin et al., 2015), while reducing Rubisco content in rice enhances nitrogen use efficiency in elevated CO_2 (Makino et al., 2000). Over-expression of electron transport components may also enhance *A* at elevated CO_2 in rice (Yamori et al., 2016). At some point, particularly in cooler conditions, P_i regeneration limitations at elevated CO_2 will become paramount as *A* is increased by over-expressing electron transport and Calvin cycle capacities (Fig. 7). When this happens, it will be necessary to boost P_i regeneration by over-expressing carbohydrate transporters, starch and sucrose synthesis enzymes, and/or genes controlling sink strength (Woodrow, 1994; Makino and Mae, 1999; Zhu et al., 2007). Such conclusions are supported by studies demonstrating that rice lines with greater sink strength have greater yield responses in elevated CO_2 (Hasegawa et al., 2013), and that transgenic plants with modest increases in sucrose-phosphate synthase activity can exhibit greater growth (Sharkey et al., 2000, 2004).

As the entry point for carbon into photosynthetic organisms, Rubisco should have been under intense selection pressure to optimize the conflicting challenges of low CO_2, inhibitory O_2 levels and variable temperatures (Sage, 2002; Tcherketz et al., 2006). In response, plants from contrasting environments have evolved Rubisco isoforms with S_{rel} and kinetic parameters that appear to be optimized for their native habitats (Galmes et al., 2005; Parry et al., 2011). In the C_3 flora, Rubisco from warm, drought-prone climates tend to have lower catalytic turnover rate (lower k_{cat}, representing a slower Rubisco) and greater S_{rel}, while cool climate plants have faster Rubisco's with greater k_{cat} values (Makino et al., 2000; Sage, 2002; Galmes et al., 2005). The fastest Rubisco's in the terrestrial flora occur in C_4 plants (Sage, 2002; Whitney et al., 2011). These observations indicate that as CO_2 levels increase, optimal forms of Rubisco in C_3 plants from warm climates will shift from low k_{cat} versions with a higher affinity (such as currently found in rice), to high k_{cat} versions with lower affinity (as found in wheat and C_4 plants). Elevated CO_2 will suppress the oxygenase reaction and thus a faster form of Rubisco would not suffer any penalty that would be apparent in low CO_2 (Makino 2011; Sharwood et al., 2016). Current research is attempting to introduce faster Rubisco's into C_3 crops, as a means to pre-adapt them to elevated CO_2 conditions (Carmo-Silva et al., 2015; RIPE, 2016). For example, Ishikawa et al. (2011) introduced the Rubisco small subunit from a C_4 plant, sorghum, into rice and obtained a more C_4-like Rubisco in terms of its kinetics. Challenges remain, however, notably in obtaining proper assembly of transgenic Rubisco's, and chloroplast transformation capabilities only exist for a few species (Parry et al., 2013; Sharwood et al., 2016).

4.3 C_4 rice engineering

Engineering the C_4 pathway into C_3 plants has long been a dream of photosynthetic biologists, since shortly after the C_4 pathway was discovered 50 years ago. This is because C_4 plants use a C_4 (four-carbon) metabolic cycle to concentrate CO_2 into an internal cellular compartment (usually the bundle sheath cells) where Rubisco is localized, thus overcoming photorespiration and providing enough CO_2 to almost saturate the enzyme (Hatch, 1987). In the C_4 cycle, PEP carboxylase first fixes CO_2 to PEP to produce the four-carbon organic compound oxaloacetate (OAA). OAA is then converted to either malate or aspartate, which diffuse into the bundle sheath where they are decarboxylated by one of three enzymes (NADP-malic enzyme, NAD-malic enzyme, or PEP carboxykinase) to release CO_2 and a three-carbon molecule, usually pyruvate. The pyruvate returns to the mesophyll cells where it is phosphorylated back to PEP using the equivalent of two ATP molecules. This represents the energy dependent step in C_4 photosynthesis. Its requirement increases the energy-expenditure of C_4 photosynthesis over C_3 photosynthesis in the absence of photorespiration. In the presence of photorespiration in warm climates, the energy requirements of the C_4 pathway are typically less than that of C_3 plants, leading to greater radiation use efficiency, by about 50% in warm environments such as where rice is grown (Sage, 2000; Zhu et al., 2008). Equally significant, the ability to concentrate CO_2 enhances the water and nitrogen use efficiencies of C_4 photosynthesis and enables C_4 plants to exhibit higher yield relative to C_3 plants while reducing resource requirements (Ghannoum et al., 2011). In warm environments of the late twentieth century, yield enhancements and resource use efficiency improvements of 40–60% in C_4 relative to C_3 plants are not

Table 3 Engineering objectives for establishing the C_4 photosynthetic pathway in rice

A) Establish the Kranz Anatomical Structure for C_4 photosynthesis
1) Reduce mesophyll cell number so each mesophyll cell contacts a bundle sheath cell
2) Increase the number of bundle sheath cells by increasing vein density
3) Increase the size of bundle sheath cells
4) Activate the bundle sheath by increasing chloroplast number and size; reduce chloroplast number in mesophyll cells
B) Establish a photosynthetically driven metabolic pump to capture CO_2
1) Boost activity of PEP carboxylase in the mesophyll cell cytosol
2) Boost the rate of oxaloacetate conversion to malate by malate dehydrogenase
3) Enhance PPDK activity in mesophyll cell chloroplasts in order to couple light energy to PEP regeneration
4) Enhance carbonic anhydrase activity to convert CO_2 to bicarbonate in mesophyll cells
5) Enhance NADP-malic enzyme capacity to decarboxylate malate in bundle sheath tissues
6) Alter regulatory set-points to coordinate C_3 and C_4 metabolic cycles
C) Establish an ability to trap CO_2 and allow its accumulation around Rubisco
1) Localize Rubisco and Calvin cycle enzymes to the BS compartment
2) Establish thick, suberized walls to provide a large diffusive resistance between the Rubisco compartment and the intercellular air spaces, to slow leakage of CO_2
D) Establish rapid metabolite transport between mesophyll and bundle sheath cells
1) Regulate metabolite flux to maintain high metabolite concentration gradients between M and BS tissues to allow for rapid diffusion
2) Create a high capacity for metabolite movement across membranes via enhanced numbers of plasmsodesmata, channels and carriers

unusual (Sheehy et al., 2007). As such, successfully engineering C_4 photosynthesis into rice represents a giant leap that could meet the world's rice demands well into the future.

C_4 engineering will not be easy, given the complexity of the genetic controls over the C_4 pathway, many of which are not understood (Sage and Zhu, 2011; Covshoff and Hibberd, 2012). To fully implement the C_4 pathway into a C_3 species such as rice, a range of changes will be required to leaf and organelle structure, metabolism and transport capabilities (Table 3). Comparisons of transcriptomes between closely related C_3 and C_4 species indicate that the necessary structure/function adjustments require alterations to hundreds of genes, at least, of which many remain undiscovered (Sage and Zhu, 2011; Brautigam et al., 2011; Gowik et al., 2011). While daunting, there are a number of developments that provide hope that C_4 engineering in rice can occur in the near future. First, genetic engineering capabilities are improving at a rapid pace, with new technologies for high-throughput sequencing, genetic transformation, and DNA editing showing particular promise for accelerating the rate of gene discovery and improving the ability to stack the required genes for C_4 function into rice (Ort et al., 2015). Second, nature has evolved

C_4 photosynthesis at least 60 times, which suggests the engineering of the C_4 pathway into rice will not be overly complex (Sage et al., 2011). These natural examples of C_4 engineering could provide multiple blueprints that can be analysed to understand how C_4 photosynthesis could be established in rice (Covshoff et al., 2014). Third, although they failed to produce C_4 rice plants, previous attempts to engineer the C_4 pathway in the 1990s created a strong foundation of experience and gene constructs to facilitate more recent efforts (Miyao et al., 2011). Because of the promise of C_4 engineering, two major funding programs for C_4 engineering have been initiated in recent years, one by the International C_4 Rice Engineering consortium (C4rice.com), which is specifically focused on creating a C_4 rice, and a second by a European consortium, which is focused on C_4 engineering in model species such as *Arabidopsis* (www.3to4.org).

A simplified summary of the C_4 rice engineering effort identifies four key stages, the first being gene discovery and development of a molecular engineering toolkit (Fig. 8). The next phase is transforming rice to express Kranz anatomy and the metabolic enzymes of the C_4 cycle, resulting in a C_4 prototype. Following this will be efforts to redesign the C_3 regulatory machinery to create an efficient C_4 cycle, while also re-engineering plants to be able to effectively utilize the additional carbohydrates flooding into the plant. Once a model C_4 rice plant is working well, it would be crossed with common varieties to create C_4 rice plants that are compatible with the local environment and produce desired grain quality. At this point, assuming regulatory approval, C_4 rice would be a commercial reality. In total, engineering a field-ready C_4 rice is assumed to require two to three decades, but this will depend upon funding, the strength of the research teams, organizational efficiency and the regulatory process.

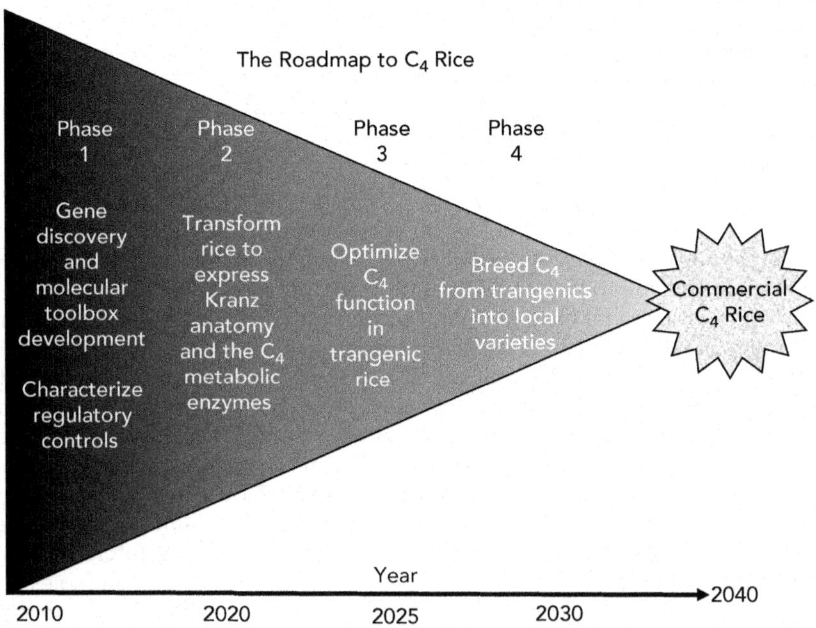

Figure 8 A simple flow diagram depicting the phases of the C_4 Rice Engineering effort (www.C4rice.org).

The first major C_4 rice effort occurred in the 1990s with the objective to engineer a functional C_4 pathway into single mesophyll cells of rice (Matsuoka et al., 2001). This mechanism would function along the lines of the single-cell C_4 mechanism present in certain aquatic plants such as *Hydrila verticilata*, which show minimal ultrastructural rearrangements (Bowes, 2011). While the research team was able to transform rice to express the enzymes of the C_4 pathway, they were unable to build an efficient C_4 metabolic cycle that concentrated CO_2 around Rubico (Miyao et al., 2011). In the current research efforts (C4rice.com; C3to4.org), the emphasis is to engineer a Kranz anatomy type of C_4 rice plant using the NADP-malic enzyme or PEP carboxykinase decarboxylating enzymes, as these subtypes exhibit greater productivity (Covshoff and Hibberd, 2012). The particular challenge of this strategy is limited understanding of the genetic controls over Kranz development in C_4 leaves (Nelson, 2011). In recognition of this limitation, identifying the genes controlling C_4 leaf development are a high priority in the first phase of the C_4 rice engineering effort (Furbank et al., 2015; Wang et al., 2016).

The strategy for gene discovery in the C_4 engineering projects is to pursue multiple options in parallel, using the latest advances in genomic sequencing and informatics to rapidly identify candidate genes (Wang et al., 2016). As diagrammed in Fig. 9, multiple types of genetic screens are being employed by the C_4 engineering consortia and collaborating labs. At IRRI, massive screens of mutagenized rice lines have been employed to find gain of function mutations, while tens-of-thousands of mutagenized sorghum lines have been examined for loss-of-function mutations (Furbank et al.,

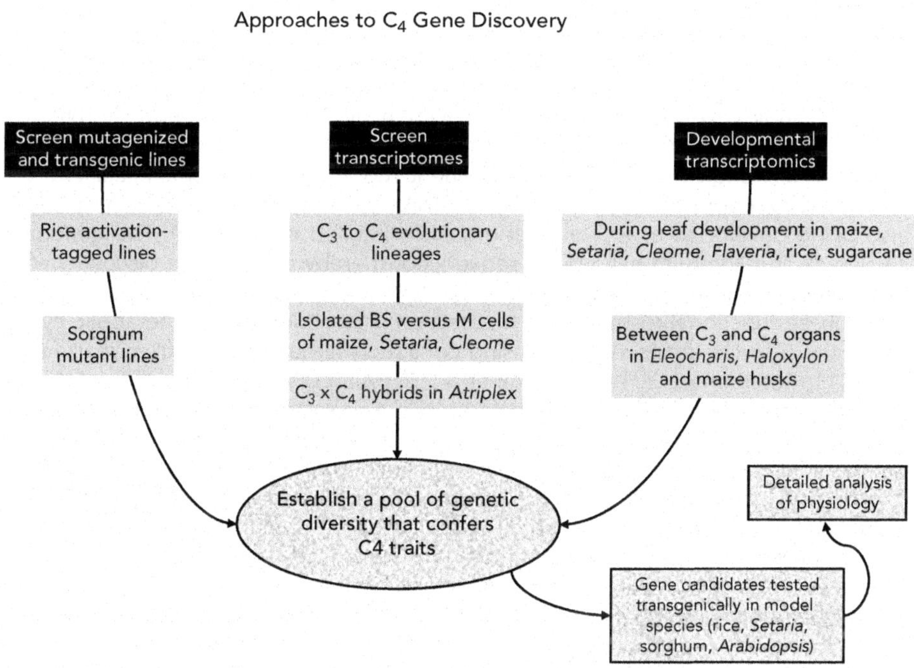

Figure 9 A schematic highlighting approaches to discovering novel genes controlling the expression of C_4 photosynthesis.

2015). These screens produced very few clear candidate genes, which may reflect the difficulties of identifying usable mutants for traits under complex genetic control. A second approach has been to compare transcriptomes of closely related C_3 and C_4 species from natural populations. Background variation in the transcriptomes will be minimized by the close relationship between species, thus allowing the C_3 to C_4 contrasts to stand out. A third approach is compare C_3 and C_4 transcriptomes in plants that exhibit C_3 and C_4 photosynthesis in distinct tissues, such as aquatic (C_3) versus aerial (C_4) culms in the marsh sedge *Eleocharis vivipara* (Chen et al., 2014). Similarly, M and BS cells in maize and *Setaria* have been dissected from C_4 leaves and their respective transcriptomes compared to identify cell-specific patterns of gene expression (Tausta et al., 2014). Fourth, gene expression along a developmental leaf gradient has been examined in rice, maize, *Setaria*, sugarcane and *Cleome* to identify when genes specific to C_3 versus C_4 leaf development are expressed (Wang et al., 2016). There is also a possibility that hybrids between C_3 and C_4 species in the Chenopod genus *Atriplex* will facilitate gene discovery, as many of the C_4 traits differentially segregate into hybrid individuals (Oakley et al., 2014). Together, these approaches have already identified dozens of gene candidates controlling critical steps in C_4 leaf development, and are providing support that certain developmental transcription factors, notably *SCARCROW* and *SHORT-ROOT*, as having a critical role in the transition from a C_3 to a C_4 pattern of leaf development (Fouracre et al., 2014; Wang et al., 2016).

Following identification of gene candidates, their function will be confirmed using genetic model species such as *Arabidopsis*, the C_3 grass *Brachypodium* and rice (Brutnell et al., 2015). Genetic models are species with easy transformability, short-generation times, small size, and prolific seed production (Covshoff et al., 2014). For rice research, the preferred model would be a fast-cycling rice line such as the *japonica* variety Kitaake; however, the rice model requires a C_4 counterpart in order to evaluate loss of function in transgenics expressing candidate genes. A key objective of the early C_4 rice engineering effort has been to develop a genetic work-horse for loss-of-function studies. The weedy *Setaria italica* was identified as a strong candidate for the C_4 genetic model, and after some initial issues with transformability it appears to be working well (Brutnell et al., 2014). Fast-cycling sorghum is also being developed as a C_4 model organism (Mullett et al., 2014). The genomes of both *Setaria* and sorghum have been sequenced, thereby aiding gene discovery efforts, particularly of promoters and other regulatory elements (Covshoff et al., 2014).

Early phase functional research has focused on assembling the C_4 metabolic cycle by stacking the C_4 cycle genes into rice, using constructs largely derived from maize (Kajaja et al., 2011; Peterhansel, 2011). Furbank et al. (2015) report that all five of the NADP-ME cycle genes have been successfully transformed into rice, and ongoing work is addressing regulatory control of these genes to ensure expression in the proper tissue compartment. Even with a functional C_4 metabolic cycle, two major hurdles remain. First, Rubisco and the C_3 biochemistry must be upregulated in the bundle sheath and silenced in the M cells. This requires activating the BS cells by enhancing chloroplast numbers (Gowik and Westhoff, 2011). Second, to establish rapid metabolite flux between M and BS cells, it is necessary to engineer the necessary mix of transporters and plasmodesmata channels between the two cell types, and between organelles and the cytosol within cells (Brautigam and Weber, 2011). A common feature of C_4 plants relative to their C_3 ancestors is the much higher frequency of plasmodesmata across the M to BS cell wall of C_4 plants, which facilitates rapid diffusive flux between these cell types. While a number of the transporters are known and their genes cloned (Furbank et al., 2015), the genes controlling plasmodesmata frequency at the M and BS interface are not known and their identification is a high priority in the gene discovery efforts.

In summary, the renewed C_4 rice engineering effort in now in its eighth year, with a number of notable achievements such as successfully stacking the C_4 cycle genes into rice, developing a working model of Kranz leaf development, putting in place a multi-pronged gene discovery effort using the latest sequencing technologies, and developing C_4 plants that can serve as genetic models. Even if the effort fails, the contribution to our understanding of C_4 photosynthesis will be immense; however, failure is not in the plan, and there is good reason to believe they will be able to engineer a C_4 rice prototype within the next decade or two (Furbank et al., 2015).

5 Future trends and conclusion

Improving *A* enhances yield in rice, as has been shown in elevated CO_2 studies, QTL approaches and genetic transformation to overcome limitations in single genes. With improved understanding of photosynthetic limitation, it is now possible to identify specific limitations and target the genes that overcome these limitations. Meanwhile, QTL approaches are enabling the pyramiding of multiple genetic elements, each of which boost photosynthetic performance. Genetic improvement, however, has to work within the context of rising atmospheric CO_2, which is altering limitation profiles by shifting control over *A* from Rubisco to RuBP regeneration. This reality requires researchers to plan for environments that do not yet exist, and engineer rice photosynthesis accordingly. Given this realization, efforts to engineer rice photosynthesis for current CO_2 conditions may be counterproductive, leading some to consider C_4 rice engineering an unwise investment. There are, however, advantages of C_4 photosynthesis in high CO_2 atmosphere that still favour engineering C_4 rice. One is the generally superior high-temperature performance exhibited by C_4 relative to C_3 crops, which is an important concern given the climate warming that accompanies CO_2 enrichment (Sage and Kubien, 2003). At high temperature, elevated CO_2 reduces the degree to which C_4 photosynthesis is superior, but does not generally erase it. Second, the high water use efficiency of C_4 photosynthesis is typically increased in high CO_2 atmospheres (Sage and Kubien, 2003; Leakey et al., 2009). In a world where fresh water use is a growing concern, the desire to maintain productivity with less water could be a critical. In rice, the C_4 pathway could provide the ability to productively cultivate rice on dry soils, thus dispensing with the fresh water needs of paddy cultivation. As such, rice cultivation could greatly expand across the globe while simultaneously freeing up water for cities, fisheries and a wider array of agricultural commodities.

6 Where to look for further information

Long S. P., Marshall-Colon A. and Zhu X.-G. (2015). Meeting the global food demand of the future by engineering crop photosynthesis and yield potential. *Cell*, 161, 56–66. An approachable review highlighting the latest developments in photosynthetic engineering.

Morton O. (2008). *Eating the Sun: How Plants Power the Planet*. Harper, New York. A popular book explaining photosynthesis and efforts to understand it.

RIPE (2016). Realizing increased photosynthetic efficiency for sustainable increases in crop yield. http://ripe.illinois.edu/. This website explains the research associated with a sophisticated funding effort by the Bill and Melinda Gates Foundation to improve photosynthesis in C_3 crop plants. Most of the options discussed in this paper, with the notable exception of C_4 rice engineering, are explained on this website.

Sheehy, J. E. and Mitchell, P. L. (2011). Developments in rice research: visions and pragmatism. *World Agriculture* 2, 7–19. This is an easily read article that documents limitations to rice yield and explains why C_4 photosynthesis will stimulate rice productivity by 50%.

Sheehy, J. E., Mitchell, P. L. and Hardy, B. (2007). *Charting New Pathways to C_4 Rice.* International Rice Research Institute. Manila. A edited volume of articles by leading C_4 researchers that comprehensively assess the rationale and strategies for engineering C_4 photosynthesis into rice.

www.c4rice.com. This is the website for the third phase C_4 Rice Engineering Project. It explains the rational and recent developments.

http://www.3to4.org/Lectures-and-Interviews. This portal on the 3 to 4 website features seven interviews and lectures by leading C_4 researchers explaining the rationale and science behind C_4 engineering.

7 References

Adachi, S., Baptista, L. Z., Sueyoshi, T., Murata, K., Yamamoto, T., Ebitani, T., Ookawa, T. and Hirasawa, T. (2014a). Introgression of two chromosome regions for leaf photosynthesis from an *indica* rice into the genetic background of a *japonica* rice. *J. Exp. Bot.* 65, 2049–56.

Adachi, M., Hasegawa, T., Fukiyama, H., Tokida, T., Sakai, H., Matsunami, T., Nakamura, H., Sameshima, R. and Okada, M. (2014b). Soil and water warming accelerates phenology and down-regulation of leaf photosynthesis of rice plants grown under free-air CO_2 enrichment (FACE). *Plant Cell Physiol.* 55, 370–80.

Adachi, S., Nakae, T., Uchida, M., Soda, K., Takai, T., Oi, T., Yamamoto, T., Ookawa, T., Miyake, H., Yano, M. and Hirasawa, T. (2013). The mesophyll anatomy enhancing CO_2 diffusion is a key trait for improving rice photosynthesis. *J. Exp. Bot.* 64, 1061–72.

Adachi, S., Nito, N., Kondo, M., Yamamoto, T., Arai-Sanoh, Y., Ando, T., Ookawa, T., Yano, M. and Hirasawa, T. (2011). Identification and characterization of genomic regions on chromosomes 4 and 8 that control the rate of photosynthesis in rice leaves. *J. Exp. Bot.* 62, 1927–38.

Ainsworth, E. A. (2008). Rice production in a changing climate: a meta-analysis of responses to elevated carbon dioxide and elevated ozone concentration. *Global Change Biol.* 14, 1642–50.

Ainsworth, E. A. and Bush, D. R. (2011). Carbohydrate export from the leaf: a highly regulated process and target to enhance photosynthesis and productivity. *Plant Physiol.* 155, 64–9.

Baker, N., Harbinson, J. and Kramer, D. M. (2007). Determining the limitations and regulation of photosynthetic energy transduction in leaves. *Plant Cell Environ.* 30, 1107–25.

Basten, C. J., Weir, B. S. and Zeng, Z. B. (2005). QTL cartographer, version 1.17. Department of Statics, North Carolina State University, Raleigh, p. 189.

Bernacchi, C. J., Pimentel, C. and Long, S. P. (2003). *In vivo* temperature response functions of parameters required to model RuBP limited photosynthesis. *Plant Cell Environ.* 26, 1419–30.

Bowes, G. (2011). Single-cell C_4 photosynthesis in aquatic plants. In A. S. Raghavendra and R. F. Sage (Eds), *C_4 Photosynthesis and Related CO_2 Concentrating Mechanisms, Advances in Photosynthesis*, vol. 32, pp. 63–80. Springer Verlag, Berlin.

Bräutigam, A., Kajala, K., Wullenweber, J., Sommer, M., Gagneul, D., Weber, K. L., Carr, K. M., Gowik, U., Mass, J., Lercher, M. J., Westhoff, P., Hibberd, J. M. and Weber, A. P. M. (2011).

An mRNA blueprint for C_4 photosynthesis derived from comparative transcriptomics of closely related C_3 and C_4 species. *Plant Physiol.* 155, 142–56.

Bräutigam, A. and Weber, A. P. M. (2011). Transport processes: connecting the reactions of C_4 photosynthesis. In A. S. Raghavendra and R. F. Sage (Eds), *C_4 Photosynthesis and Related CO_2 Concentrating Mechanisms, Advances in Photosynthesis*, vol. 32, pp. 199–219. Springer Verlag, Berlin.

Broman, K. W. and Sen, S. (2009). *A Guide to QTL Mapping with R/qtl.* Springer, New York.

Brutnell, T. P., Bennetzen, J. L. and Vogel, J. P. (2015). *Brachypodium distachyon* and *Setaria viridis*: Model genetic systems for the grasses. *Annu. Rev. Plant Biol.* 66, 465–85.

Carmo-Silva, E., Scales, J. C., Madgwick, P. J. and Parry, M. A. (2015). Optimizing Rubisco and its regulation for greater resource use efficiency. *Plant Cell Environ.* 38, 1817–32.

Chen, G. Y., Yong, Z. H., Liao, Y., Zhang, D. Y., Chen, Y., Zhang, H. B., Chen, J., Zhu, J. G. and Xu, D. Q. (2005). Photosynthetic acclimation in rice leaves to free-air CO_2 enrichment related to both ribulose-1,5-bisphosphate carboxylation limitation and ribulose-1,5-bisphosphate regeneration limitation. *Plant Cell Physiol.* 46, 1036–45.

Chen, T., Zhu, X. G. and Lin, Y. (2014). Major alterations in transcript profiles between C_3–C_4 and C_4 photosynthesis of an amphibious species *Eleocharis baldwinii. Plant Mol. Biol.* 86, 93–110.

CO2now.org (2016). www.co2.earth (Accessed 16 March 2016).

Collins, M. R., Knutti, J., Arblaster, J. L., Dufresne, T., Fichefet, P., et al. (2013). Long-term climate Change: Projections, Commitments and Irreversibility. In T. Stocker, D. Qin, G. K. Plattner, M. Tignor, S. K. Allen, J. Boschung, A., Nauels, Y. Xia, V. Bex and P. M. Midgley (Eds), *Climate Change 2013: The Physical Science Basis. Contribution of Working Group I to the Fifth Assessment Report of the Intergovernmental Panel on Climate Change*, pp. 1029–136. Cambridge University Press, Cambridge, UK..

Cook, M. and Evans, L. (1983). Some physiological aspects of the domestication and improvement of rice (*Oryza* spp.). *Field Crops Res.* 6, 219–38.

Covshoff, S., Burgess, S. J., Kneřová, J. and Kümpers, B. M.C. (2014). Getting the most out of natural variation in C_4 photosynthesis. *Photosynth. Res.* 119, 157–67.

Covshoff, S. and Hibberd, J. M. (2012). Integrating C_4 photosynthesis into C_3 crops to increase yield potential. *Curr. Opin. Biotech.* 23, 209–14.

Czyczyło-Mysza, I., Marcińska, I., Skrzypek, E., Chrupek, M., Grzesiak, S., Hura, T., Stojałowski, S., Myśków, B., Milczarski, P. and Quarrie, S. (2011). Mapping QTLs for yield components and chlorophyll a fluorescence parameters in wheat under three levels of water availability. *Plant Genetic Res.* 9, 291–5.

Dalal, J., Lopez, H., Vasani, N. B., Hu, Z., Swift, J. E., Yalamanchili, R., Dvora, M., Lin, X., Xie, D. and Qu, R. (2015). A photorespiratory bypass increases plant growth and seed yield in biofuel crop *Camelina sativa. Biotech. Biofuels* 8, 175.

Evans, J. R. (1989). Photosynthesis and nitrogen relationships in leaves of C_3 plants. *Oecologia* 78, 9–19.

Evans, J. R., Kaldenhoff, R., Genty, B. and Terashima, I. (2009). Resistances along the CO_2 diffusion pathway inside leaves. *J. Exp. Bot.* 60, 2235–48.

Evans, J. R. and Loreto, F. (2000). Acquisition and diffusion of CO_2 in higher plant leaves. In Leegood, R. C., Sharkey, T. D. and von Caemmerer, S. (Eds), *Photosynthesis: Physiology and Metabolism*, pp. 321–51. Kluwer Academic, Dordrecht, The Netherlands.

Farquhar, G. D., von Caemmerer, S. and Berry, J. A. (1980). A biochemical model of photosynthetic CO_2 assimilation in leaves of C_3 species. *Planta* 149, 78–90.

Farquhar, G. D. and Sharkey, T. D. (1982). Stomatal conductance and photosynthesis. *Annu. Rev. Plant Physiol.* 33, 317–45.

Flexas, J., Diaz-Espejo, A., Conesa, M., A., Coopman, R. E., Douthe, C., Gago, J., Galmes, J., Medrano, H., Ribas-Carbo, M., Tomas, M. and Niinemets, U. (2016). Mesophyll conductance to CO_2 and Rubisco as targets for improving intrinsic water use efficiency in C_3 plants. *Plant Cell Environ.* doi:10.1111/pce.12622.

Flood, P. J., Harbinson, J. and Aarts, M. G. M. (2011). Natural genetic variation in plant photosynthesis. *Trend Plant Sci.* 16, 327–35.

Fouracre, J. P., Ando, S. and Langdale, J. A. (2014). Cracking the Kranz enigma with systems biology. *J. Exp. Bot.* 65, 3327–39.

Fukayama, H., Ueguchi, C., Nishikawa, K., Katoh, N., Ishikawa, C., Masumoto, C., Hatanaka, T. and Misoo, S. (2012). Overexpression of Rubisco activase decreases the photosynthetic CO_2 assimilation rate by reducing Rubisco content in rice leaves. *Plant Cell Physiol.* 53, 976–86.

Furbank, R. T., Quick, W. P. and Sirault, X. R. R. (2015). Improving photosynthesis and yield potential in cereal crops by targeted genetic manipulation: prospects, progress and challenges. *Field Crops Res.* 182, 19–29.

Furbank, R. T. and Tester, M. (2011). Phenomics-technologies to relieve the phenotyping bottleneck. *Trends Plant Sci.*16, 635–44.

Galmés, J., Flexas, J., Keys, A. J., Cifre, J., Mitchell, R. A. C., Madgwick, P. J., Haslam, R. P., Medrano, H. and Parry, M. A. J. (2005). Rubisco specificity factor tends to be larger in plant species from drier habitats and in species with persistent leaves. *Plant Cell Environ.* 28, 571–9.

Ghannoum, O., Evans, J. R. and von Caemmerer, S. (2011). Nitrogen and water use efficiency in C_4 plants. In A. S. Raghavendra and R. F. Sage (Eds), *C_4 Photosynthesis and Related CO_2 Concentrating Mechanisms, Advances in Photosynthesis*, vol. 32, pp. 129–46. Springer Verlag, Berlin.

Gowik, U., Bräutigam, A., Weber, K. L., Weber, A. P. and Westhoff, P. (2011). Evolution of C_4 photosynthesis in the genus *Flaveria*: how many and which genes does it take to make C_4? *Plant Cell.* 23, 2087–105.

Gowik, U. and Westhoff, P. (2011). The path from C_3 to C_4 photosynthesis. *Plant Physiol.* 155, 56–63.

Gu, J., Yin, X., Struik, P. C., Stomph, T. J. and Wang, H. (2012). Using chromosome introgression lines to map quantitative trait loci for photosynthesis parameters in rice (*Oryza sativa* L.) leaves under drought and well-watered field conditions. *J. Exp. Bot.* 63, 455–69.

Guo, P. G., Baum, M., Varshney, R. K., Graner, A., Grando, S. and Ceccarelli, S. (2008). QTLs for chlorophyll and chlorophyll fluorescence parameters in barley under post-flowering drought. *Euphytica* 163, 203–14.

Hanba, Y. T., Shibasaka, M., Hayashi, Y., Hayakawa, T., Kasamo, K., Terashima, I. and Katsuhara, M. (2004). Overexpression of the barley aquaporin HvPIP2;1 increases internal CO_2 conductance and CO_2 assimillation in the leaves of transgenic rice plants. *Plant Cell Physiol.* 45, 521–9.

Hasegawa, T., Sakai, H., Tokida, T., Nakamura, N., Zhu, C., Usui, Y., Yoshimoto, M., Fukuoka, M., Wakatsuki, H., Katayanagi, N., Matsunami, T., Kaneta, Y., Sato, T., Takakai, F., Sameshima, R., Okada, M., Mae, T. and Makino, A. (2013). Rice cultivar responses to elevated CO_2 at two free-air CO_2 enrichment (FACE) sites in Japan. *Funct. Plant Biol.* 40, 148–59.

Hasegawa, T., Sakai, H., Tokida, T., Usui, Y., Yoshimoto, M., Fukuoka, M., Nakamura, H., Shimono, H. and Okada, M. (2015). Rice free-air carbon dioxide enrichment studies to improve assessment of climate change effects on rice agriculture. In J. L. Hatfield and D. Fleisher (Eds), *Improving Modeling Tools to Assess Climate Change Effects on Crop Response. Advances in Agricultural Systems Modeling*, vol. 7, pp. 1–24. American Society of Agronomy, Madison, WI, USA.

Hatch, M. D. (1987). C_4 photosynthesis – a unique blend of modified biochemistry, anatomy and ultrastructure. *Biochim. Biophys. Acta* 895, 81–106.

Hirasawa, T., Ozawa, S., Taylaran, R. D. and Ookawa, T. (2010). Varietal differences in photosynthetic rates in rice plants with special reference to the nitrogen content of leaves. *Plant Prod. Sci.* 13, 53–7

Hu, S. P., Zhou, Y., Zhang, L., Zhu, X. D., Li, L., Luo, L. J., Liu, G. L. and Zhou, Q. M. (2009). Correlation and quantitative trait loci analyses of total chlorophyll content and photosynthetic rate of rice (*Oryza sativa*) under water stress and well-watered conditions. *J. Integrative Plant Biol.* 51, 879–88.

International Rice Genome Sequencing Project (2005). The map-based sequence of the rice genome. *Nature* 436, 793–800.

Ishikawa, C., Hatanaka, T., Misoo, S., Miyake, C. and Fukayama, H. (2011). Functional incorporation of sorghum small subunit increases the catalytic turnover rate of Rubisco in transgenic rice. *Plant Physiol.* 156, 1603–11.

Ishimaru, K., Kobayashi, N., Ono, K., Yano, M. and Ohsugi, R. (2001). Are contents of Rubisco, soluble protein and nitrogen in flag leaves of rice controlled by the same genetics? *J. Exp. Bot,* 52, 1827–33.

Kajaja, K., Covshoff, S., Karki, S., Woodfield, H., Tolley, B. J., Dionora, M. J., Mogul, R. T., Mabilangan, A. E., Danila, F. R., Hibberd, J. M. and Quick, W. P. (2011). Strategies for engineering a two-celled C_4 photosynthetic pathway into rice. *J. Exp. Bot.* 62, 3001–10.

Kanemura, T., Homma, K., Ohsumi, A., Shiraiwa, T. and Horie, T. (2007). Evaluation of genotypic variation in leaf photosynthetic rate and its associated factors by using rice diversity research set of germplasm. *Photosynth. Res.* 94, 23–30.

Kasajima, I., Ebana, K., Yamamoto, T., Takahara, K., Yano, M., Kawai-Yamada, M. and Uchimiya, H. (2011). Molecular distinction in genetic regulation of nonphotochemical quenching in rice. *Proc. Nat. Acad. Sci. (USA)* 138, 13835–40.

Kebeish, R., Niessen, M., Thiruveedhi, K., Bari, R., Hirsch, H. J., Rosenkranz, R. and Peterhänsel, C. (2007). Chloroplastic photorespiratory bypass increases photosynthesis and biomass production in *Arabidopsis thaliana*. *Nat. Biotech.* 25, 593–9.

Laza, M. R., Kondo, M., Ideta, O., Barlaan, E. and Imbe, T. (2006). Identification of quantitative trait loci for $\delta^{13}C$ and productivity in irrigated lowland rice. *Crop Sci.* 46, 763–73.

Leakey, A. D.B., Ainsworth, E. A., Bernacchi, C. J., Rogers, A., Long, S. P. and Ort, D. R. (2009). Elevated CO_2 effects on plant carbon, nitrogen, and water relations: six important lessons from FACE. *J. Exp. Bot.* 60, 2859–76.

Liang, Y., Zhang, K. P., Zhao, L., Liu, B., Meng, Q. W., Tian, J. C. and Zhao, S. J. (2010). Identification of chromosome regions conferring dry matter accumulation and photosynthesis in wheat (*Triticum aestivum* L.). *Euphytica* 171, 145–56.

Long, S. P., Ainsworth, E. A., Rogers, A. and Ort, D. R. (2004). Rising atmospheric carbon dioxide: plants FACE the future. *Annu. Rev. Plant Biol.* 55, 591–628.

Long, S. P. and Bernacchi C. J. (2003). Gas exchange measurements, what can they tell us about the underlying limitations to photosynthesis? Procedures and sources of error. *J. Exp. Bot.* 54, 2393–401.

Long, S. P., Marshall-Colon, A. and Zhu, X. G. (2015). Meeting the global food demand of the future by engineering crop photosynthesis and yield potential. *Cell* 161, 56–66.

Makino A. (2003). Rubisco and nitrogen relationships in rice: Leaf photosynthesis and plant growth. *Soil Sci. Plant Nut.* 49, 319–27.

Makino, A. (2011). Photosynthesis, grain yield, and nitrogen utilization in rice and wheat. *Plant Physiol.* 155: 125–9.

Makino, A., Harada, M., Kaneko, K., Mae, T., Shimada, T. and Yamamoto, N. (2000). Whole-plant growth and N allocation in transgenic rice plants with decreased content of ribulose-1,5-bisphosphate carboxylase under different CO_2 partial pressures. *Aust. J. Plant Physiol.* 27, 1–12.

Makino, A. and Mae, T. (1999). Photosynthesis and plant growth at elevated CO_2. *Plant Cell Physiol.* 40, 999–1006.

Makino, A., Shimada, T., Takumi, S., Kaneko, K., Matsuoka, M., Shimamoto, K., Nakano, H., Miyao-Tokutomi, M., Mae, T. and Yamamoto, N. (1997). Does decrease in ribulose-1,5-bisphosphate carboxylase by antisense *RbcS* lead to a higher N-use efficiency of photosynthesis under conditions of saturating CO_2 and light in rice plants? *Plant Physiol.* 114, 483–91.

Matsuoka, M., Furbank, R. T., Fukayama, H. and Miyao, M. (2001). Molecular engineering of C_4 photosynthesis. *Annu. Rev. Plant Physiol. Plant Mol. Biol.* 52, 297–314.

Miyao, M., Masumoto, C., Miyazawa, S. and Fukuyama, H. (2011). Lessons from engineering a single-cell C_4 photosynthetic pathway into rice. *J. Exp. Bot.* 62, 3021–9.

Monteith, J. L. (1977). Climate and the efficiency of crop production in Britain. *Philos. Trans. R. Soc. Lond. Ser. B.* 281, 277–94.

Moore, B., Cheng, S., Sims, D. and Seemann, J. R. (1999). The biochemical and molecular basis for photosynthetic acclimation to elevated atmospheric CO_2. *Plant Cell Environ.* 22, 567–82.
Mullett, J., Morishige, D., McCormick, R., Truong, S., Hilley, J., McKinley, B., Anderson, R., Olson, S. N. and Rooney, W. (2014). Energy sorghum – a genetic model for the design of C_4 grass bioenergy crops. *J. Exp. Bot.* 65, 3479–89.
Myles, S., Peiffer, J., Brown, P. J., Ersoz, E. S., Zhang, Z., Costich, D. E. and Buckler, E. S. (2009). Association mapping: critical considerations shift from genotyping to experimental design. *Plant Cell* 21, 2194–202.
Nelson, T. (2011). The grass leaf developmental gradient as a platform for a systems understanding of the anatomical specialization of C_4 leaves. *J. Exp. Bot.* 62, 3039–48.
Oakley, J. C., Sultmanis, S., Stinson, C. R., Sage, T. L. and Sage, R. F. (2014). Comparative studies of C_3 and C_4 *Atriplex* hybrids in the genomics era: physiological assessments. *J. Exp. Bot.* 65, 3637–47.
Ort, D. R., Merchant, S. S., Alric, J., Barkan, A., Blankenship, R. E., Bock, R., Croce, R., Hanson, M. R., Hibberd, J. M. and Long, S. P. (2015). Redesigning photosynthesis to sustainably meet global food demand and bioenergy demand. *Proc. Nat. Acad. Sci. (USA)* 112, 8529–36,
Parry, M. A. J., Andralojc, P. J., Scales, J. C., Salvucci, M. E., Carmo-Silva, A. E., Alonso, H. and Whitney, S. M. (2013). Rubisco activity and regulation as targets for crop improvement. *J. Exp. Bot.* 64, 717–30.
Pearcy, R. W. (1990). Sunflecks and photosynthesis in plant canopies. *Annu. Rev. Plant Physio. Molec. Biol.* 41, 421–53.
Peng, F. S., Huang, J., Sheehy, J. E., Laza, R. C., Visperas, R. M., Zhong, X., Centeneno, G. S., Kush, G. S. and Kassman, K. G. (2004). Rice yields decline with higher night temperature from global warming. *Proc. Natl. Acad. Sci. USA* 101, 997–9975.
Peterhansel, C. (2011). Best practice procedures for the establishment of a C_4 cycle in transgenic C_3 plants. *J. Exp. Bot.* 62, 2011–3020.
Peterhansel, C. and Maurino, V. G. (2011). Photorespiration redesigned. *Plant Physiol.* 155, 49–55.
Raines, C. A. (2011). Increasing photosynthetic carbon assimilation in C_3 plants to improve crop yield: current and future strategies. *Plant Physiol.* 155, 36–42.
Ray, D. K., Mueller, N. D., West, P. C. and Foley, J. A. (2013). Yield trends are insufficient to double global crop production by 2050. *PLoS ONE* 8, e66428.
Ray, D. K., Ramankutty, N., Mueller, N. D., West, P. C. and Foley, J. A. (2012). Recent patterns of crop yield growth and stagnation *Nat. Commun.* 3, 1293
Reynolds, M., and Langridge, P. (2016). Physiological breeding. *Curr. Opinion Plant Sci.* 31, 162–71.
RIPE (2016). Realizing Increased Photosynthetic Efficiency for sustainable increases in crop yield. http://ripe.illinois.edu/(Accessed 16 March 2016).
Rosenthal, D. M., Locke, A. M., Khozaei, M., Raines, C. A., Long, S. P. and Ort, D. R. (2011). Over-expressing the C_3 photosynthesis cycle enzyme sedoheptulose-1–7 bisphosphatase improves photosynthetic carbon gain and yield under fully open air CO_2 fumigation (FACE). *BMC Plant Biology* 11, 123.
Ruuska, S. A., Andrews, T. J., Badger, M. R., Price, G. D. and von Caemmerer, S. (2000). The role of chloroplast electron transport and metabolites in modulating rubisco activity in tobacco. Insights from transgenic plants with reduced amounts of cytochrome *b/f* complex or glyceraldehyde 3-phosphate dehydrogenase. *Plant Physiol.* 122, 491–504.
Sage, R. F. (1990). A model describing the regulation of ribulose-1,5-bisphosphate carboxylase, electron transport, and triose phosphate use in response to light intensity and CO_2 in C_3 Plants. *Plant Physiol.* 94, 1728–34.
Sage, R. F. (1995). Was low atmospheric CO_2 during the Pleistocene a limiting factor for the origin of agriculture? *Global Change Biol.* 1, 93–106.
Sage, R. F. (2000). C_3 versus C_4 photosynthesis in rice: ecophysiological perspectives. In J. E. Sheehy, P. L. Mitchell and B. Hardy (Eds), *Redesigning Rice Photosynthesis to Increase Yield*, pp. 13–38. International Rice Research Institute, Manila, The Philippines.
Sage, R. F. (2002). Variation in the molar activity of Rubisco in C_3 and C_4 plants with some implications for photosynthetic performance at high and low temperature. *J. Exp. Bot.* 369, 609–20.

Sage, R. F. (2013). Photorespiratory compensation: a driver for biological diversity. *Plant Biol.* 15, 624–38.
Sage, R. F., Christin, P. A. and Edwards, E. J. (2011). The C_4 lineages of planet Earth. *J. Exp. Bot.* 62, 3155–69.
Sage, R. F. and Coleman, J. R. (2001). Effects of low atmospheric CO_2 on plants: more than a thing of the past. *Trends Plant Sci.* 6, 18–24.
Sage, R. F. and Kubien, D. S. (2003). *Quo Vadis* C_4? An ecophysiological perspective on global change and the future of C_4 plants. *Photosynth. Res.* 77, 209–25.
Sage, R. F. and Kubien, D. S. (2007). The temperature response of C_3 and C_4 photosynthesis. *Plant, Cell Environ.* 30, 1086–107.
Sage, R. F., Pearcy, R. W. and Seemann, J. R. (1987). The nitrogen use efficiency of C_3 and C_4 plants. III. Leaf nitrogen effects on the activity of carboxylating enzymes in *Chenopodium album* L. and *Amaranthus retroflexus* L. *Plant Physiology* 85, 355–9.
Sage, R. F., Sharkey, T. D. and Seemann, J. R. (1988). The *in vivo* response of ribulose-1,5-bisphosphate carboxylase activation state and pool sizes of photosynthetic metabolites to elevated CO_2 in *Phaseolus vulgarius* L. *Planta* 174, 407–16.
Sage, T. L. and Sage, R. F. (2009). The functional anatomy of rice leaves: implications for refixation of photorespiratory CO_2 and efforts to engineer C_4 photosynthesis into rice. *Plant Cell Physiol.* 50, 756–72.
Sage, R. F. and Stata, M. (2014). Photosynthetic diversity meets biodiversity: the C_4 plant example. *J. Plant Physiol.* 172, 104–19.
Sage, R. F. and Zhu, X. G. (2011). Exploiting the engine of C_4 photosynthesis. *J. Exp. Bot.* 62, 2989–3000.
Salvucci, M. E. and Crafts-Brandner, S. J. (2004). Mechanism for deactivation of Rubisco under moderate heat stress. *Physiologia Plant.* 122, 513–19.
Scafaro, A. P., von Caemmerer, S., Evans, J. R. and Atwell, B. J. (2011). Temperature response of mesophyll conductance in cultivated and wild *Oryza* species with contrasting mesophyll cell wall thickness. *Plant, Cell Environ.* 34, 1999–2008.
Sharkey, T. D. (1985). Photosynthesis in intact leaves of C_3 plants: physics, physiology and rate limitations. *Bot. Rev.* 51, 53–105.
Sharkey, T. D. (1988). Estimating the rate of photorespiration in leaves. *Physiol. Plant.* 73, 147–52.
Sharkey, T. D., Bernacchi, C. J., Farquhar, G. D. and Singsaas, E. L. (2007). Fitting photosynthetic carbon dioxide response curves for C_3 leaves. *Plant, Cell Environ.* 30, 1035–40.
Sharkey, T. D., Laporte, M. M. and Kruger, E. L. (2000). Will increased photosynthetic efficiency lead to increased yield in rice. In Sheehy, J. E., Mitchell, P. L. and Hardy, B. (Eds), *Redesigning Rice Photosynthesis to Increase Yield*, pp. 73–86. International Rice Research Institute, Manila, The Philippines.
Sharkey, T. D., Laporte, M. M., Lu, Y., Weise, S. E. and Weber, A. P.M. (2004). Engineering plants for elevated CO_2: A relationship between sugar sensing and starch degradation. *Plant Biol.* 6, 280–8.
Sharkey, T. D., Stitt, M., Heineke, D., Gerhardt, R., Raschke, K. and Heldt, H. W. (1986). Limitation of photosynthesis by carbon metabolism II. O_2 insensitive CO_2 assimilation results from triose phosphate utilization limitations. *Plant Physiol.* 81, 1123–9.
Sharwood, R. E., Ghannoum, O. and Whitney, S. (2016). Prospects for improving CO_2 fixation in C_3 crops through understanding C_4 Rubisco biogenesis and catalytic diversity. *Curr. Opin. Plant Biol.* 31, 135–42.
Sheehy, J. E., Ferrer, A. B., Mitchell, P. L., Elmido-Mablilangen, A., Pablico, P. and Dionroa, M. J. A. (2007). How the rice crop works and why it need a new engine. In J. Sheehy, A. K. Mitchell and B. Hardy (Eds), *Reconfiguring the Rice Plant's Photosynthetic Pathway*, pp. 3–26. International Rice Research Institute, Manila, The Philippines.
Sheehy, J. E. and Mitchell, P. L. (2011). Developments in rice research: visions and pragmatism. *World Agric.* 2, 7–19.
Simkin, A. J., NcAusland, L., Headland, L. R., Lawson, T. and Raines, C. A. (2015). Multigene manipulation of photosynthetic carbon assimilation increases CO_2 fixation and biomass yield in tobacco. *J. Exp. Bot.* 66, 4075–90.

Spreitzer, R. J. and Salvucci, M. E. (2002). Rubisco: structure, regulatory interactions, and possibilities for a better enzyme. *Annu. Rev. Plant Biol.* 53, 449–75.

Sudo, E., Suzuki, Y. and Makino, A. (2014). Whole-plant growth and N utilization in transgenic rice plants with increased or decreased rubisco content under different CO_2 partial pressures. *Plant Cell Physiol.* 55, 1905–11.

Suzuki, Y., Ohkubo, M., Hatakeyama, H., Ohashi, K., Yoshizawa, R., Kojima, S., Hayakawa, T., Yamaya, T., Mae, T. and Makino, A. (2007). Increased Rubisco content in transgenic rice transformed with 'sense' *rbcS* gene. *Plant Cell Physiol.* 48, 626–37.

Takai, T., Adachi, S., Taguchi-Shiobara, F., Sanoh-Arai, Y., Iwasawa, N., Yoshinaga, S., Hirose, S., Taniguchi, Y., Yamanouchi, U., Wu, J., Matsumoto, T., Sugimoto, K., Kondo, K., Ikka, T., Ando, T., Kono, I., Ito, S., Shomura, A., Ookawa, T., Hirasawa, T., Yano, M., Kondo, M. and Yamamoto, T. (2013a). A natural variant of *NAL1*, selected in high-yield rice breeding programs, pleiotropically increases photosynthesis rate. *Sci. Report* 3, 2149.

Takai, T., Ohsumi, A., Arai, Y., Iwasawa, N., Yano, M., Yamamoto, T., Yoshinaga, S. and Kondo, M. (2013b). QTL analysis of leaf photosynthesis in rice, *JARQ* 47, 227–35.

Takai, T., Ohsumi, A., San-oh, Y., Laza, M. R.C., Kondo, M., Yamamoto, T. and Yano, M. (2009). Detection of a quantitative trait locus controlling carbon isotope discrimination and its contribution to stomatal conductance in *japonica* rice. *Theor. Applied Gen.* 118, 1401–10.

Tamoi, M., Nagaoka, M., Miyagawa, Y. and Shigeoka, S. (2006). Contribution of fructose-1,6-bisphosphatase and sedoheptulose-1,7-bisphosphatase to the photosynthetic rate and carbon flow in the Calvin cycle in transgenic plants. *Plant Cell Physiol.* 47, 380–90.

Tanksley, S. D. (1993). Mapping polygenes. *Annu. Rev. Gen.* 27, 205–33.

Tausta, S. L., Li, P., Si, Y., Gandotra, N., Liu, P., Sun, Q., Brutnell, T. P. and Nelson, T. (2014). Developmental dynamics of kranz cell transcriptional specificity in maize leaf reveals early onset of C_4-related processes. *J. Exp. Bot.* 65, 3543–55.

Tcherkez, G. G. B., Farquhar, G. D. and Andrews, T. J. (2006). Despite slow catalysis and confused substrate specificity, all ribulose bisphosphate carboxylases may be nearly perfectly optimized. *Proc. Natl. Acad. Sci. USA* 103, 7246–51.

Teng, S., Qian, Q., Zeng, D., Kunihiro, Y., Fujimoto, K., Huang, D. and Zhu, L. (2004). QTL analysis of leaf photosynthetic rate and related physiological traits in rice (*Oryza sativa* L.) *Euphytica* 135, 1–7.

Terashima, I., Hanba, Y. T., Tholen, D. and Niinemets, U. (2011). Leaf functional anatomy in relation to photosynthesis. *Plant Physiol.* 155, 108–16.

Tholen, D., Boom, C. and Zhu, X. G. (2012). Opinion: prospects for improving photosynthesis by altering leaf anatomy. *Plant Science* 197, 92–101.

Tilman, D. Balzer, C., Hill, J. and Befort, B. L. (2011). Global food demand and the sustainable intensification of agriculture. *Proc. Natl. Acad. Sci. USA* 108, 20260–4

Uematsu, K., Suzuki, N., Iwamae, T., Inui, M. and Yukawa, H. (2012). Increased fructose 1,6-bisphosphate aldolase in plastids enhances growth and photosynthesis of tobacco plants. *J. Exp. Bot.* 63, 3001–9.

von Caemmerer, S. (2013). Steady-state models of photosynthesis. *Plant Cell Environ.* 36, 1617–30.

von Caemmerer, S. and Evans, J. R. (1991). Determination of the average partial pressure of CO_2 in chloroplasts from leaves of several C_3 plants. *Aust. J. Plant Physiol.* 18, 287–305.

von Caemmerer, S. and Evans, J. R. (2010). Enhancing C_3 photosynthesis. *Plant Physiol.* 154, 589–92.

von Caemmerer, S. and Farquhar, G. D. (1981). Some relationships between the biochemistry of photosynthesis and the gas exchange of leaves. *Planta* 153, 376–87.

von Caemmerer, S. and Quick, W. P. (2000). Rubisco: physiology *in vivo*. *Photosynthesis*: Springer, 85–113.

Wang, P., Vlad, D. and Langdale, J. A. (2016). Finding the genes to build C_4 rice. *Curr. Opin. Plant Biol.* 31, 44–50.

White, A., Rogers, A., Rees, M. and Osborne, C. P. (2016). How can we make plants grow faster? A source-sink perspective on growth rate. *J. Exp. Bot.* 67, 31–45.

Whitney, S. M., Sharwood, R. E., Orr, D., White, S. J., Alonso, H. and Galmés, J. (2011). Isoleucine 309 acts as a C_4 catalytic switch that increases ribulose-1,5-bisphosphate carboxylase/oxygenase (Rubisco) carboxylation rate in *Flaveria*. *Proc. Natl. Acad. Sci. USA* 108, 14688–93.

Woodrow, I. (1994). Optimal acclimation of the C_3 photosynthetic system to enhanced CO_2. *Photosynth. Res.* 39, 351–68.

Yamamoto, T., Uga, Y. and Yano, M. (2014). Genomics-assisted allele mining and its integration into rice breeding. In R. Tuberosa, A. Graner and E. Frison (Eds), *Genomics of Plant Genetic Resources*, vol. 2, pp. 251–65. *Crop Productivity, Food security and Nutritional Quality.* Springer, Berlin.

Yamamoto, T. and Yano, M. (2008). Detection and molecular cloning of genes underlying quantitative phenotypic variations in rice. In H. Y. Hirano, Y. Sano, A. Hirai and T. Sasaki (Eds), *Rice Biology in the Genomics Era*, pp. 295–308. Springer, Heidelberg, Berlin.

Yamori, W. (2013). Improving photosynthesis to increase food and fuel production by biotechnological strategies in crops. *J. Plant Biochem. Physiol.* 1, 113.

Yamori, W., Kondo, E., Suguira, D., Terashima, I., Suzuki, Y. and Makino, A. (2016). Enhanced leaf photosynthesis as a target to increase grain yield: insights from transgenic rice lines with variable Rieske FeS protein content in the cytochrome b_6f complex. *Plant, Cell Environ.* 39, 80–7.

Yamori, W., Matsumoto, C., Fukayama, H. and Makino, A. (2012). Rubisco activase is a key regulator of non steady-state photosynthesis at any leaf temperature and, to a lesser extent, of steady-state photosynthesis at high temperature. *Plant J.* 71, 871–80.

Yang, W., Duan, L., Chen, G., Xiong, X. and Liu, Q. (2013). Plant phenomics and high-throughput phenotyping: accelerating rice functional genomics using multidisciplinary technologies *Curr. Opin. Plant Biol.* 16, 180–7.

Yano, M. (2001). Genetic and molecular dissection of naturally occurring variation. *Curr. Opin. Plant Biol.* 4, 130–5.

Zhu, X. G., Long, S. P. and Ort, D. R. (2008). What is the maximum efficiency with which photosynthesis can convert solar energy into biomass? *Curr. Opin. Biotechnol.* 19, 153–9.

Zhu, X. G., Long, S. P. and Ort, D. R. (2010). Improving photosynthetic efficiency for greater yield. *Annu. Rev. Plant Biol.* 61, 235–61.

Zhu, X. G., Portis, A. R. and Long, S. P. (2004). Would transformation of C_3 crop plants with foreign Rubisco increase productivity? A computational analysis extrapolating from kinetic properties to canopy photosynthesis. *Plant, Cell Environ.* 27, 155–65.

Zhu, X. G., de Sturler, E. and Long, S. P. (2007). Optimizing the distribution of resources between enzymes of carbon metabolism can dramatically increase photosynthetic rate: a numerical simulation using an evolutionary algorithm. *Plant Physiol.* 145, 513–26.

Chapter 5

Breeding green super rice (GSR) varieties for sustainable rice cultivation

Z. Li, Chinese Academy of Agricultural Sciences, China; and J. Ali, International Rice Research Institute, The Philippines

1 Introduction

Rice is the major cereal crop in Asia where 90% of the world's rice is produced and consumed. As a C3 crop, rice productivity has reached a very high level of up to >8 t/ha in the irrigated areas of many countries, resulting primarily from improvement in breeding in previous decades, including the 'Green Revolution' since 1950s and the development of hybrid rice technology in China since late 1970s. Rice production and productivity need to keep pace with a growing global population likely to reach 9 billion by 2050 in order to have a hunger-free world and to ensure sustainable production in the face of depleting resources such as land, water and nutrients as well as changing climatic conditions. The greatest challenge we face is how to sustain the current productivity of rice, or further increase it, without undermining the sustainability of rice production for future generations. Achieving this depends primarily on improved rice cultivars and improved rice farming technologies.

From a breeder's point of view, there are a number of challenges to achieve sustainable rice production. The first is to greatly increase rice productivity in the rainfed lowland rice areas that occupy approximately 35% of global rice cultivation and which remain the poorest and often the most neglected areas (Mackill et al., 1996; Pandey, 1997). Rainfed rice areas are characterized by multiple abiotic and biotic stresses; therefore, farmers do not have resources to buy expensive inputs for rice crop to get better and stable yields. Most modern super rice cultivars could not make inroads in the rainfed areas because of lack of traits for high nutrient- and water-use efficiency and tolerance/resistance to multiple abiotic and biotic stresses. Recently released rice varieties with improved tolerance to

http://dx.doi.org/10.19103/AS.2016.0003.05

single abiotic stresses such as drought, salinity and submergence have limited comparative advantages over landraces in such conditions. Secondly, improving yield potential has been the highest priority in almost all rice breeding programmes worldwide in the post-Green Revolution era. In China, large numbers of 'super' inbred and hybrid rice cultivars have been developed by 'ideotype' breeding and by exploiting both intra- and inter-subspecific heterosis using conventional cross-making and phenotypic selection (Li and Zhang, 2013). These 'super inbred' or 'super hybrid' cultivars normally perform well under very high inputs. However, to get the fantastic yields of these super hybrids or inbreds, farmers have had to use more and more fertilizer and irrigation water inputs, leading to unsustainable farm practices and damaging rice ecosystems (Zhang, 2007). Current rice farming practices, especially under irrigated conditions, use huge volumes of water, which is becoming an increasing luxury in many rice-growing countries that have gone through rapid urbanization. Paddy cultivation practices, especially in irrigated rice conditions, are becoming increasingly expensive and less profitable for farmers. Labour-intensive operations of transplanting, weed management and the need for regular chemical spray protection of crops are dissuading farmers from paddy cultivation. This trend has already been noticed in the traditional paddy areas of post-Green Revolution era, especially in the Indian states of Punjab, Haryana and Uttar Pradesh. Higher labour wages around the industrial cities in China has already resulted in farmers moving away from farming. Labour shortages have led to mechanization in China and in many parts of Asia as well. These problems mean that water- and nutrient-efficient rice cultivars are needed. In addition, salinity and alkalinity are key factors limiting rice productivity in coastal and inland areas of most rice-growing countries, where millions of ha lands have been, or can potentially be, used for rice production. Rice varieties suitable for these areas require not only a high level of tolerance to salinity/alkalinity, but also good tolerance/resistance to other associated abiotic stresses (submergence, drought and specific mineral deficiencies or toxicities), in addition to high yield potential. Finally, the global climate changes with El Niño effects have intensified the frequency and extent of drought and flooding in many rice production systems of Asia and Africa. As a result, many traditional irrigated areas are becoming more salinized and drought prone. Sustainable rice production therefore requires new types of rice cultivars for specific target environments (TPE) that have high yield potential with greatly improved nutrient- and water-use efficiency and tolerance/resistance to multiple abiotic/biotic stresses specific to each TPE.

2 The concept of green super rice (GSR)

Green super rice (GSR) is defined as rice cultivars that can produce high and stable yield under less input (Zhang, 2007). Since firstly proposed by a group of Chinese scientists, GSR has become an important concept to achieve sustainable rice production in China. The early breeding efforts to develop GSR cultivars in China focused primarily on drought tolerance/water-use efficiency and N/P use efficiency (Luo, 2010). This concept has now also been extended to promoting high resource use efficiency in rice production, to support more efficient and environmental-friendly crop management technologies and ecological crop production systems, labour-saving and mechanized crop cultivation technologies, and improved adaptability to future climate change. Global rice production requirements in the future need to come from increased productivity per se of improved rice varieties and requires major shifts in breeding goals and objectives (Li and Zhang, 2013). In other words, breeders will have to improve many 'green' traits in addition to high yield potential and desirable quality. These green traits include sufficient levels of tolerance/resistance to multiple abiotic and biotic stresses, high nutrient- and water-use efficiencies in different target environments in the developing world.

These problems facing the global rice production have stimulated tremendous efforts in basic and breeding research for abiotic stress tolerances in rice. These efforts have largely focused on single abiotic stresses and related traits. Progress has been made in improving tolerances of mega-varieties to single abiotic stresses (Guan et al., 2010; Witcombe, 2008). However, varieties with good tolerance to single abiotic stresses have not been able to attain their maximum yield potentials in irrigated ecosystems, nor have they been able to achieve stable yields in most rainfed areas, where multiple abiotic and biotic stresses coexist in farmer's fields. These conditions require varieties able to cope with periodic abiotic stresses such as drought and flood whilst also being able to perform well under favourable conditions. Breeding multiple stress tolerant rice varieties could have a significant impact on the livelihood of resource-poor farmers. Many of these varieties are expected to provide a crop insurance against more unpredictable climatic conditions that would otherwise wreck their livelihoods. In addition, developing cultivars with improved nitrogen (N)- and phosphate (P)-use efficiency are essential for the sustainability of rice production in most rice-growing areas, as inefficient utilization of N and P fertilizer by rice plants increases the cost of rice production and causes environmental pollution.

Recent progress in rice genomic/genetic research, as well as the availability of high-throughput genotyping technology, has provided huge amounts of genetic information and tremendous opportunities for rice improvement. However, limitations in molecular tools have so far limited applications in rice breeding, and the classical pedigree breeding based on crossing and phenotypic selection remains the predominant approach in most rice breeding programmes worldwide. The lack of connectivity of these modern tools to plant breeders is at least partially responsible for the slow progress in breeding superior rice varieties for complex rainfed conditions. Traditional pedigree breeding practised by most breeders in developing countries is ineffective as a way to combine multiple abiotic and biotic stress tolerance traits with high yield potential and does not normally generate any genetic information on target traits during breeding. Recent efforts to apply marker-assisted selection (MAS) to introgress major gene/QTL into elite adapted mega-varieties to improve their tolerances to single abiotic stresses have proven to be effective. One successful example is the development of Swarna *Sub1* in the Indian subcontinent, which has helped farmers to tackle the flooding that often destroyed their rice crops. However, rainfed lowlands environments are highly complex and need series of genetic interventions to develop resilient varieties to meet the challenge. MAS breeding schemes may take many generations since there is a need to handle very large populations to combine many QTLs for different target traits. The risk of failure is high as differing types of abiotic stress tolerance are genetically complex. For example, to combine a set of 10 genes/QTLs into a single elite genetic background, the required minimum population size to obtain a single desirable homozygous genotype that carries them all in each generation of a MAS experiment would be 4^{10}, which is too large to be handled in most breeding programmes. It would also be even more problematic to break the possible linkage drags associated with the target genes/QTL to ensure most progenies carry all the desirable features or traits of the adapted recipient. With the availability of virtually unlimited numbers of single nucleotide polymorphism (SNP) markers and the rapid progress in the global efforts in rice functional genomics research (Alexandrov et al., 2015; Chen et al., 2013), we expect increasing applications of DNA markers in rice breeding programmes. However, the requirement for any successful MAS breeding programme is the accuracy of genetic information of target QTLs in parental lines, as QTL effects are highly dependent on genetic backgrounds which are often difficult to obtain (Li, 2001; Li et al., 1997).

Now the question is, 'Do we have a simple, feasible holistic approach that is easily adoptable by small breeding programmes and produces good genetic gains?' In the following section, we will show a modified backcross (BC) breeding procedure, as part of our GSR breeding technology, which can be easily adopted by small breeding programmes. Using this approach, we were able to develop from a small number of breeding populations, within a short period of 5 years, large numbers of promising GSR lines with good tolerances to multiple abiotic/biotic stresses.

3 GSR breeding techniques

Traditionally, BC breeding is widely adopted to improve highly heritable traits controlled by single major genes. It was also proven equally effective for improving single complex traits including abiotic/biotic stress tolerances/resistances and yield potential, as there is sufficient genetic diversity in the primary gene pool for almost all complex traits (Ali et al., 2006, 2012; He et al., 2010; Lafitte et al., 2006; Meng et al., 2010). However, there are few successful cases reported for developing rice varieties with improved grain yield as well as tolerances to two or more abiotic stresses that are suitable to both rainfed and irrigated conditions. We have recently demonstrated an improved BC breeding approach that results in faster breeding cycle to fix the breeding lines tolerant to multiple abiotic stresses (drought and saline) while performing well under normal irrigated conditions (Ali et al., 2016).

Figure 1 shows a shuttle breeding and molecular recurrent selection breeding strategy we have currently adopted for developing high yielding GSR varieties with multiple stress tolerances. This strategy involves four major activities: (1) developing segregating BC populations by crossing elite widely adaptable recipients with a diverse set of donors; (2) strong phenotypic selection and progeny testing of the BCF_2 populations, tested for multiple stresses and selected for target traits in different target countries; (3) QTL/QTL network discovery by genetic characterization of selected introgression lines (ILs) using high-throughput genotyping; and (4) development of superior GSR varieties by designed QTL pyramiding (DQP, Fig. 2) or by molecular recurrent selection (MRS) procedure.

3.1 Segregating backcross (BC) populations

In activity 1, random BC_1F_2 or BC_2F_2 populations were derived from crosses between elite widely adaptable recurrent parents (RPs) and a set of core donors of diverse origins. Then, the BCF_2 populations were screened for different target traits, including abiotic (drought, salinity, submergence and low fertilizer input) stress tolerances and biotic stress (blast, bacterial leaf blight, Tungro, etc.) resistances, and normal irrigated conditions (for high yield potential) in different target countries. Following this, the selected (typically ~5%) BCF_3 and BCF_4 progenies were progeny tested simultaneously under all stress and non-stress testing conditions for two consecutive rounds. Only the 2–3 best plants in the best performing

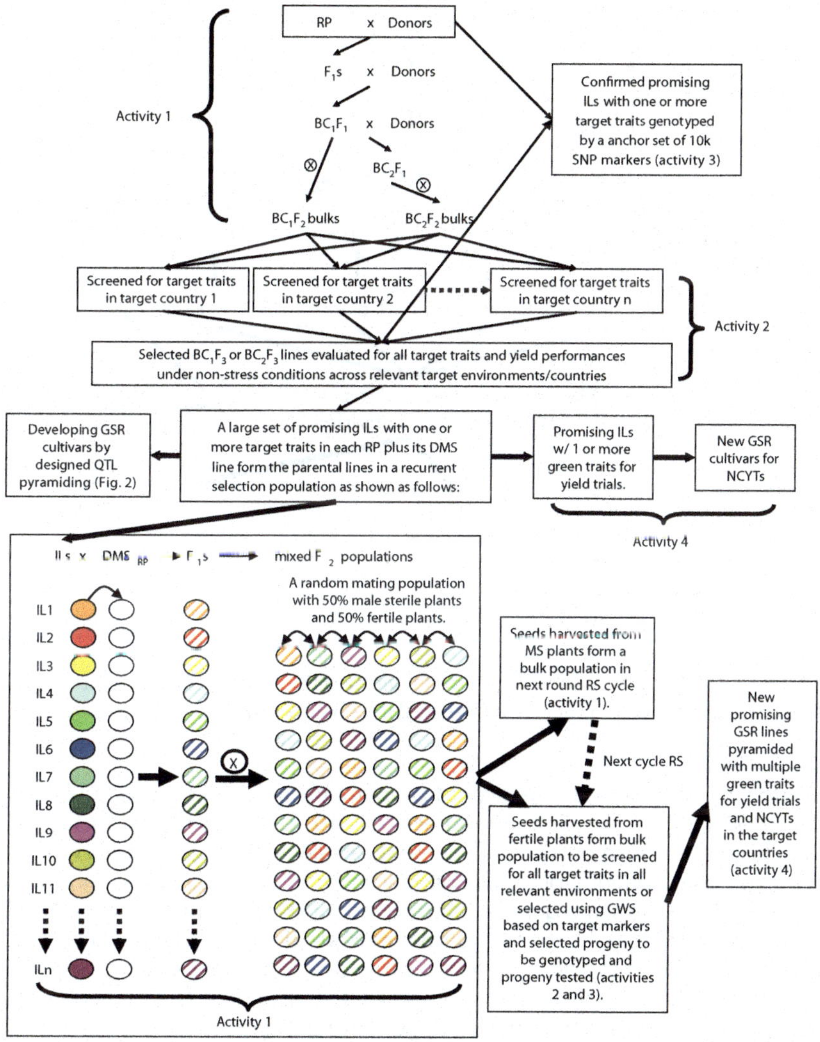

Figure 1 A shuttle breeding and molecular recurrent selection breeding strategy to develop 2nd generation GSR cultivars for multiple target countries (environments); DMS = dominant male sterility line in the RP genetic background.

BCF_3 and BCF_4 lines under each set of conditions were selected and were advanced to the preliminary yield trials. Next, the promising BCF_5 introgression lines (ILs) performing significantly better than the checks under one or more stress/non-stress conditions were identified and advanced to preliminary yield trials in target environments. Many new GSR varieties were developed in step 1, as shown in the next section. The selected BCF_5 could be used for discovering QTLs and QTL networks affecting both target traits and non-target traits using the approach of selective introgression based on genotypic data of ILs and

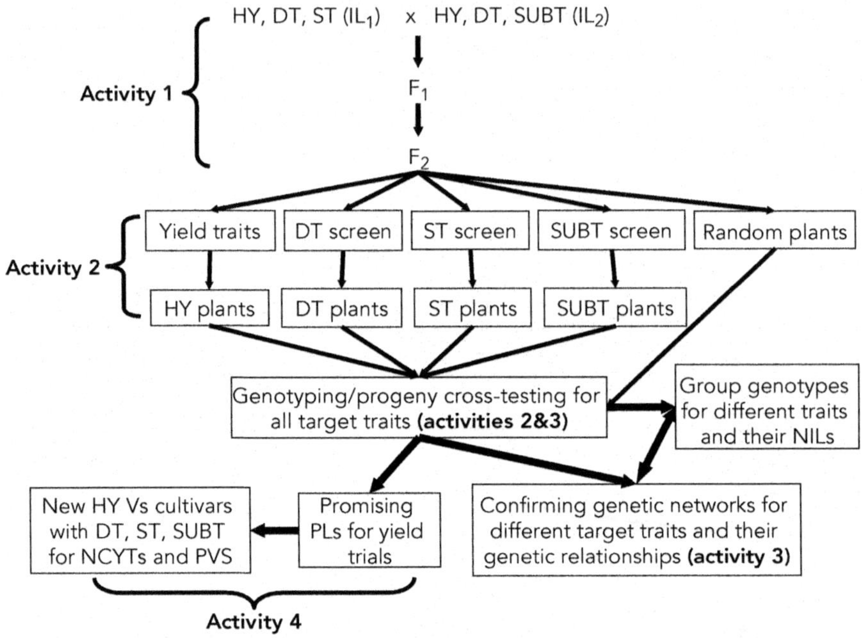

Figure 2 DQP strategy for combining high yield (HY), drought tolerance (DT), salinity tolerance (ST) and submergence tolerance (SUBT) using promising ILs as parental lines.

their phenotypic data from the replicated progeny testing experiments (Cui et al., 2015; Li et al., 2005; Wang et al., 2015; Zhang et al., 2010, 2012). Using ILs for QTL discovery is highly efficient, because only a small number of selected informative ILs will be used in both genotyping and phenotyping. The genetic information obtained is highly informative because all ILs share the same genetic background, but each carries a different set of QTLs affecting different combinations of traits from multiple donors. The genetic information of selected ILs can be used to select parental ILs that carry QTLs for target traits from different donors as parental lines in the MRS (Fig. 1) and DQP (Fig. 2) breeding procedures based on specific breeding objectives for different target environments.

3.2 Phenotypic selection and progeny testing of populations

Table 1 shows the time periods and outputs from our GSR breeding cycle undertaken at the IRRI since the 2007 wet season (WS), which resulted in 3 sets of 4973 ILs developed from 71 BC_1 populations between three recipients (Huanghuazhan – HHZ, Weed Tolerant Rice 1 – WTR1, and TME50518) and 71 donors. Specifically, it took only 4 years (8 seasons) to complete the breeding cycle of the GSR breeding step 1 from cross-making to final product development for each set of ILs, including two rounds of replicated yield trials across different stress and non-stress conditions. These ILs showed relative yield superiority over the checks under one or more of testing conditions (irrigated, saline, drought and flooding). From these, 530 HHZ ILs were sent for testing the different levels of yield trials

Table 1 The total numbers of introgression lines (ILs) in three genetic backgrounds of Huanghuazhan (HHZ), weed-tolerant rice 1 (WTR1) and TME50518 developed at IRRI

Recipient	Number of donors	Number of ILs	Number of DQP crosses	Number of PLs	Breeding cycle
HHZ	16	1,333	179	2,023	2006WS-2010WS
WTR1	16	2,232	45	661	2010WS-2014WS
TME50518	39	1,408	73	–	2011WS-2015WS
Total	71	4,973	297	2,684	

*Breeding cycle was from cross-making to final product development, including one round of replicated preliminary yield trial and one round of advance yield trial.

Table 2 Green super rice (GSR) varieties released in Asia

S.no.	Country	GSR inbreds	Released name
1	Pakistan	GSR IR1-5-S10-D3-Y2	NIBGE GSR-I
2	Pakistan	GSR IR1-5-S12-D3-Y2	NIBGE GSR-II
3	Philippines	IR83140-B-36-B	NSIC Rc 29 (Katihan 4)
4	Philippines	IR83140-B-28-B	NSIC Rc 390 (Salinas 19)
5	Philippines	IR84675-58-4-1-B-B	NSIC Rc 392 (Salinas 20)
6	Philippines	GSR IR1-8-S6-S3-Y2	NSIC Rc 480 (GSR 8)
7	Philippines	IR 83142-B-19-B	NSIC Rc 434 (GSR 2)
8	Philippines	GSR IR1-12-D10-S1-D1	GSR 12
9	Philippines	GSR IR1-12-S2-Y3-Y2	GSR 12a
10	Philippines	GSR IR1-5-S8-D3-SU1	GSR 5a
11	Philippines	IR 83140-B-11-B	GSR 11
12	Philippines	GSR IR1-1-Y4-Y1	GSR 1
13	Philippines	GSR IR1-5-S14-S2-Y2	GSR 5

in different Asian and African countries, from which 13 GSR varieties were released in the Philippines and Pakistan (Table 2), 77 promising GSR lines are currently being tested in national cooperative yield trials and are nearing release, and another 18 GSR inbreds have reached the final stage of testing through participatory varietal selection (PVS) in East and Southern Africa (Table 3).

The superior performances of many GSR varieties have been well demonstrated in the field trials under both irrigated and drought conditions at IRRI and in many target countries across Asia and Africa (Ali et al., 2012). For example, the resulting WTR1 ILs demonstrated many features related to high yield, including higher biomass; a robust plant stature with a thicker culm/stem with lodging resistance; moderate tillering ability; broader, thicker,

Table 3 Seventy-seven promising GSR varieties nominated into the national cooperative yield trials in Asia and Africa

S.no.	Country	GSR line	Nominated category – national trials	S.no.	Country	GSR line	Nominated category – national trials
I South Asia (21)				41	Philippines	GSR IR1-3-S13-Y2-D1	Submergence
1	Bangladesh	GSR IR1-15-D4-D1-Y1	Irrigated lowland	42	Philippines	GSR IR1-18-D4-Y1-SU1-L2	Submergence
2	Bangladesh	GSR IR1-6-S3-Y1-SU2	Irrigated lowland	43	Philippines	GSR IR1-5-S14-S2-Y2	Submergence
3	Bangladesh	GSR IR1-23-D16-D1-D1	Irrigated lowland	44	Philippines	GSR IR1-8-S14-S3-Y2	Submergence
4	Bangladesh	GSR IR1-11-Y10-D3-Y3	Irrigated lowland	45	Philippines	GSR IR2-5-L10-U1-R2	Rainfed upland
5	Bangladesh	GSR IR1-12-D10-S1-D1	Irrigated lowland	46	Philippines	GSR IR1-6-Y2-Y1-D1	Rainfed (dry seeded)
6	Bangladesh	GSR IR1-15-D7-S4-S1	Irrigated lowland	47	Philippines	GSR IR1-3-S6-Y1-Y1	Rainfed (dry seeded)
7	India	GSR IR1-11-Y6-Y1-Y1	Irrigated lowland	48	Philippines	IR83140-B-11-B	Irrigated lowland, Saline
8	India	GSR IR1-5-S9-Y3-Y1	Irrigated lowland	49	Philippines	IR83140-B-28-B	Saline
9	India	IR 82858-B-B-1	Direct seeded rice	50	Philippines	GSR IR1-6-S9-L2-Y1-Y1	Cool elevated
10	India	IR 83108-B-B-1 3	Direct seeded rice	51	Philippines	GSR IR1-18-D4-Y1-L1-L1	Cool elevated
11	India	GSR IR1-17-D6-S3-D1	Irrigated lowland	52	Philippines	GSR IR1-1-D1-D1-Y1-D3	Irrigated
12	India	IR 83141-B-17-B	Early direct seeded	53	Philippines	GSR IR1-2-Y13-D1-D1	Irrigated
13	India	IR 83142-B-57-B	Irrigated lowland (medium)	54	Philippines	GSR IR1-8-S6-S3-D1	Irrigated lowland
14	India	GSR IR1-5-S10-D1-D1	Alkaline and inland saline	55	Indonesia	GSR IR1-5-S9-Y3-Y1	Irrigated lowland
15	India	GSR IR1-5-Y4-S1-Y1-3953-1	Coastal Saline	56	Indonesia	GSR IR1-5-S10-D2-D1	Rainfed lowland
16	India	GSR IR1-17-D6-S3-D1	IVT-Aerobic	57	Indonesia	GSR IR1-5-D1-D1	Rainfed lowland
17	India	GSR IR1-9-D 12-D1-SU-1	IVT-Rainfed shallow lowland	58	Vietnam	GSR IR1-11-Y10-D3-Y3	Rainfed lowland
18	India	GSR IR1-8-Y7-D2-S-1	IVT-Irrigated (medium)	59	Vietnam	GSR IR1-5-D1-D1	Rainfed lowland
19	India	GSR IR1-5-D 20-D 3-Y2	IVT-Irrigated (medium)				
20	India	GSR IR1-5-Y4-S 1-Y1	IVT-Irrigated lowland (late)				
21	India	IIRRH – 107 (HYBRID)	IVT-Aerobic				

II Southeast Asia (38)				**III East and Southern Africa (18)**			
22	Philippines	GSR IR1-3-S13-Y1-S1	Irrigated lowland (Early-TPL)	60	Tanzania	GSR IR1-8-S14-S1-SU1	Irrigated and rainfed lowland
23	Philippines	GSR IR2-5-L10-Y1-Y2	Irrigated lowland (TPR)	61	Tanzania	GSR IR1-12-D10-S1-D1	Irrigated and rainfed lowland
24	Philippines	GSR IR2-5-L10-U1-L2	Irrigated lowland (DSR)	62	Tanzania	GSR IR1-8-S12-Y2-D1	Irrigated and rainfed lowland
25	Philippines	GSR IR2-9-L1-L3-L2	Irrigated lowland (DSR)	63	Tanzania	GSR IR1-5-D20-D2-D1	Irrigated and rainfed lowland
26	Philippines	GSR IR1-21-Y4-Y2-Y1	Irrigated lowland (DSR)	64	Tanzania	GSR IR1-5-D1-D1	Irrigated and rainfed lowland
27	Philippines	GSR IR1-24-S6-Y1-L1-D1	Irrigated lowland (TPR)	65	Tanzania	GSR IR1-5-Y3-S2-SU1	Irrigated and rainfed lowland
28	Philippines	GSR IR1-S5-Y2-Y1	Special (aromatic)	66	Tanzania	GSR IR1-5-Y4-S1-Y1	Irrigated and rainfed lowland
29	Philippines	GSR IR1-14-Y7-Y1-D2	Special (aromatic)	67	Tanzania	GSR IR1-5-S12-D3-Y2	Irrigated and rainfed lowland
30	Philippines	GSR IR1-6-S3-Y1-SU2	Special (aromatic)	68	Mozambique	GSR IR1-12-S2-Y3-Y2	Irrigated and rainfed lowland
31	Philippines	GSR IR1-6-S3-Y1-N1	Special (aromatic)	69	Mozambique	GSR IR1-12-Y4-D1-Y1	Irrigated and rainfed lowland
32	Philippines	GSR IR1-12-D10-S1-D1	Saline	70	Mozambique	GSR IR1-5-S14-S2-Y1	Irrigated and rainfed lowland
33	Philippines	GSR IR2-6-R4-S3-Y2	Saline	71	Mozambique	GSR IR1-5-S14-S2-Y2	Irrigated and rainfed lowland
34	Philippines	GSR IR2-1-Y11-S1-R2	Saline	72	Mozambique	GSR IR1-5-S8-D2-S1	Irrigated and rainfed lowland
35	Philippines	GSR IR2-19-R2-SU2-R2	Saline	73	Mozambique	GSR IR1-5-Y3-Y1-D1	Irrigated and rainfed lowland
36	Philippines	GSR IR1-8-S6-S3-Y2	Saline	74	Mozambique	GSR IR1-8-S14-S3-Y2	Irrigated and rainfed lowland
37	Philippines	GSR IR1-2-Y3-D1-SU1-L1	Rainfed upland	75	Mozambique	GSR IR1-8-S6-S3-S1	Irrigated and rainfed lowland
38	Philippines	GSR IR1-15-S13-Y1	Rainfed upland	76	Mozambique	GSR IR1-8-S6-S3-Y1	Irrigated and rainfed lowland
39	Philippines	GSR IR1-19-S14-Y1	Rainfed upland	77	Mozambique	GSR IR1-8-S6-S3-Y2	Irrigated and rainfed lowland
40	Philippines	GSR IR2-1-L1-L1-L2	Rainfed upland				

Table 4 Yield performances of some promising WTR1 GSR ILs in multi-environment trials (MET) in the Philippines

	BLUEs (t/ha) at specified test sites											Across mean (t/ha)	%Yield advantage over best inbred check	%Yield advantage over best hybrid check
	Munoz		San Mateo		Los Baños		Ubay		RTR	Average				
Designation	DS	WS	DS	WS	DS	WS	DS	WS	WS	DS	WS			
GSR IR2-8-Y14-SU3-R2	8.81	3.56	8.03	6.25	8.31	4.61	7.19	4.3	6.67	8.09	5.08	6.58	20.24	7.85
GSR IR2-5-L10-Y1-Y2	8.43	3.79	7.65	6.62	7.92	5.0	6.8	4.33	6.84	7.7	5.32	6.51	18.88	6.49
GSR IR2-1-R5-N1-Y3	8.87	3.08	8.09	6.02	8.36	3.42	7.24	4.09	6.02	8.14	4.52	6.33	15.69	3.3
GSR IR2-7-Y11-SU3-Y2	8.25	3.33	7.47	6.32	7.75	4.58	6.62	4.32	6.55	7.52	5.02	6.27	14.57	2.18
GSR IR2-1-Y16-S1-R2	NA	2.5	7.89	5.68	8.17	4.38	7.05	4.04	6.21	7.95	4.56	6.25	14.26	1.87
GSR IR2-5-L10-U1-R2	8.04	3.91	7.26	6.3	7.53	4.48	6.41	4.44	6.61	7.31	5.14	6.23	13.77	1.38
GSR IR1-21-S16-Y1-SU1-D3	8.37	3.7	7.59	6.11	7.86	3.78	6.74	4.11	6.35	7.64	4.81	6.22	13.71	1.32
GSR IR2-19-Y14-L2-L2	8.86	2.52	8.09	5.48	8.36	3.32	7.24	3.59	6.23	8.14	4.23	6.18	12.94	0.55
NSIC Rc132H (Mestizo 6)	8.56	2.16	7.78	6.08	8.05	3.84	6.93	4.06	6.24	7.83	4.48	6.15	12.39	0
IRRI 156 (NSIC Rc238)	6.91	3.69	6.13	6.07	6.41	3.02	5.29	4.33	6.71	6.18	4.77	5.47	0	
IRRI 123 (PSB Rc82)	6.79	3.22	6.01	5.88	6.28	4.14	5.16	4.02	6.29	6.06	4.71	5.39	–1.6	
IRRI 104 (PSB Rc10)	6.56	1.54	5.78	3.56	6.06	2.45	4.93	2.66	4.32	5.83	2.91	4.37	–20.2	

DS and WS = dry season and wet season, respectively.

longer, more erect leaves; stay-green flag leaves; and larger panicle size. These WTR1 ILs lines performed exceedingly well in the replicated multi-environment trials, with relative yield advantages of 7.8% and 20% over the best yielding hybrid check (Meztiso 6) and over the inbred check variety (NSIC Rc 238) across seasons at 5 locations across the Philippines (Table 4), and yield advantages over the checks by 31–36% ranging from fully irrigated to drought-stressed environments (lii et al., 2014). One of these lines, GSR IR1-12-D10-S1-D1 performed extremely well across many sites globally and showed the highest yield among 12 elite cultivars tested globally with multiple abiotic stress tolerances against drought and salinity (Dingkuhn et al., 2015).

The resulting ILs from this modified BC breeding procedure have the same, elite, uniform genetic backgrounds, but together they have high levels of genetic diversity for both target traits and non-target traits as the donors are highly diverse. This property makes these ILs extremely useful materials for breeding and genetic/genomic research (Li et al., 2005). In breeding, one important concept is the customized set of ILs for

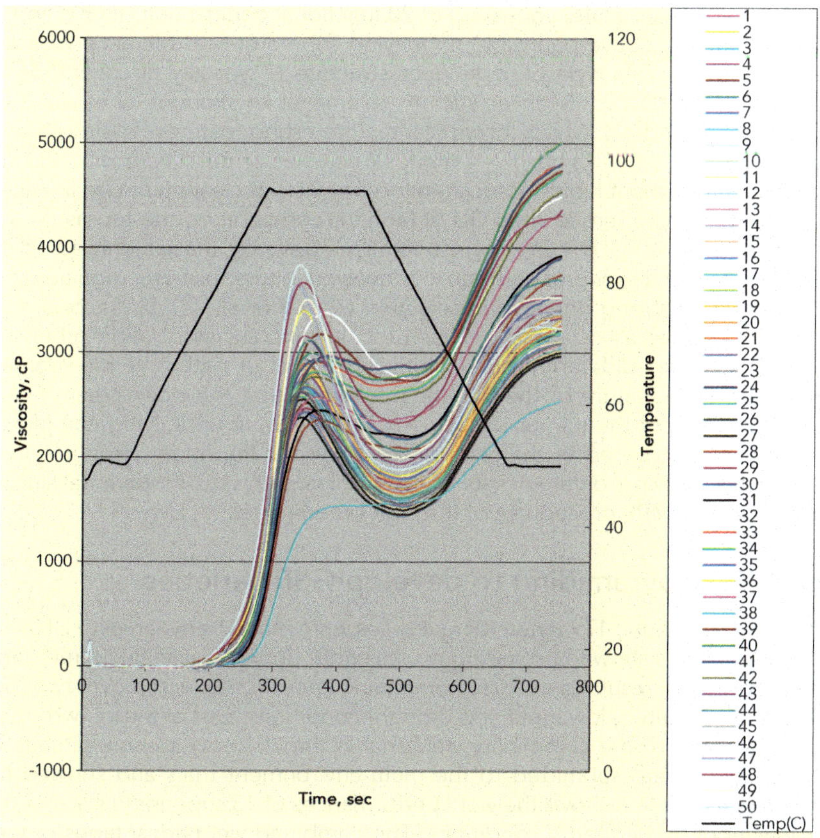

Figure 3 Rapid visco analyzer (RVA) pasting properties of 50 HHZ ILs suitable for varied consumers with different taste preferences.

grain quality for different target countries. Figure 3 shows the diversity of 50 HHZ ILs in grain quality by measuring pasting properties using a Rapid Visco Analyser. The major grain quality parameters of these 50 HHZ ILs are amylose contents of 14.5–31.6%, gel consistency from low to intermediate and high, and protein contents of 7.8–11.2%. As the grain quality requirements vary considerably in different countries, we were able to send different customized sets of GSR lines with multiple stress tolerances, but with desirable grain quality parameters, to different target countries through our GSR project funded by the Bill & Melinda Gates Foundation.

3.3 Quantitative/Qualitative Trait Locus (QTL/QTL) network discovery

Activity 2 of the GSR breeding strategy involves QTL/QTL network discovery by genetic characterization of introgression lines using DNA markers, which is well demonstrated in Figure 4 and Table 5. Based on 10 ILs selected from 855 BC_3F_2 plants of the cross between IR64 (recipient) and TKM9 (donor) under 14 days of complete submergence, a genetic network (multi-locus structure) consisting of 23 functional genetic units (FGUs and 54 loci) in 3 major groups (putative pathways) underlying submergence tolerance was detected (Wang et al., 2015). This type of multi-locus structure is typically discovered in rice ILs selected for abiotic stress tolerances such as tolerances to drought (Li et al., 2005) and salinity (unpublished data), which shows several interesting features. The most important feature is that most networks have 2–4 groups (branches or putative positively regulated pathways) of unlinked but highly associated loci or FGUs of clear hierarchy. In each group of the associated loci, there is one FGU of high introgression on the top as the putative regulator and multiple FGUs in the downstream. Theoretically, these highly associated loci most likely represent a group of epistatic loci involved in the positively regulated pathway that co-respond to strong phenotypic selection (Zhang et al., 2010). Second, there are often strong negative associations between loci in different association groups (Fig. 4) possibly representing differently but negatively regulated pathways for submergence tolerance. The third feature of the genetic network is that the downstream FGUs (loci) accounted for most of the identified loci. The QTL network underlying complex abiotic stress tolerance discovered in the selected ILs provides important genetic information in selecting ILs carrying different sets of loci/alleles/pathways as parental lines in the following DQP and MRS procedures of the GSR breeding technology.

3.4 QTL/QTL pyramiding to develop GSR varieties

In our DQP experiments, 179 pyramiding crosses were made between elite HHZ ILs, from which 76 populations showed promising results under drought, salinity, submergence and low input conditions, resulting in 2,023 pyramiding lines (PLs) selected over two rounds of selection under irrigated, low input and drought conditions. Out of these, two sets of 125 PLs were advanced to the preliminary yield trial in the 2015 dry season. Simultaneously, 33 elite HHZ PLs were submitted to the multi-environment trials and showed multiple stress tolerances. Table 6 shows the yield performances of 13 superior PLs across drought, salinity, low input and irrigated conditions. The combined yield advantages of these PLs over the best check SL8 (a hybrid) range from 16 to 54%, validating the power of the GSR breeding technology. The MRS component of the GSR breeding technology is designed to take advantage of CMS-facilitated breakdown of undesirable linkage by more efficient

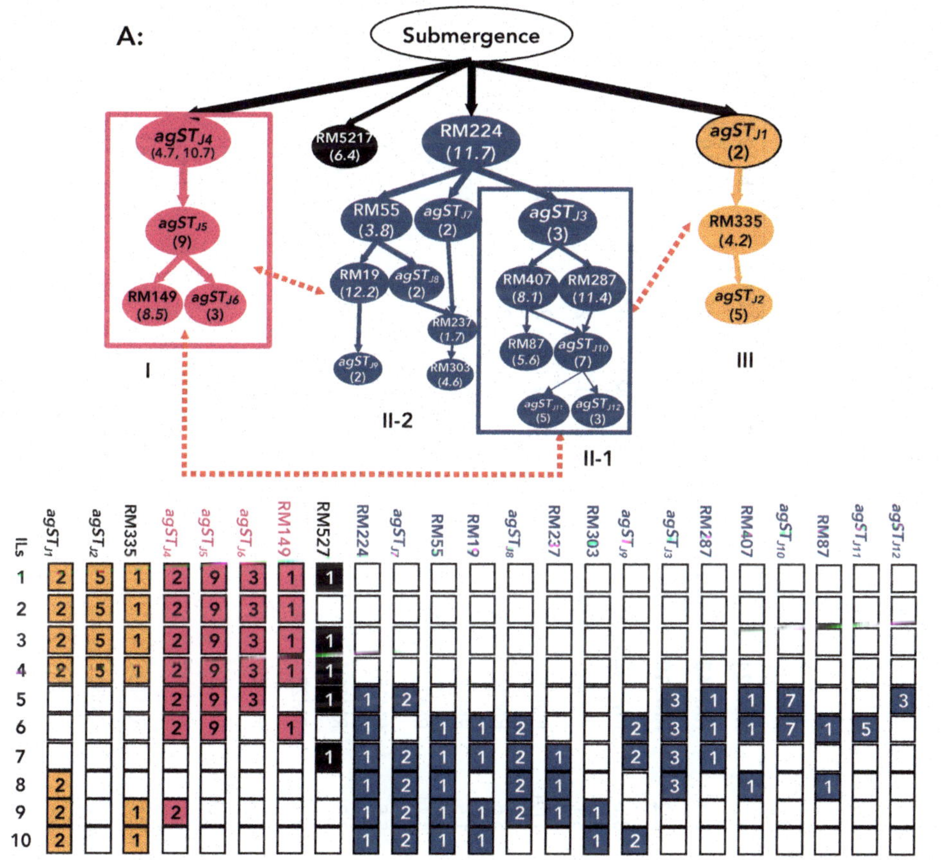

Figure 4 The putative genetic network (multi-locus structure) consisting of 23 detected functional genetic units (FGUs) in 3 major groups (putative pathways) and their graphic genotypes underlying submergence tolerance (ST) of rice detected in 10 ST introgression lines (ILs) the NPT/TKM9 BC3 population. In the corresponding graphical genotypes, the unfilled and fully colored cells represent the recipient homozygote and donor homozygote genotypes. The numbers in the cells are the number of loci included in each of the FGUs. The loci (markers) included in each of the association groups (AGs) are shown in Table 5.

recombination and to promote more efficient pyramiding of multiple target traits (Li and Zhang, 2013), which is in progress.

Several key findings can be summarized from this breeding strategy. First, there is sufficient hidden diversity in the primary gene pool for abiotic stress tolerances, and BC breeding plus appropriate phenotypic selection is a powerful way to exploit it. Compared to most conventional pedigree breeding approach, the modified BC breeding procedure is highly efficient. Second, selecting the appropriate recurrent parent(s) with high yield potential and wide adaptability to the target environments is the key to success for improving complex traits in a BC breeding programme. This is because a widely adaptable variety must have key adaptive traits and pathways to most environments, in

Table 5 Genomic information of 23 functional genetic units (FGUs or 54 loci) associated with submergence tolerance (ST) detected by X^2 tests (single loci) and multi-locus probability tests (association groups) in 10 ST introgression lines (ILs) selected from BC3F2 populations derived from crosses between IR64 and TKM9

Population												
Code	Gen.	FGU[1]	Donor	Branches	Bin	Marker	B	H	IF	X^2_2	P(1)	P(2)
J	**BC_3F_2**	qST11.7	**TKM9**		**11.7**	**RM224**	**6**	**0**	**0.600**	**107.0**	**$5.8x10^{-24}$**	
J	**BC_3F_2**	$agST_{J3}$	**TKM9**	**J-1**	**5.4**	**RM163**	**4**	**0**	**0.400**	**45.2**	**$1.3x10^{-30}$**	
J	**BC_3F_2**	$agST_{J3}$	**TKM9**	**J-1**	**6.2**	**RM510**	**4**	**0**	**0.400**	**45.2**	**$1.3x10^{-30}$**	
J	**BC_3F_2**	$agST_{J3}$	**TKM9**	**J-1**	**8.6**	**RM502**	**4**	**0**	**0.400**	**45.2**	**$1.3x10^{-30}$**	
J	BC3F2	*qST8.1*	TKM9	J-1	8.1	RM407	3	0	0.300	24.2	$5.6x10^{-6}$	
J	BC3F2	*qST11.4*	TKM9	J-1	11.4	RM287	3	0	0.300	24.2	$5.6x10^{-6}$	
J	BC3F2	*qST5.6*	TKM9	J-1	5.6	RM87	3	0	0.300	24.2	$5.6x10^{-6}$	
J	BC3F2	$agST_{J10}$	TKM9	J-1	1.5	RM583	2	0	0.200		$1.9x10^{-19}$	
J	BC3F2	$agST_{J10}$	TKM9	J-1	2.11	RM207	2	0	0.200		$1.9x10^{-19}$	
J	BC3F2	$agST_{J10}$	TKM9	J-1	5.5	RM161	2	0	0.200		$1.9x10^{-19}$	
J	BC3F2	$agST_{J10}$	TKM9	J-1	7.3	RM11	2	0	0.200		$1.9x10^{-19}$	
J	BC3F2	$agST_{J10}$	TKM9	J-1	8.3	RM25	2	0	0.200		$1.9x10^{-19}$	
J	BC3F2	$agST_{J10}$	TKM9	J-1	12.5	RM235	2	0	0.200		$1.9x10^{-19}$	
J	BC3F2	$agST_{J11}$	TKM9	J-1	10.6	RM228	1	0	0.100		$7.1x10^{-9}$	
J	BC3F2	$agST_{J11}$	TKM9	J-1	12.1	RM415	1	0	0.100		$7.1x10^{-9}$	
J	BC3F2	$agST_{J11}$	TKM9	J-1	7.1	RM481	1	0	0.100		$7.1x10^{-9}$	
J	BC3F2	$agST_{J11}$	TKM9	J-1	8.5	RM531	1	0	0.100		$7.1x10^{-9}$	
J	BC3F2	$agST_{J11}$	TKM9	J-1	4.4	RM564	1	0	0.100		$7.1x10^{-9}$	
J	BC3F2	$agST_{J12}$	TKM9	J-1	2.3	RM8	1	0	0.100		$1.3x10^{-5}$	

J	BC3F2	*$agST_{J12}$*	TKM9	J-1	2.7	RM263	1	0	0.100		$1.3x10^{-5}$
J	BC3F2	*$agST_{J12}$*	TKM9	J-1	11.6	RM206	1	0	0.100		$1.3x10^{-5}$
J	**BC_3F_2**	$agST_{J4}$	**TKM9**	J-2	**4.7**	**RM348**	**6**	**1**	**0.650**	**119.9**	**$2.0x10^{-21}$**
J	**BC_3F_2**	$agST_{J4}$	**TKM9**	J-2	**10.7**	**RM6**	**7**	**0**	**0.700**	**147.8**	**$2.0x10^{-21}$**
J	**BC_3F_2**	$agST_{J5}$	**TKM9**	J-2	**1.3**	**RM220**	**6**	**0**	**0.600**	**107.0**	**$9.8x10^{-81}$**
J	**BC_3F_2**	$agST_{J5}$	**TKM9**	J-2	**2.5**	**RM27**	**6**	**0**	**0.600**	**107.0**	**$9.8x10^{-81}$**
J	**BC_3F_2**	$agST_{J5}$	**TKM9**	J-2	**2.11**	**RM166**	**6**	**0**	**0.600**	**107.0**	**$9.8x10^{-81}$**
J	**BC_3F_2**	$agST_{J5}$	**TKM9**	J-2	**6.6**	**RM412**	**6**	**0**	**0.600**	**107.0**	**$9.8x10^{-81}$**
J	**BC_3F_2**	$agST_{J5}$	**TKM9**	J-2	**7.5**	**RM234**	**6**	**0**	**0.600**	**107.0**	**$9.8x10^{-81}$**
J	**BC_3F_2**	$agST_{J5}$	**TKM9**	J-2	**9.3**	**RM321**	**6**	**0**	**0.600**	**107.0**	**$9.8x10^{-81}$**
J	**BC_3F_2**	$agST_{J5}$	**TKM9**	J-2	**9.6**	**RM278**	**6**	**0**	**0.600**	**107.0**	**$9.8x10^{-81}$**
J	**BC_3F_2**	$agST_{J5}$	**TKM9**	J-2	**10.3**	**RM467**	**6**	**0**	**0.600**	**107.0**	**$9.8x10^{-81}$**
J	**BC_3F_2**	$agST_{J5}$	**TKM9**	J-2	**12.6**	**RM17**	**6**	**0**	**0.600**	**107.0**	**$9.8x10^{-81}$**
J	BC3F2	*$agST_{J6}$*	TKM9	J-2	4.2	RM261	4	0	0.400	45.2	$1.1x10^{-12}$
J	BC3F2	*$agST_{J6}$*	TKM9	J-2	8.3	RM544	4	0	0.400	45.2	$1.1x10^{-12}$
J	BC3F2	*qST8.5*	TKM9	J-2	8.5	RM149	5	0	0.500	72.8	$1.6x10^{-16}$
J	**BC_3F_2**	$agST_{J1}$	**TKM9**	**J-3**	**5.2**	**RM413**	**7**	**0**	**0.700**	**147.8**	**$8.0x10^{-33}$**
J	**BC_3F_2**	$agST_{J1}$	**TKM9**	**J-3**	**10.2**	**RM311**	**8**	**0**	**0.800**	**195.2**	**$4.1x10^{-43}$**
J	**BC_3F_2**	qST4.2	**TKM9**	**J-3**	**4.2**	**RM335**	**1**	**6**	**0.400**	**138.5**	**$8.4x10^{-31}$**
J	BC3F2	*$agST_{J2}$*	TKM9	J-3	1.5	RM23	4	0	0.400	45.2	$1.3x10^{-30}$
J	BC3F2	*$agST_{J2}$*	TKM9	J-3	3.9	RM571	4	0	0.400	45.2	$1.3x10^{-30}$
J	BC3F2	*$agST_{J2}$*	TKM9	J-3	5.5	RM421	4	0	0.400	45.2	$1.3x10^{-30}$
J	BC3F2	*$agST_{J2}$*	TKM9	J-3	7.6	RM248	4	0	0.400	45.2	$1.3x10^{-30}$

Continued

Table 5 Continued

Population												
Code	Gen.	FGU[1]	Donor	Branches	Bin	Marker	B	H	IF	X^2_2	P(1)	P(2)
J	BC3F2	*agST*$_{J2}$	TKM9	J-3	12.4	RM463	4	0	0.400	45.2	1.3×10^{-30}	
J	**BC_3F_2**	qST3.8	**TKM9**	**J-4**	**3.8**	**RM55**	**5**	**0**	**0.500**	**72.8**	**1.6×10^{-16}**	
J	**BC_3F_2**	agST$_{J7}$	**TKM9**	**J-4**	**6.2**	**RM314**	**5**	**0**	**0.500**	**72.8**	**1.4×10^{-15}**	
J	**BC_3F_2**	agST$_{J7}$	**TKM9**	**J-4**	**11.5**	**RM21**	**5**	**0**	**0.500**	**72.8**	**1.4×10^{-15}**	
J	BC3F2	*agST*$_{J8}$	TKM9	J-4	4.5	RM252	4	0	0.400	45.2	1.1×10^{-12}	
J	BC3F2	*agST*$_{J8}$	TKM9	J-4	6.4	RM541	4	0	0.400	45.2	1.1×10^{-12}	
J	BC3F2	*agST*$_{J9}$	TKM9	J-4	1.1	RM200	3	0	0.300	24.2	9.1×10^{-10}	
J	BC3F2	*agST*$_{J9}$	TKM9	J-4	9.4	RM566	3	0	0.300	24.2	9.1×10^{-10}	
J	BC3F2	*qST1.7*	TKM9	J-4	1.7	RM237	3	0	0.300	24.2	1.6×10^{-16}	
J	BC3F2	*qST4.6*	TKM9	J-4	4.6	RM303	2	0	0.200	9.9	0.0072	
J	BC3F2	*qST12.2*	TKM9	J-4	12.2	RM19	3	0	0.333	27.4	1.1×10^{-6}	
J	BC3F2	*qST6.4*	TKM9		6.4	RM527	5	0	0.500	72.8	1.6×10^{-16}	

[1] AGs are defined as a group unlinked but perfectly associated loci of equal introgression in the selected ST ILs from each BC population, detected by multi-locus probability tests. P(1) or P(2) are the probabilities for the null hypothesis that the genotypic frequencies fit the Mendelian segregation based either on single locus X^2 tests or the multilocus probability tests for association groups (AGs), respectively (see the materials and methods). The bold ones are putative regulators of high introgression frequency in the ST ILs (Wang et al. 2015).

Table 6 Grain yield performances of some promising elite Huang-hua-zhan pyramiding lines under 4 different conditions in 2015 wet season

						Mean yield advantage (%)		
Pyramiding Lines	Irrigated	Low Input	Rainfed	Salinity	Combined	Over SL8	Over IRRI123	Over HHZ
GSR IR1-DQ142-Y1-Y1	7.54	4.99	4.78	2.38	4.92	54.8	58.3	106.8
GSR IR1-DQ123-Y3-Y2	6.75	4.99	4.62	2.39	4.69	47.3	50.7	96.9
GSR IR1-DQ140-LI4-Y1	7.25	5.02	4.68	0.89	4.46	40.3	43.4	87.4
GSR IR1-DQ62-D7-D1	7.40	4.63	4.33	1.12	4.37	37.4	40.5	83.6
GSR IR1-DQ121-Y6-D2	6.96	4.76	5.03	0.57	4.33	36.1	39.2	81.9
GSR IR1-DQ120-LI3-Y1	6.86	4.12	4.52	1.53	4.26	33.9	36.9	78.9
GSR IR1-DQ122-Y7-D1	7.58	3.19	3.98	1.25	4.00	25.8	28.6	68.1
GSR IR1-DQ122-Y4-D4	6.04	3.85	3.91	2.05	3.96	24.6	27.4	66.5
GSR IR1-DQ129-Y3-L1	6.31	4.09	3.64	1.66	3.92	23.4	26.2	64.9
GSR IR1-DQ123-Y1-L1	6.19	4.14	3.51	1.54	3.84	20.9	23.6	61.6
GSR IR1-DQ130-Y5-Y1	6.61	4.08	3.56	0.85	3.77	18.7	21.4	58.6
GSR IR1-DQ130-LI3-D2	6.35	3.71	3.96	1.02	3.76	18.2	20.9	58.0
GSR IR1-DQ136-Y8-Y1	6.25	3.72	3.75	1.04	3.69	16.1	18.7	55.1
SL 8 (Hybrid Check)	4.97	4.40	2.14	1.24	3.18	0.2	2.4	33.8
IRRI 123(Inbred check)	5.42	3.53	2.55	0.95	3.11	-2.2	0.0	30.7
HHZ (Recipient check)	4.02	2.59	2.27	0.64	2.38	-25.2	-23.5	0.0
IRRI 148 (Inbred Check)	2.90	1.94	1.76	0.85	1.86	-41.4	-40.1	-21.8
LSD (5%)	2.48	1.10	1.33	1.26	0.89			

Low Input=No fertilizers ever added over 13 consecutive seasons to the plots; Salinity=8-14 dSm^{-1}(seed to seed); Rainfed=21 d old single seedling transplanted and irrigated for 30 d and later, no irrigation until harvest (110-120d crop duration). During wet season at Los Baños has regular showers and all bunds are kept open.

addition to its high yield potential, and require minimum numbers of additional alleles for improving specific target traits. It is more desirable for a BC breeding programme to have 2–3 recipients with different plant types and many lines of diverse origins (but distantly related to the recipients as donors) if the breeding is aiming at broad and diverse target environments. This is because favourable loci/alleles for target traits from different donors are unlikely to be the same. Uses of diverse donors are expected to broaden the genetic diversity for target traits in selected ILs, providing the genetic basis for further improving the target traits by DQP. Third, introgression or segregation in BC progenies of rice line crosses is characteristic of specific crosses, which may or may not follow the Mendelian expectations. Fourth, we observed generally reduced heterozygosity in the BC progenies selected under drought and salinity (Li and Zhang, 2013; Wang et al., 2015).

4 Assessing the impact of GSR varieties

Pandey et al. (2012) has studied the use of improved rice varieties and assessed farm-level impacts in stress-prone rainfed areas in South Asia. One important conclusion they reached is that technological development and dissemination strategies in stress-prone rainfed environments should consider the heterogeneity and complexity of the stresses in those environments. Putting multiple types of abiotic stress tolerance into single widely-adaptable varieties is the most efficient way to deal with these conditions. The newly released GSR varieties should serve this purpose. In a recent social economic assessment of some newly released GSR varieties in the rainfed and irrigated ecosystems in the Philippines and Bangladesh, Yorobe et al. (2016) reported positive and significant effects of GSR rice varieties on yield and net farm income particularly at high rainfall levels. Using submergence-tolerant GSR varieties results in significantly higher yield compared to conventional rice varieties. The average net farm income difference between GSR and non-GSR varieties was $230.9 per hectare. Their results suggest a stronger positive impact of GSR varieties for rice farmers in areas experiencing extreme climatic events and during the wet season of the Philippines with high occurrence of flooding. Currently, several GSR varieties with multiple abiotic stress tolerances are quickly expanding to replace conventional varieties in the Philippines under the support of the Department of Agriculture. Thus, GSR varieties offer a real opportunity for food security and poverty alleviation in the Philippines (Yorobe et al., 2016). In 2015, a total of 54 613 tons of certified seeds were produced for the 13 GSR varieties released in the Philippines and Vietnam, which are sufficient for more than 1.5 million ha of more than 1.6 million small-hold rice farmers in the two countries (Table 7). With many more GSR varieties, as shown in Table 3,

Table 7 GSR varietal releases, seed production in Asia and Africa (Source:)

Country	No. of released varieties	Seeds produced (t)	Estimated Area (ha) (SR = 35kg/ha)	Average household area (ha)	No. of households reached
Philippines	11	18,827	537,914	2.01	267,619
Vietnam	2	34,733	992,364	0.71	1,397,696
Total	**13**	**54,613**	**1,560,373**		**1,665,315**

SR = Seeding rate

to be released for diverse rice ecologies and taken up by small-hold farmers in different Asian and African countries, we expect significant impact of these GSR varieties on food security and poverty alleviation in these developing countries.

5 Linking genomic tools and information to breeding

Rice is the model species for monocot crops and tremendous progress has been made in the areas of rice functional genomics with over 2000 rice genes (affecting many traits of agronomic importance) cloned and characterized (Zhang et al., 2008). Recently, a core collection of 3024 rice accessions from 89 countries has been sequenced with an average 14X sequencing depth, and over 42 million of SNP markers have been identified from the 3K rice genomes (Alexandrov et al., 2015; Li and al, 2014; Li, 2001; The 3000 Rice Genome Project, 2014; Zheng TianQing, 2015; Zheng et al., 2015). The complete genome sequence data of the 3K sequenced rice genomes are freely available in the public domain. Meanwhile, tremendous global efforts are being undertaken to systematically phenotype the sequenced 3K accessions to identify important genes/alleles associated with a wide range of rice traits by GWAS (Mau et al., 2016; The 3000 Rice Genome Project, 2014). Deep analyses of this huge resource of rice genome/genetic data are in progress and are expected to result in several comprehensive rice genetic information databases and platforms for future rice improvement (3KRGP). Based on the SNPs discovered in the 3K RGP, an Affymetrix genotyping chip containing more than 56 000 carefully selected informative SNP markers has been developed, which would enable high-throughput genotyping of rice breeding and genetic materials in a highly cost-effective manner.

As briefly described above, the real power of the GSR breeding technology is the full integration of the trait-specific ILs with the genetic information and new genomic technologies (Li and Zhang, 2013). The large set of the trait-specific ILs with significantly improved tolerances to more abiotic stresses developed in step 1 BC breeding procedure forms the material platform for large-scale discovery of genes/QTL underlying both target and non-target traits and for highly efficient development of better GSR varieties by DQP and MRS. In fact, all parental lines in our GSR breeding programmes are from the core collection of rice germplasm and are included in the re-sequenced 3K lines. More of the sequenced accessions are being systematically used as donors in our GSR breeding programmes so that the available genomics information in the 3K rice genomes can be used in important steps of the GSR breeding, including parental selection, number of crosses made and selection schemes.

6 Future trends and conclusion

Sustainable rice production is the key to food security and poverty alleviation of many Asian and African countries, particularly under changing climatic conditions. Rice productivity is low in the rainfed areas of Asia and Africa, particularly in the face of the global climate change. Raising the low rice productivity of most rainfed areas of Asia and Africa and sustaining rice productivity in global irrigated rice areas are the biggest challenge in the coming decades for the global rice research community. 'Green Super Rice (GSR)', defined as rice cultivars that can produce high and stable yields under less input, is an important concept proposed to meet the challenge. Breeding programmes such as GSR signal a major shift in breeding objectives from emphasizing yield improvement in the past to high yield potential plus resilience to more frequent and extreme environmental disturbances from climatic changes in the future. Here, we have briefly described and demonstrated a novel breeding strategy, 'GSR breeding technology'. This integrates strong phenotypic selection in a modified BC breeding procedure with highly efficient QTL network discovery by selective introgression and DNA markers, followed by efficient development of GSR varieties by designed QTL pyramiding (DQP) and molecular recurrent selection (MRS). This breeding strategy has been demonstrated by developing large numbers of promising GSR lines with good tolerances to multiple abiotic stresses from a small number of breeding populations in a short period of 6 years and by releases and up-scaling of 13 GSR varieties in Asia. It is believed that wide adoption of the GSR concept and the GSR BC breeding strategy will help achieve the sustainable rice production and food security in the future.

7 References

Alexandrov N, Tai S, Wang W, Mansueto L, Palis K, Fuentes RR, Ulat VJ, Chebotarov D, Zhang G and Li Z (2015). SNP-Seek database of SNPs derived from 3000 rice genomes. *Nucleic Acids Research* 43: D1023–D1027.

Ali AJ, Xu JL, Ismail AM, Fu BY, Vijaykumar CHM, Gao YM, Domingo J, Maghirang R, Yu SB and Gregorio G (2006). Hidden diversity for abiotic and biotic stress tolerances in the primary gene pool of rice revealed by a large backcross breeding program. *Field Crops Research* 97: 66–76.

Ali J, Xu J, Gao Y, Ma X, Meng L, Wang Y, Pang Y, Guan Y, Xu M, Revilleza J, Neil J, Zhou S and Li Z (2016). Harnessing the hidden genetic diversity for improving multiple abiotic stress tolerances in rice (*Oryza sativa* L.) *PloS ONE* 12(3): e0172515. doi:10,1371/journal.pone.0172515.

Ali J, Xu JL, Gao YM, Fontanillo MA and Marfori CM (2012). Innovative Green Super Rice (GSR) molecular breeding strategy: achievements and advances. *Philippine Journal of Crop Science* 37(1): 49.

Chen H, Hang H, Zhou F, Yu H and Xing WD (2013). Development of genomics-based genotyping platforms and their applications in rice breeding. *Current Opinion in Plant Biology* 16: 247–54.

Cui Y, Zhang F, Xu J, Li Z and Xu S (2015). Mapping quantitative trait loci in selected breeding populations: A segregation distortion approach. *Heredity* 115(6): 538–46.

Dingkuhn M, Laza MRC, Kumar U, Mendez KS, Collard B, Jagadish K, Singh RK, Padolina T, Malabayabas M and Torres E (2015). Improving yield potential of tropical rice: Achieved levels and perspectives through improved ideotypes. *Field Crops Research* 182: 43–59.

Guan YS, Serraj R, Liu SH, Xu JL, Ali J, Wang WS, Venus E, Zhu LH and Li ZK (2010). Simultaneously improving yield under drought stress and non-stress conditions: a case study of rice (*Oryza sativa* L.). *Journal of Experimental Botany* 61: 4145–56.

He YX, Zheng TQ, Hao XB, Wang LF, Gao YM, Hua ZT, Zhai HQ, Xu JL, Xu ZJ and Zhu LH (2010). Yield performances of japonica introgression lines selected for drought tolerance in a BC breeding programme. *Plant Breeding* 129: 167–75.

Iii MM, Li T, Angeles O, Evangelista GK, Fontanilla MA, Xu J, Gao Y, Li Z and Ali J (2014). Biomass accumulation and partitioning of newly developed Green Super Rice (GSR) cultivars under drought stress during the reproductive stage. *Field Crops Research* 162: 30–8.

Lafitte HR, Li ZK, Vijayakumar CHM, Gao YM, Shi Y, Xu JL, Fu BY, Yu SB, Ali AJ and Domingo J (2006). Improvement of rice drought tolerance through backcross breeding: Evaluation of donors and selection in drought nurseries. *Field Crops Research* 97: 77–86.

Li ZK (2001). QTL mapping in rice: a few critical considerations. *Rice Genetics IV*, 153–71.

Li ZK, Fu BY, Gao YM, Xu JL, Ali J, Lafitte HR, Jiang YZ, Rey JD, Vijayakumar CHM and Maghirang R (2005). Genome-wide introgression lines and their use in genetic and molecular dissection of complex phenotypes in rice (*Oryza sativa* L.). *Plant Molecular Biology* 59: 33–52.

Li ZK, Pinson SRM, Park WD, Paterson AH and Stansel JW (1997). Epistasis for three yield components in rice *Oryza sativa* L. *Genetics* 145: 453–65.

Li ZK and Zhang F (2013). Rice breeding in the post-genomics era: from concept to practice. *Current Opinion in Plant Biology* 16: 261–9.

Luo LJ (2010). Breeding for water-saving and drought-resistance rice (WDR) in China. *Journal of Experimental Botany* 61: 3509–17.

Mackill DJ, Coffman WR and Garrity DP (1996). Rainfed lowland rice improvement. International Rice Research Institute.

Meng L, Xu J, Li Z, Lin X, Sun H, Ma X, Cui Y, Chen K, Sun Y and Zhu L (2010). Identification and screening of salt and alkaline tolerance in rice using advanced backcross introgression populations. *Molecular Plant Breeding* 18: 72–80.

Pandey S (1997). Rainfed lowland rice research: challenges and priorities for the 21st century: Breeding strategies for rainfed lowland rice in drought-prone environments. *Proceedings of an International Workshop*, Held at Ubon Ratchathani, Thailand, 5–8 November.

Pandey S, Gauchan D, Malabayabas M, Bool-Emerick M and Hardy B (2012). *Patterns of Adoption of Improved Rice Varieties and Farm-Level Impacts in Stress-Prone Rainfed Areas in South Asia.* Los Baños (Philippines): International Rice Research Institute, p. 318.

The 3,000 Rice Genomes (2014). The 3,000 rice genomes project. *GigaScience* 3: 1–6.

Wang W, Fu B, Ali J, Xu J, Gao Y, Zheng T, Zhang F and Li Z (2015). Genome-wide responses to selection and genetic networks underlying submergence tolerance in rice. *Plant Genome* 8.

Witcombe JR (2008). Breeding for abiotic stresses for sustainable agriculture. *Philosophical Transactions of the Royal Society of London* 363: 703–16.

Yorobe JM, Ali J, Pede VO, Rejesus RM, Velarde OP and Wang H (2016). Yield and income effects of rice varieties with tolerance of multiple abiotic stresses: the case of green super rice (GSR) and flooding in the Philippines. *Agricultural Economics* 47: 261–71.

Zhang F, Zhai HQ, Paterson AH, Xu JL, Gao YM, Zheng TQ, Wu RL, Fu BY, Ali J and Li ZK (2010). Dissecting genetic networks underlying complex phenotypes: the theoretical framework. *PloS ONE* 6: e14541.

Zhang H, Wang H, Qian Y, Xia J, Li Z, Shi Y, Zhu L, Ali J, Gao Y and Li Z (2012). Simultaneous improvement and genetic dissection of grain yield and its related traits in a backbone parent of hybrid rice (*Oryza sativa* L.) using selective introgression. *Molecular Breeding* 31: 181–94.

Zhang Q (2007). Strategies for developing green super rice. *Proceedings of the National Academy of Sciences of the United States of America* 104: 16402–9.

Zhang Q, Li J, Xue Y, Han B and Deng XW (2008). Rice 2020: A call for an international coordinated effort in rice functional genomics. *Molecular Plant* 1: 715–19.

Zheng TQ, Hong YU, Zhang HL, Zhichao WU, Wang WS, Tai SS, Lu C, Jue R, Wei CC, Shi JX, Gao YM, Fu BY, Zhou YL, Zhao XQ, Zhang F, McNALLY K, Li ZC, Zhang GY, Li JY, Zhang DB, Xu JL and Li ZK (2015). Rice functional genomics and breeding database (RFGB):3K-rice SNP and InD el sub-database. *Chinese Science Bulletin* 60: 367–71.

Chapter 6

Mechanisms of drought tolerance in rice

Anuj Kumar, Supratim Basu, Venkategowda Ramegowda and Andy Pereira, University of Arkansas, USA

1 Introduction

Rice (*Oryza sativa*) is one of the most important cereal crops belonging to the grass family (Poaceae) and has the second largest area under production following maize worldwide. The tremendous growth of human population worldwide has increased the demand for rice production (Liang et al., 2010), requiring an improvement of 50% by the year 2025 (Khush, 2001). Rainfed ecosystems contribute to only 25% of the total water supply, thereby making rice more vulnerable to increased frequency of drought stress under the ensuing threat of global climate change. Rice, being a water-loving crop, is severely affected by drought stress that depresses yield by 15–50% depending on the vigour and period of stress (Kumar et al., 2008; Srividhya et al., 2011). Rice is especially sensitive to drought stress during reproductive growth and even moderate stress can result in drastic reduction in grain yield (GY) (Hsiao et al., 1984; O'Toole, 1982; Venuprasad et al., 2008).

Drought mitigation, through development of drought-resistant varieties with higher yields suitable for water-limiting environments, will be the key factor to improve stable rice production. The development of drought-tolerant and high-yielding varieties requires a good knowledge of the physiological mechanisms, yield under drought-related

http://dx.doi.org/10.19103/AS.2106.0003.08

components and the genetic control of traits contributing to drought resistance. Moreover, efforts to increase yield under drought also focus on improving secondary traits such as root architecture, leaf water potential, panicle water potential, osmotic adjustment and relative water content (RWC). Rice can be classified into three subspecies namely *indica*, *japonica* and *javanica* also known as 'tropical japonica' (Mae, 1997), which is the type of rice commonly grown in the US (Garris et al., 2005). Due to its wide distribution and adaptability, rice serves as an excellent model system for studying plant evolutionary genomics with a broad range of morphological, physiological and developmental diversity found in both *Oryza sativa* and its widely distributed wild ancestors, *Oryza rufipogon* and *Oryza nivara* (Zhao et al., 2010).

Classical breeding for drought tolerance, involving the screening of landraces and varieties under controlled drought or field drought conditions, followed by incorporation of the tolerant genotypes in breeding programmes, has helped make steady progress in breeding programmes (Kumar et al., 2008). In parallel, molecular genetic studies of drought-responsive candidate genes, such as transcription factors and other regulatory genes, and their expression in transgenic plants have produced examples of drought tolerance tested under laboratory or controlled conditions (Todaka et al., 2015).

2 Mechanisms of drought resistance

Drought resistance is defined as the ability of a plant to produce its maximum economic product under water-deficit environmental conditions relative to the water-constraint-free management. Drought resistance is a complex trait whose effect depends on the action and interaction of different morphological, biochemical and physiological responses (Mitra, 2001). The rice crop mainly responds to drought stress conditions by leaf rolling, stomatal closure and increased abscisic acid (ABA) synthesis to minimize water deficit (Price et al., 2002). Physiological research suggests that drought resistance in rice mainly relies on water use efficiency (WUE) that allows minimum water usage for maximum production and osmotic adjustment that enables plants to maintain turgor and protect the meristem (Nguyen et al., 2004).

There are many mechanisms by which plants can adapt to drought stress, namely drought escape, drought avoidance and drought tolerance; they adapt on the basis of their molecular responses and morpho-physiological changes (Fukai and Cooper, 1995).

2.1 Drought escape

Drought escape is defined as the ability of a plant to complete its life cycle before the development of serious soil water deficits. It essentially comprises two different mechanisms such as rapid phenological development and developmental plasticity. In rapid phenological development, plants are able to produce flowers with minimum vegetative growth enabling them to produce seeds on a limited water supply. In developmental plasticity, plants are able to produce an abundance of vegetative growth, flowers and seeds in seasons of abundant rain. This enables the desert ephemerals to both escape drought and survive long periods without rain. In drought-prone areas of the southern US, drought escape is an important mechanism that allows rice to produce grains despite limited water availability (Kumar et al., 2008).

2.2 Drought avoidance

Drought avoidance is defined as the ability of plants to maintain relatively high tissue water potential despite a shortage of soil moisture. The plant tissue has two options to maintain a high water level during a period of high evaporative demand and increasing soil water deficit, to either reduce water loss or maintain the supply of water. Rice genotypes that can maintain water status through ABA biosynthesis, other biochemical mechanisms or adapted root systems come under drought avoidance mechanism category. These genotypes are able to minimize the yield losses caused by drought (Singh et al., 2012). Mechanisms for improving WUE and reducing water loss also confer drought avoidance. Rice genotypes that avoid drought usually have deep, coarse roots with a high ability for branching and soil penetration, higher root to shoot ratio, elasticity in leaf rolling, early stomatal closure and high cuticular resistance (Blum et al., 1989; Samson et al., 2002; Wang et al., 2006).

2.3 Drought tolerance

Drought tolerance is the ability of plants to survive under low tissue water content (Turner, 1979; Delphine et al., 2010). Genetically, drought tolerance in rice is a complex trait which is controlled by polygenic effect and involves complex morpho-physiological mechanisms (Li and Xu, 2007), such as maintenance of turgor through osmotic adjustment, increased elasticity in cell, decreased cell size and desiccation tolerance by protoplasmic resistance (Sullivan and Ross, 1979). The responses of plants to tissue water potential determine their level of drought tolerance (Mitra, 2001) and the traits associated with such phenomena are considered as secondary traits. Secondary traits such as osmotic adjustment, RWC, leaf rolling and stomatal conductivity have been used for selection (Nguyen et al., 1997; Babu et al., 2001; Kato et al., 2006).

3 Assessing drought tolerance performance

Plant productivity is a key component that needs to be considered when abiotic stress resistant rice is developed. Plant productivity is immensely affected by drought stress under field conditions that varies in severity and duration and is also accompanied by concurrent stresses like high or low temperature. Thus, drought tolerance analysis obtained from greenhouse conditions is not necessarily comparable to observations made under field conditions. Field experiments are thus critical for the proper evaluation of drought stress resistance.

Drought stress has been studied in upland and lowland conditions at different stress levels such as mild, moderate and severe stress environments. Drought stress evaluation in the field is done by imposing continuous stress from ten days before heading to ten days after heading, following IRRI's standard procedures that involve monitoring changes in soil moisture and penetration resistance in stressed field with gravimetric measurements and a penetrometer (Bengough et al., 2011). RWC is evaluated when leaf rolling and dying have started (Raman et al., 2012; Jeong et al., 2013).

The response of plants to drought stress is further classified into mild, moderate and severe based on the percentage of yield reduction compared with GY from non-stress

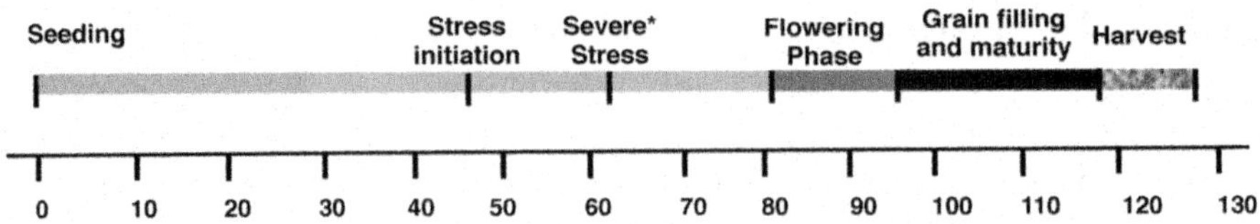

Figure 1 Drought screening under reproductive stage in direct-seeded upland conditions. The different stages of plant growth are shaded, and the time of stress application is shown with stress affecting the flowering phase. The field is irrigated when the soil tensiometer shows a reading of –70 kPa at 30-cm depth. Drought at the R3 stage of flowering has been shown to be important in reducing the yield of rice plants (adapted from Kumar et al., 2014).

experiments (Kumar et al., 2008). Under lowland conditions, stress experiments exhibiting a yield reduction of 30% or less are termed mild stress, reductions of 31–65% are termed moderate stress and yield reductions above 65% are referred to as severe stress treatments. In the absence of an upland non-stress counterpart of this experiment, the percentage of yield reduction compared with the mean yield of a lowland non-stress experiment was used to classify the experiment as upland mild stress treatment (Dixit et al., 2014b).

For drought evaluation in upland conditions in the field, rice populations are usually screened using sprinkler-irrigated dry direct-seeded trials (Fig. 1). The field plots are irrigated by sprinklers twice a week during establishment and early vegetative stage (Kumar et al., 2014). Drought stress is initiated after this duration by withholding irrigation and irrigated only when the soil water tension goes below –50 kPa at 30-cm soil depth. At this soil water potential, most of the rice plants wilt and exhibit leaf drying. This type of cyclic stress is considered to be efficient in screening for drought tolerance in rice populations consisting of genotypes with a broad range of growth duration (Lafitte et al., 2004) and it ensures that all plants experience adequate stress during the reproductive development. The same agricultural practices are applied to upland non-stress field plots, except that irrigation is continued twice a week up to ten days before harvest. Under lowland conditions in the field, transplanted field plots are drained out after 30 days of transplanting and irrigation is withheld to impose drought stress at the reproductive stage (Kumar et al., 2014). Drought stress is continued until severe leaf rolling. Irrigation for plant survival is provided thereafter through flash flooding, and water is drained after 24 h to impose a second cycle of drought stress. The depth of water table is measured by inserting a 1.1-m polyvinyl chloride (PVC) pipe in the field plots at regular time intervals. Pipes are usually inserted to 1.0-m depth and 10 cm of pipe remains above the soil surface. Depletion in the water table is always measured through a metre scale daily after the onset of the stress (Xiao et al., 2009; Wang et al., 2014).

4 Identifying genes promoting drought tolerance: quantitative trait loci (QTL) analysis

The effectiveness of yield-related traits under water-limited environments has been useful in other cereals crops such as wheat, maize and sorghum, and molecular tools facilitate the identification and genomic locations of genes controlling traits to drought tolerance using QTL analysis.

After identifying the components contributing to GY, the relationship of yield components to morpho-physiological traits will help in understanding the mechanisms of drought tolerance. These components include plant height, days to flowering, delay in flowering, GY per panicle, biomass/biological yield, harvest index, number of panicles per plant, panicle length, panicle excretion, spikelet fertility, total number of spikelets, peduncle length, 1000 grain weight and seed setting. They are all affected by different intensities of drought stress and flowering stages of rice evaluated under field conditions. For higher and stable yield under water-limited conditions, each component has its effect on yield. Although plant height is not directly related to yield, it has a constructive effect on yield by modifying the expression of secondary and integrated traits by affecting the transpiration demand. Rice genotypes with greater plant height are often larger in overall plant size, intercept more light and lose water faster by transpiration, leading to lower plant water status (Kamoshita et al., 2004), higher leaf death scores and more spikelet sterility (Pantuwan et al., 2002; Kato et al., 2007).

4.1 Major QTLs for grain yield under drought

Several studies have been reported for major and independent QTLs related to GY and its components under water-limited conditions in rice (Table 1). Most mapped QTLs have positive effect on GY but some have also shown negative effect on GY because of negative alleles from recipient parents.

qDTY12.1, the first large-effect QTL reported for GY under drought, was identified in a population of 436 random F3-derived lines from a cross between upland rice cultivars Vandana and Way Rarem (Mishra et al., 2013). Although this QTL is contributed by the susceptible parent Way Rarem, it also exhibits digenic interaction QTLs (qDTY2.3 and qDTY3.2) contributed by the tolerant parent Vandana. It was thus observed that the lines possessing a combination of these QTLs showed more yield under drought than those containing the individual QTL (Dixit et al., 2012). qDTY3.1 was identified in a backcross 1 (BC_1) progeny derived from a cross of high-yielding lowland rice variety Swarna and upland rice variety Apo, while qDTY6.1 was identified from a cross of high-yielding lowland rice variety Swarna and upland rice variety Apo. Both these QTLs have a complementary effect under drought stress conditions (Dixit et al., 2014a). A large-scale experiment conducted using F_3 populations developed from crosses between drought-tolerant donor N22 and high-yielding mega-varieties Swarna, IR64 and MTU1010 identified one of the most consistent QTLs, qDTY1.1, that showed a significant effect on GY under drought (Vikram et al., 2011). Two consistent trait effects, QTL (qDTY8.1) and (qDTY10.1), were identified in a population derived from a cross between Basmati 334/Swarna (Vikram et al, 2012) and an Aday Sel/*4 IR64 BIL population (Swamy et al., 2013).

4.2 Effect of grain yield QTLs on other traits

A major proportion of the QTLs identified for GY under drought affect other traits as well, which can be due to three probable reasons: (i) tolerance contributed by the QTL can affect all these traits, (ii) tolerance can be contributed by the cumulative effect of all the traits influenced by the QTL and (iii) the tolerance mechanism can be a combined effect of the aforementioned causes. The qDTY12.1 locus showed its effect on almost all traits such as days to flowering, plant height, biomass, harvest index, drought response index and panicle number, which affect GY under drought (Bernier et al., 2006). Similarly, the

Table 1 Major QTLs for grain yield and component traits in rice

QTL	Ch	Marker interval	LOD	Trait	Parents	Populations	Associated traits	References
sd1	1	RZ730–RZ801	2.53	PH	IR64 × Azucena	DH	TN	Hemamalini et al., 2000
phs4.1	4	RG476–RG214	7.82	PH	CT9993 × IR62266	DH	GY	Babu et al., 2003
gpps4.1	4	RG214–RG620	3.58	G/P	CT9993 × IR62266	DH	–	Babu et al., 2003
qgy3.1	3	RG348/ RM22–RZ329	2.6	GY	CT9993-5-10-1-M × IR62266-42-6-2	DH	PN	Lanceras et al., 2004
qgy4.3	4	RM120–RZ675/ RM274	2.57	GY	CT9993-5-10-1-M × IR62266-42-6-2	DH	PN	Lanceras et al., 2004
qgy3.2	3	RG348/ RM22–RZ329	2.84	GY	CT9993-5-10-1-M × IR62266-42-6-2	DH	PN	Lanceras et al., 2004
qGY-2a	2	RM561–RM341	4.26	GY	Zhenshan 97B × IRAT109	RILs(F9)	SF, GW	Zou et al., 2005
qGY-2b	2	RM526–RM525	4.23	GY	Zhenshan 97B × IRAT109	RILs(F9)	SF, GW	Zou et al., 2005
qTGW-2b	2	RM525–RM318	3.07	TGW	Zhenshan 97B × IRAT109	RILs(F9)	GY	Zou et al., 2005
qPN-4a	4	RM451–RM317	4.61	PN	Zhenshan 97B × IRAT109	RILs(F9)	–	Zou et al., 2005
qPN-8	8	RM404–RM339	3.62	PN	Zhenshan 97B × IRAT109	RILs(F9)	SF, GY	Zou et al., 2005
qSN-4b	4	RM255–RM349	13.2	SN	Zhenshan 97B × IRAT109	RILs(F9)	–	Zou et al., 2005
qSN-8	8	RM80–RM149	6.66	SN	Zhenshan 97B × IRAT109	RILs(F9)	–	Zou et al., 2005
qSN-9	9	RM215–RM245	12.88	SN	Zhenshan 97B × IRAT109	RILs(F9)	–	Zou et al., 2005
qtl1.1	1	RM3103–RM511	5.1	PH	Way Rarem x Vandana	F3 derived lines	–	Bernier et al., 2007
qDTY1.1	1	EM11–RG109	43.5	GY	CT9993-5-10-1-M × IR62266-42-6-2	DH	SDW	Kumar et al., 2007
qDTY12.1	12	RM201–RG667	34	GY	Way Rarem × Vandana	F3 derived lines	HI, PN, PH	Bernier et al., 2007
qtl3.1	3	RM277–RM260	4.8	BIO	Way Rarem × Vandana	F3 derived lines	HI, PN, DTH	Bernier et al., 2007
qtl6.1	6	RM523–RM545	4.8	BIO	Way Rarem × Vandana	F3 derived lines	HI, PN, DTH	Bernier et al., 2007

qtl2.1	2	RM7195–RM28166	4	PN	Way Rarem × Vandana	F3 derived lines	–	Bernier et al., 2007
qtl6.1	6	RM1367–RM250	4	PN	Way Rarem × Vandana	F3 derived lines	–	Bernier et al., 2007
qGy10	10	RM242–RM278/ RZ12	8.78	GY	Tequing × Lemont	BILs	PN, GW	Zhao et al., 2008
qGy7	7	RM458C–RM139	12.17	GY	Tequing × Lemont	BILs	PN, GW	Zhao et al., 2008
qGy5	5	RM126–RM483	7.82	GY	Tequing × Lemont	BILs	PN, GW	Zhao et al., 2008
qGy8	8	RM509–RM163	13.24	GY	Tequing × Lemont	BILs	PN, GW	Zhao et al., 2008
qPL-4a	4	RM470–RM451	10.02	PL	Zhenshan 97B × IRAT109	RILs(F10)	–	Liu et al., 2008
qPL-8	8	RM149–RM477	7.04	PL	Zhenshan 97B × IRAT109	RILs(F10)	SBN	Liu et al., 2008
qPL-9	9	RM553–RM160	4.72	PL	Zhenshan 97B × IRAT109	RILs(F10)	–	Liu et al., 2008
qdw2.1	2	RM495–RM1	2.63	SDW	IR 64 × INRC10192	RILs(F7)	–	Srividhya et al., 2011
qsdw2.2	2	RM262–RM106	2.51	SDW	IR 64 × INRC10192	RILs(F7)	–	Srividhya et al., 2011
qsdw9.1	9	RM561–RM341	3.03	SDW	IR 64 × INRC10192	RILs(F7)	–	Srividhya et al., 2011
qDTY8.1	8	RM339–RM210	5.1	GY	Swarna × Basmati 334	RILs(F3:4)	GN	Vikram et al., 2012
qDTY10.1	10	RM223–RM210	3.9	GY	Swarna × Basmati 334	RILs(F3:4)	GN	Vikram et al., 2012
QRF6.1	6	Pgi-2–RM527/ RM238B	9	HI	IR62266 × Norungan	RILs	–	Suji et al., 2012
QRF6.2	6	RM510–RZ398/ RM225/RM584	10.2	HI	IR62266 × Norungan	RILs	–	Suji et al., 2012
QRF6.3	6	RM510–RZ398/ RM225/RM584	9.6	HI	IR62266 × Norungan	RILs	–	Suji et al., 2012
QDS_9.1	9	RM453–RM491	5.5	GY	IR64 × IR77298-5-6-B-18(Aday Sel)	BILs(BC2F2:3)	PL, GW	Swamy et al., 2013
QDS_9.2	9	RM460/RM409–RZ422/RM434	11.7	GY	IR64 × IR77298-5-6-B-18 and -11(Aday Sel)	BILs(BC2F2:3)	PL, GW	Swamy et al., 2013

(Continued)

Table 1 (*Continued*)

QTL	Ch	Marker interval	LOD	Trait	Parents	Populations	Associated traits	References
QDS_2.1	2	RM460/RM409–RZ422/RM434	3.4	GY	IR64 × IR77298-5-6-B-18(Aday Sel)	BILs(BC2F2:3)	PL, GW	Swamy et al., 2013
QDS_10.1	10	RM236–RM279	10	GY	IR64 × IR77298-14-1-2-B-10(Aday Sel)	BILs(BC2F3:4)	PL, GW	Swamy et al., 2013
QDS_4.1	4	RM269–RM304	6.5	GY	IR64 × IR77298-14-1-2(Aday Sel)	BILs(BC2F3:4)	PL, GW	Swamy et al., 2013
QGyp1	1	RM286–RM332	9.66	GY	IR 64 × Tarom Molaei	BILs(BC2F6)	TN, SF	Wang et al., 2013
QGyp11	11	RM13–RM434	2.8	GY	IR 64 × Tarom Molaei	BILs(BC2F6)	TN, SF	Wang et al., 2013
QGyp2	2	RM5–RM237	3.72	GY	Teqing × Tarom Molaei	BILs(BC2F6)	TN, SF	Wang et al., 2013
QGyp10	10	RM16–RM319	3.12	GY	Teqing × Tarom Molaei	BILs(BC2F6)	TN, SF	Wang et al., 2013
QFgp1	1	RM5–RM237	4.43	FG/P	IR 64 × Tarom Molaei	BILs(BC2F6)	SF, GY	Wang et al., 2013
QFgp6	6	RM136–RM527B	5.06	FG/P	IR 64 × Tarom Molaei	BILs(BC2F6)	GW	Wang et al., 2013
QFgp11	11	RM286–RM332	2.62	FG/P	IR 64 × Tarom Molaei	BILs(BC2F6)	SF, GY	Wang et al., 2013
QFgp9	9	RM242–RM201	9.3	FG/P	Teqing × Tarom Molaei	BILs(BC2F6)	–	Wang et al., 2013
QPn9a	9	OSR28–RM189	2.54	PN/P	IR 64 × Tarom Molaei	BILs(BC2F6)	–	Wang et al., 2013
QSf6	6	RM261–RM471	1.94	SF	IR 64 × Tarom Molaei	BILs(BC2F6)	GY, HI, PF	Wang et al., 2013
QSf1	1	RM286–RM332	6.59	SF	Teqing × Tarom Molaei	BILs(BC2F6)	GY, FG	Wang et al., 2013
QSf7	7	RM421–RM87	3.08	SF	Teqing × Tarom Molaei	BILs(BC2F6)	–	Wang et al., 2013
qDTH6.1	6	RM586–RM217	6.4	PH	TDK1 × IR55419-04	BILs(BC1F3:4)	–	Dixit et al., 2014b
qDTH3.1	3	RM275–RM528	3.4	PH	TDK1 × IR55419-05	BILs(BC1F3:4)		Dixit et al., 2014b
qPH1.1	1	RM315–RM568	16.3	PH	IR64 × IRAT177	RILs(F6)	–	Trijatmiko et al., 2014
qYP-12-1	12	RZ488–RG477	2.92	YP	Xiaobaijingzi × Kongyu131	RILs(F2:7)	–	Xing et al., 2014

qDTY 3.4	3	RG191–RM546	4.29	GY	Danteshwari × Dagaddeshi	RILs (F11)	–	Verma et al., 2014
qDTY 1.3	1	RM411–RM55	3.23	GY	Danteshwari × Dagaddeshi	RILs (F11)	–	Verma et al., 2014
qDTY 11.1	11	RM24–RM449	7.39	GY	Danteshwari × Dagaddeshi	RILs (F11)	–	Verma et al., 2014
qDTY 3.3	3	RG191–RM546	4.38	GY	Danteshwari × Dagaddeshi	RILs (F11)	–	Verma et al., 2014
qDTY3.1	3	RM275–RM340	6.3	GY	TDK1 × IR55419-04	BILs(BC1F3:4)	–	Dixit et al., 2014a
qDTY6.1	6	RM168–RM468	7.5	GY	TDK1 × IR55419-04	BILs(BC1F3:4)	–	Dixit et al., 2014a
qDTY6.2	6	RM586–RM217	7	GY	TDK1 × IR55419-04	BILs(BC1F3:4)	–	Dixit et al., 2014a
qDTY6.3	6	RM168–RM468	4.3	GY	TDK1 × IR55419-04	BILs(BC1F3:4)	–	Dixit et al., 2014a
qGY1.1	1	RM212–RM414	3.2	YP	IR64 × IRAT177	RILs(F6)	–	Trijatmiko et al., 2014
qTGW1	1	RM315–RM431	5.66	1,000 Gw	Swarna × WAB 450	BILs(BC1F6)	–	Sangodele et al., 2014
qGW3.1	3	RM489–RM36	3.09	GW/P	Swarna × WAB 450	BILs(BC1F6)	–	Sangodele et al., 2014
qGW3.2	3	RM563–RM16	3.11	GW/P	Swarna × WAB 450	BILs(BC1F6)	–	Sangodele et al., 2014
qDTF3.1	3	RM121–M541	9.5	DTF	TDK1 × IR55419-04	BILs(BC1F3:4)	GY, HI	Dixit et al., 2014a
qDTF3.1	3	RM586–RM217	10	DTF	TDK1 × IR55419-04	BILs(BC1F3:4)	GY, HI	Dixit et al., 2014a
qDTF6.1	6	RM168–RM468	3.7	DTF	TDK1 × IR55419-04	BILs(BC1F3:4)	GY, HI	Dixit et al., 2014a
qDTH8.1	8	RM547–RM978	3.1	DTH	IR64 × IRAT177	RILs(F6)	GY, HI	Trijatmiko et al., 2014
qDTH10.1	10	RM596–RM333	3.9 1	DTH	IR64 × IRAT177	RILs(F6)	GY, HI	Trijatmiko et al., 2014
qGN7	7	RM180–RM455	3.52	G/P	Swarna × WAB 450	BILs(BC1F6)	PT	Sangodele et al., 2014
qGPP2.1	2	RM279–RM452	3.1	G/P	IR64 × IRAT177	RILs(F6)	–	Trijatmiko et al., 2014
qGPP8.1	8	RM547–RM978	4	G/P	IR64 × IRAT177	RILs(F6)	–	Trijatmiko et al., 2014
qGPP8.2	8	RM339–RM223	6	G/P	IR64 × IRAT177	RILs(F6)	GY	Trijatmiko et al., 2014
qPN-6-2	6	RM589–RM402	4.25	PN/P	Xiaobaijingzi × Kongyu131	RILs(F2:7)	–	Xing et al., 2014

(Continued)

Table 1 (*Continued*)

QTL	Ch	Marker interval	LOD	Trait	Parents	Populations	Associated traits	References
qPN-8	8	RM515–RM547	3.25	PN/P	Xiaobaijingzi × Kongyu131	RILs(F2:7)	–	Xing et al., 2014
qSPP4.1	4	RM451–RM348	3.7	SN/PA	IR64 × IRAT177	RILs(F6)	PN	Trijatmiko et al., 2014
qPSS8.1	8	RM547–RM978	5.3	PSS	IR64 × IRAT177	RILs(F6)	DTH, G/P, PN	Trijatmiko et al., 2014
qPSS8.2	8	RM339–RM223	7.7	PSS	IR64 × IRAT177	RILs(F6)	GY	Trijatmiko et al., 2014
qYLD1.2	1	RM237–RM128	3.92	GY	Appo × Moroberekan	BILs(BC1F3:4 andBC1F3:6)	–	Sellamuthu et al., 2015
qYLD2.1	2	RM145–RM300	3.03	GY	Appo × Moroberekan	BILs(BC1F3:4 andBC1F3:6)	–	Sellamuthu et al., 2015
qYLD4.2	4	RM252–RM303	3.56	GY	Appo × Moroberekan	BILs(BC1F3:4 andBC1F3:6)	–	Sellamuthu et al., 2015
qYLD6.3	6	RM527–RM3	3.12	GY	Appo × Moroberekan	BILs(BC1F3:4 andBC1F3:6)	–	Sellamuthu et al., 2015
qYLD3	3	RM545–RM186	3.83	GY	Appo × Moroberekan	BILs(BC1F3:4 andBC1F3:6)	–	Sellamuthu et al., 2015
qYLD4.1	4	RM261–RM273	4.71	GY	Appo × Moroberekan	BILs(BC1F3:4 andBC1F3:6)	–	Sellamuthu et al., 2015
qYLD6.3	6	RM3–RM340	4.38	GY	Appo × Moroberekan	BILs(BC1F3:4 andBC1F3:6)	–	Sellamuthu et al., 2015
qYLD12	12	RM519–RM17	3.58	GY	Appo × Moroberekan	BILs(BC1F3:4 andBC1F3:6)	–	Sellamuthu et al., 2015
qHI8.1	8	RM404–RM233	3.26	HI	Appo × Moroberekan	BILs(BC1F3:4 andBC1F3:6)	–	Sellamuthu et al., 2015
qHI3	3	RM545–RM186	4.28	HI	Appo × Moroberekan	BILs(BC1F3:4 andBC1F3:6)	–	Sellamuthu et al., 2015

qHI4	4	RM261–RM273	3.56	HI	Appo × Moroberekan	BILs(BC1F3:4 andBC1F3:6)	–	Sellamuthu et al., 2015
qHI6.3	6	RM3–RM340	6.22	HI	Appo × Moroberekan	BILs(BC1F3:4 andBC1F3:6)	–	Sellamuthu et al., 2015
qHI12.1	12	RM519–RM17	3.1	HI	Appo × Moroberekan	BILs(BC1F3:4 andBC1F3:6)	–	Sellamuthu et al., 2015
qDMY1.3	1	RM246–RM128	3.11	DMY	Appo × Moroberekan	BILs(BC1F3:4 andBC1F3:6)	–	Sellamuthu et al., 2015
qDMY9	9	RM257–RM108	3.04	DMY	Appo × Moroberekan	BILs(BC1F3:4 andBC1F3:6)	–	Sellamuthu et al., 2015
qDMY1.1	1	RM220–RM600	4.12	DMY	Appo × Moroberekan	BILs(BC1F3:4 andBC1F3:6)	–	Sellamuthu et al., 2015
qDMY12	12	RM257–RM108	3.04	DMY	Appo × Moroberekan	BILs(BC1F3:4 andBC1F3:6)	–	Sellamuthu et al., 2015
qHGW6.1	6	RM204–RM253	5.43	1,000 Gw	Appo × Moroberekan	BILs(BC1F3:4 andBC1F3:6)	–	Sellamuthu et al., 2015
qHGW10.1	10	RM222–S10001	4.38	1,000 Gw	Appo × Moroberekan	BILs(BC1F3:4 andBC1F3:6)	–	Sellamuthu et al., 2015
qHGW1	1	RM220–RM600	4.3	1,000 Gw	Appo × Moroberekan	BILs(BC1F3:4 andBC1F3:6)	DMY	Sellamuthu et al., 2015
qHGW2.2	2	RM425–RM166	4.45	1,000 Gw	Appo × Moroberekan	BILs(BC1F3:4 andBC1F3:6)	FD	Sellamuthu et al., 2015
qHGW4	4	RM348–RM127	4.12	1,000 Gw	Appo × Moroberekan	BILs(BC1F3:4 andBC1F3:6)	–	Sellamuthu et al., 2015
qHGW8	8	RM506–RM38	5.61	1,000 Gw	Appo × Moroberekan	BILs(BC1F3:4 andBC1F3:6)	–	Sellamuthu et al., 2015
qFGP3	3	RM130–RM514	3.54	FG/PA	Appo × Moroberekan	BILs(BC1F3:4 andBC1F3:6)	–	Sellamuthu et al., 2015

(Continued)

Table 1 *(Continued)*

QTL	Ch	Marker interval	LOD	Trait	Parents	Populations	Associated traits	References
qFGP10	10	RM147–RM496	3.17	FG/PA	Appo × Moroberekan	BILs(BC1F3:4 andBC1F3:6)	–	Sellamuthu et al., 2015
qNOP1.2	1	RM302–RM212	3.01	PN	Appo × Moroberekan	BILs(BC1F3:4 andBC1F3:6)	–	Sellamuthu et al., 2015
qNSP4	4	RM252–RM303	4.55	SN/PA	Appo × Moroberekan	BILs(BC1F3:4 andBC1F3:6)	–	Sellamuthu et al., 2015
qSRP4.1	4	RM252–RM348	5.04	SN/PA	Appo × Moroberekan	BILs(BC1F3:4 andBC1F3:6)	–	Sellamuthu et al., 2015
qSRP9	9	RM410–RM257	3.1	SN/PA	Appo × Moroberekan	BILs(BC1F3:4 andBC1F3:6)	–	Sellamuthu et al., 2015
qPRP4	4	RM303–RM348	3.99	SN/PA	Appo × Moroberekan	BILs(BC1F3:4 andBC1F3:6)	–	Sellamuthu et al., 2015
qPNX2	2	RM145–RM300	3.41	PE	Appo × Moroberekan	BILs(BC1F3:4 andBC1F3:6)	GY	Sellamuthu et al., 2015
qPNX5.2	5	RM164–RM430	3.6	PE	Appo × Moroberekan	BILs(BC1F3:4 andBC1F3:6)	–	Sellamuthu et al., 2015
qPNX6	6	RM176–RM494	3.43	PE	Appo × Moroberekan	BILs(BC1F3:4 andBC1F3:6)	–	Sellamuthu et al., 2015
qPNX8	8	RM223–RM210	3.4	PE	Appo × Moroberekan	BILs(BC1F3:4 andBC1F3:6)	–	Sellamuthu et al., 2015
qPNL11	11	RM209–RM224	3.32	PL	Appo × Moroberekan	BILs(BC1F3:4 andBC1F3:6)	–	Sellamuthu et al., 2015
qPNL2	2	RM525–RM318	4.24	PL	Appo × Moroberekan	BILs(BC1F3:4 andBC1F3:6)	–	Sellamuthu et al., 2015
qPNL5	5	RM163–RM440	5.65	PL	Appo × Moroberekan	BILs(BC1F3:4 andBC1F3:6)	–	Sellamuthu et al., 2015

qSPF8	8	RM404–RM223	3.36	SF	Appo × Moroberekan	BILs(BC1F3:4 andBC1F3:6)	HI	Sellamuthu et al., 2015
qSPF3	3	RM545–RM186	3.22	SF	Appo × Moroberekan	BILs(BC1F3:4 andBC1F3:6)	GY, HI	Sellamuthu et al., 2015
qPNF3.1	3	RM545–RM186	3.58	PF	Appo × Moroberekan	BILs(BC1F3:4 andBC1F3:6)	GY, HI	Sellamuthu et al., 2015
qPNF4.2	4	RM261–RM273	3.54	PF	Appo × Moroberekan	BILs(BC1F3:4 andBC1F3:6)	GY, HI	Sellamuthu et al., 2015
qPNF9	9	RM257–RM108	3.28	PF	Appo × Moroberekan	BILs(BC1F3:4 andBC1F3:6)	DMY	Sellamuthu et al., 2015
qPNF3.1	3	RM545–RM186	3.04	PF	Appo × Moroberekan	BILs(BC1F3:4 andBC1F3:6)	GY, HI	Sellamuthu et al., 2015
qPNF6.3	6	RM3–RM340	3.23	PF	Appo × Moroberekan	BILs(BC1F3:4 andBC1F3:6)	GY, DTF, HI	Sellamuthu et al., 2015
qPDL1.2	1	RM488–RM237	3.02	PEL	Appo × Moroberekan	BILs(BC1F3:4 andBC1F3:6)	GY	Sellamuthu et al., 2015
qPDL6	6	RM527–RM3	3.23	PEL	Appo × Moroberekan	BILs(BC1F3:4 andBC1F3:6)	GY, HI	Sellamuthu et al., 2015

Abbreviations: DTF, days to flowering; DMY, DTH-total dry matter yield; Days to heading; FD, flowering days; FG, filled grain; FG/P, filled grain per plant; GN, grain number; G/P, grains per panicle; GW, grain weight; GW/P, grain weight per plant; GY, grain yield; HI, Harvest Index; PE, panicle exertion; PEL, peduncle length; PF, panicle fertility; PH, plant height; PL, panicle length; PN, panicle number; PN/P, panicle number/plant; PT, productive tiller; SBN, secondary branch number; SDW, shoot dry weight; SF, spikelet fertility; SN, spikelet number; SN/PA, number of spikelets/panicle; SPP, spikelets per plant; TN, tiller number.

large-effect QTL, qDTY1.1, co-localizes with the semi-dwarfing locus *sd1* and also has a variety of effects on other traits (Vikram et al., 2011). Moreover, other secondary traits such as lower canopy temperature and leaf drying co-localize with yield under drought QTLs. Lower canopy temperature under drought stress indicates favourable water status that enables rice plants to survive drought by avoidance mechanism. Previous research studies conducted using mapping population obtained from Zhenshan97B/IRAT109 reported epistatic QTL interactions for canopy temperature, leaf water potential and spikelet fertility (Tuberosa and Salvi, 2006). Phenotypic correlations and co-localization of these QTLs reflect on the existence of genetic relationships between physiological traits and GY under drought stress. Hence, a clear insight into understanding the mechanism governing drought resistance and identification of alleles that could be used for breeding programmes will hasten the development of drought-resistant cultivars (Prince et al., 2015).

4.3 Potential solutions for enhanced yield stability under drought stress

The primary aim of rice breeders is to increase the yield potential (YP). Recent research in the last decade have shown that the continuous selection for high YP has sometimes resulted in higher yields under drought conditions, thereby suggesting a positive relationship between YP and drought tolerance (Guan et al., 2010). However, the major constraint for selection under drought stress is the prevalence of different strategies (mechanisms) to respond to different types of drought stress with stress-adaptive mechanisms differing from vegetative to reproductive stage (Kamoshita et al., 2008). Thus, in the light of specific adaptation of plants to prevailing water supply it has been proposed that YP should be selected under conditions like drought escape (e.g. accelerated flowering) for terminal stress (Blum, 2005). Previous reports have shown that the direct selection for yield under drought stress results in plants exhibiting dehydration-avoidance phenotype with reduced biomass (early flowering, smaller plant size, small leaf area or limited tillering) or the ability of plants to maintain water status under drought stress (Blum, 2005; Gu et al., 2012). The progress in breeding programmes in increasing YP is slow due to the complexity of the genetic control of GY along with other major traits in determining factors like number of tillers, panicles and fertile grains (Dixit et al., 2014a).

4.4 Molecular analysis of a gene *DRO1* for improved root traits

The ability of rice plants to survive under drought stress is the establishment of a deep rooting system that enables accessing water moisture reserves, an avoidance mechanism. *DEEPER ROOTING 1* (*DRO1*) QTL was identified from a breeding population derived from the cross between widely cultivated shallow rooting IR64 rice and deep rooting Kinandang Patong landrace that provided yield stability under drought (Uga et al., 2011, 2013). A single 1-bp deletion within the exon 4 of the IR64 *DRO1* gene results in a premature stop codon consequently forming a truncated protein and a reduced root angle in response to gravity while on the contrary, Kinandang Patong allele, *DRO1–kp* promotes cell elongation at the root tip causing downward growth. Introgression of *DRO1–kp* into IR64 results in a deep rooting phenotype with improved drought avoidance and increased yield performance (number of filled grains) under prolonged or severe water-deficit conditions in comparison to the recipient parent. *DRO1*, though negatively regulated by auxin, is thus involved in increased root growth and enhanced root angle consequently promoting gravitropism and higher YP.

5 Genetic engineering for drought tolerance

Genetic engineering by the transformation of rice plants using modified gene constructs has provided genetic evidence of drought tolerance components that can serve as models for the development of cultivars with improved YP. A selected list of transgenic rice plants with information on transgenes and promoters used for rice transformation as well as their yield performance is given in Table 2. The candidate genes were identified by virtue of their role in different drought tolerance mechanisms or in their response to drought stress at flowering, identifying them as potential agents for inducing stress tolerance pathways during flowering.

5.1 AP2/ERF transcription factors

Yield in rice plants under drought stress conditions is greatly affected by the photosynthetic capacity of rice plants that is reliant on three major resistance steps, one at the stomata and the other at another mesophyll layer (Lauteri et al., 2014), as well as on the ability of plants to harness solar energy. An AP2/ERF transcription factor, *HYR* (HIGHER YIELD RICE) associated with photosynthetic carbon metabolism, was identified from a gene regulatory network. The overexpression of HYR in Nipponbare under the control of *CaMV35S* promoter shows enhanced photosynthesis, determining a morpho-physiological programme contributing to higher GY both under control and drought-stressed conditions. *HYR* is identified as a switch activating genes related to light harvesting and PSII efficiency as well as cascade of genes related to carbon metabolism (Ambavaram et al., 2014). Transgenic rice plants overexpressing the AP2/ERF gene *OsAP37* showed enhanced drought tolerance at the vegetative stage. These transgenic rice plants when exposed to drought stress in field conditions exhibited an increased yield under drought in comparison to the control, achieved primarily due to improved grain filling. Thus, it can be suggested that *OsAP37* can cause significant increase in GY under drought without causing undesirable growth phenotypes (Oh et al., 2009). *DREBs/CBFs* are transcription factors that regulate various stress-induced genes by binding to the CRT/DRE elements and their overexpression results in drought tolerance with less photo-oxidative damage, and more accumulation of osmolytes (Datta et al., 2012). Transgenic rice plants overexpressing *DREB1A* from rice and *Arabidopsis* showed enhanced drought tolerance at the vegetative stage and relatively high spikelet fertility and increase in yield in comparison to the non-transgenic controls (Datta et al., 2012; Ravikumar et al., 2014). *AtDREB1C* transgenic rice plants showed improved drought tolerance at a vegetative stage, and at the reproductive stage, they had larger sink capacity with an improved yield under drought compared to the untransformed drought-treated plants (Ishizaki et al., 2013).

5.2 bZIP transcription factors

Previous research has shown that the subfamily of *bZIP* transcription factors, ABA responsive element binding factors (ABF/AREB/ABRE), plays a crucial role in regulating

Table 2 Genes transformed into rice affecting grain yield under drought and other drought response parameters

Gene function	Gene name	Gene source	Promoter	Transgene construct	Growth trait	Reference
NAC	SNAC1	Rice	CamV35S	COE	>SF	Hu et al., 2006
Isopentenyltransferase	IPT	Agrobacterium	SARK	SMIE	GY>160%	Rivero et al., 2007
LEA protein	OsLEA3-1	Rice	HVA1/ CamV35S/actin	COE	>GY	Xiao et al., 2007
bzip	OsbZIP23	Rice	Ubiquitin	COE	>GY	Xiang etal., 2008
CBL-interacting protein kinase	OsCIPK23	Rice	Ubiquitin	CRNAi/ hybrid	Seed set reduced in the RNAi lines	Yang et al., 2008
Stress-associated protein	OsiSAP8	Rice	CamV35S	COE	No yield penalty	Kanneganti et al., 2008
AP2/ERF	AP37	Rice	OsCc1	COE	>Seed filling, GW	Oh et al., 2009
AP2/ERF	AP59	Rice	OsCc1	COE	<GW	Oh et al., 2009
Jasmonic acid carboxyl methyltransferase gene	AtJMT	Arabidopsis	Ubiquitin	COE	<GY, SPP, FR	Kim et al., 2009
MAP kinase kinase kinase	NPK1	Arabidopsis	Actin	COE	>GY, SF	Xiao et al., 2009
AP2/ERF	CBF3	Arabidopsis	Actin	COE	>GY, SF	Xiao et al., 2009
Molybdenum cofactor sulphurase	LOS5	Arabidopsis	Actin	COE	>GY, SF	Xiao et al., 2009
Vacuolar Na+/H+ antiporter	NHX1	Arabidopsis	Actin	COE	>GY, SF	Xiao et al., 2009
C2H2-type zinc finger transcription factor	ZAT10	Arabidopsis	Actin	COE	>GY, SF	Xiao et al., 2009
Serine/threonine protein kinase	SOS2	Arabidopsis	Actin	COE	>GY, SF	Xiao et al., 2009

Human Ski-interacting protein	OsSKIPa	Rice	CamV35S	COE	>GY, SF	Hou et al., 2009
Mitogen-activated protein kinase	DSM1	Rice	CamV35S	COE	<GY, SF	Ning et al., 2010
b-Carotene hydroxylase	DSM2	Rice	CamV35S	COE	>SF, seed setting	Du et al., 2010
NAC	OsNAC10	Rice	RCc5	RSE	>GY	Jeong et al., 2010
Isopentenyltransferase	IPT	Agrobacterium	SARK	SMIE	>GY	Peleg et al., 2011
Inositol 1,3,4-trisphosphate 5/6-kinase	OsITPK2/DSM3	Rice	CamV35S	COE	<Seed setting, biomass	Du et al., 2011
AP2/ERF	AtDREB1A	Arabidopsis	rd29A	SIE	>GY	Datta et al., 2012
AP2/ERF	AtDREB1C	Arabidopsis	lip9	SIE	>SN, FG	Ishizaki et al., 2012
bzip	OsbZIP46	Rice	Ubiquitin	COE	>GY	Tang et al., 2012
NAC	OsNAC9	Rice	RCc4	RSE	>GY	Redillas et al., 2012
Ornithine δ-aminotransferase	OsOAT	Rice	Ubiquitin	COE	> Seed setting rate	You et al., 2012
Squalene synthase	SQS1	Maize	Ubiquitin	COE	>SF, GY	Manavalan et al., 2012
DEEPER ROOTING 1	DRO1	Rice	Ubiquitin	CRNAi/ Hybrid	>FG	Uga et al., 2013
Homeodomain	EDT1/HDG11	Arabidopsis	Ubiquitin	COE	>Biomass, seed setting	Yu et al., 2013
Ascorbate peroxidases	APX2	Rice	Actin	COE	>SF	Zhang et al., 2013
Lipid transfer protein	OsDIL	Rice	Ubiquitin	COE	Less degeneration of tapetum and abnormal anthers	Guo et al., 2013

(Continued)

Table 2 (*Continued*)

Gene function	Gene name	Gene source	Promoter	Transgene construct	Growth trait	Reference
NAC	OsNAC5	Rice	RCc3	RSE	>GY	Jeong et al., 2013
AP2/ERF	AtDREB1A	Arabidopsis	rd29A	SIE	>SF, GY	Ravikumar et al., 2014
RNA-binding protein	AtGRP2 / AtGRP7	Arabidopsis	Ubiquitin	COE	>FG, GY	Yang et al., 2014
Receptor-like kinase	GUDK	Rice		T-DNA mutant	<SF	Ramegowda et al., 2014
bzip	OsbZIP71	Rice	CamV35S	COE	>Seed setting	Liu et al., 2014
NAC	OsNAP	Rice	CamV35S	COE	>GY, SF	Chen et al., 2014
AP2/ERF	HYR	Rice	CamV35S	COE	>PN,GN,GW,GY	Ambavaram et al., 2014
GRAS transcription factor	OsGRAS23	Rice	CamV35S	COE	>Seed setting, GY	Xu et al., 2015
WRKY	OsWRKY47	Rice	Ubiquitin	COE	Relatively low yield reduction	Raineri et al., 2015
AP2/ERF	OsEREBP1	Rice	Ubiquitin	COE	More new tillers that mature to grain	Jisha et al., 2015

Abbreviations: COE, constitutively overexpression; CRNAi, constitutively triggering RNA interference; FG, filled grain; GN, grain number; GW, grain weight; GY, grain yield; PN, panicle number; RSE, root-specific expression; SMIE, stress- and maturation-induced promoter; SF, spikelet fertility; SN, spikelet number; SPP, spikelets per plant; >, increased; <, reduced.

drought tolerance in rice in an ABA-dependent manner (Fukao and Xiong, 2013). Transgenic rice plants overexpressing *OsbZIP23* showed an increased ABA sensitivity but enhanced drought tolerance both at the vegetative and reproductive stage, where the relative GY per plant was higher than the untransformed plants or the mutants with no detrimental phenotypes under the control conditions. Microarray analysis revealed many downstream genes with diverse functions (~30) as putative targets that include stress-inducible transcription factors and dehydrins (Xiang et al., 2008).

OsbZIP46 is another member of the subfamily that has an internal amino acid sequence (residues 122–219), with a negative effect on transactivation assays. Transgenic rice plants overexpressing *OsbZIP46CA1*, bearing a deletion of the internal residues described above, showed increased drought tolerance with a relatively high GY in comparison to rice plants overexpressing native *OsbZIP46* and WT plants. *OsbZIP46* can interact with homologs of SnRK2 protein kinases that phosphorylate ABFs in *Arabidopsis* (Tang et al., 2012; Todaka et al., 2015). These results suggest that *OsbZIP46* and *OsbZIP23* are positive regulators of drought stress signalling in rice plants, mediated by the ABA-dependent pathway.

OsbZIP71, when overexpressed under the constitutive CaMV35S promoter or inducible rd29A promoter, showed an increased tolerance to drought with higher seed setting rate and increased yield per plant under drought. Tolerance to drought is given by an interaction with *OsMYB4*, which interacts with *OsNHX1* and *COR413-TM1* (Liu et al., 2014).

These reports suggest that *bZIP* transcription factors are potential candidates for genetic engineering to increase yield under drought stress. The tolerance mechanism is regulated via the ABA-signalling pathway that is indicated by an increased ABA content and enriched ABRE motifs in the promoter regions of responding genes (Maruyama et al., 2012, 2014).

5.3 NAC transcription factors

NAC transcription factors are a highly conserved group of transcription factors with a variable C-terminal domain that are known to play a major role in drought stress tolerance of rice by controlling biochemical and molecular pathways (Olsen et al., 2005; Nuruzzaman et al., 2010). Constitutive overexpression of *SNAC1* in transgenic rice showed enhanced drought tolerance at the vegetative as well as reproductive stage, with a higher seed setting rate and spikelet fertility in comparison to the non-transformed plants but with no yield penalty. Drought resistance of the transgenic plants was achieved through the expression of *OsSRO1c* causing reduction in water loss by closing stomatal pores, but with no reduction in photosynthesis (Hu et al., 2006; You et al., 2013). Rice plants overexpressing *OsNAC10*, *OsNAC9* and *OsNAC5* under the control of a root-specific promoter showed thickened roots and increased tolerance to drought at the reproductive stage with a higher yield than the control under field conditions. Microarray analysis revealed various downstream genes encoding *NCED, calcium-transporting ATPase, P450, Zn finger* and *HAK5* were upregulated (Jeong et al., 2010, 2013; Redillas et al., 2012). *OsNAP* is another member of the NAC family of transcription factors that is highly induced by ABA, and abiotic stresses like salinity and drought. Constitutive overexpression in rice showed no growth penalty, reduced water loss, enhanced drought tolerance at the vegetative stage and improved yield and spikelet fertility at the reproductive stage. *OsNAP* acts as a transcriptional activator of various stress regulatory transcription factors like *OsAP37, OsMYB2, OsDREB1A* and other genes like *OsPP2C06/OsABI2, OsPP2C09 and OsPP2C68* (Chen et al., 2014).

5.4 Other transcription factors

The analysis of other transcription factors in transformed rice plants has also shown enhanced drought tolerance as well as increased yield components. The *Arabidopsis EDT1/HDG11* gene encodes a homeodomain-leucine zipper transcription factor, which when overexpressed in rice gives enhanced drought tolerance. This drought tolerance has been attributed to an efficient root system, reduced stomatal pores and improved WUE. Improved GY under drought in transgenic rice was mediated through improved seed setting, an increase in the number of panicles and photosynthetic efficiency (Yu et al., 2013). *OsGRAS23* belonging to the family of *GRAS* transcription factors was genetically identified in a drought-resistant QTL on chromosome 4, from the cross between upland rice IRAT109 and paddy rice Zhenshan97B (Zou et al., 2005). The expression of *OsGRAS23* was induced by the abiotic stresses such as drought, cold and salinity, and its overexpression in rice displayed drought resistance accompanied by reduction in the accumulation of H_2O_2. The transgenic rice plants also exhibited higher seed setting rate and improved yield under drought (Xu et al., 2015). *OsWRKY47*, a transcription factor from group II of the WRKY family, when overexpressed in rice, exhibits improved drought tolerance and higher yield under drought. Drought tolerance can also be achieved by controlling the expression of *OsWRKY47* regulated genes *cysteine-rich repeat secretory protein 55 precursor (CRRSP) and calmodulin-binding protein (CaMBP)* (Raineri et al., 2015).

6 Drought-protective genes

6.1 Protein kinases

OsCIPK23, a member of CBL (calcineurin B-like) interacting protein kinase (CIPK) family, expressed primarily in the pistil and anther, is upregulated by abiotic stresses and phytohormones like ABA (Shi et al., 1999; Pandey et al., 2015). RNAi mediated the repression of *OsCIPK23* in rice and showed a hypersensitive response to drought at the vegetative stage and reduced seed set at the flowering stage under drought stress, indicating that it has a key role in pollination and drought tolerance (Yang et al., 2008). A receptor-like cytoplasmic kinase, *GUDK*, was identified from the screening of mutants of drought-responsive genes that showed hypersensitive phenotype to different abiotic stresses. The loss of function of *GUDK* showed a reduction in photosynthesis and biomass at the vegetative stage and reduced yield under drought stress at the reproductive stage. Phosphoproteome analysis identified *GUDK* as the primary regulator of *OsAP37* that is responsible for controlling GY in rice (Oh et al., 2009; Ramegowda et al., 2014). Mitogen-activated protein kinases have been identified as key factors in regulating plant development and stress response (Sinha et al., 2011). *DSM1*-RNA interference lines were found sensitive to drought stress at the panicle development stage with reduced spikelet fertility and reduced yield. Overexpression lines showed drought tolerance that is achieved by reactive oxygen species (ROS) scavenging (Ning et al., 2010). The screening of T-DNA mutants identified a drought- and salt-hypersensitive gene *dsm3* encoding inositol 1,3,4-trisphosphate 5/6-kinase. This mutant showed a reduction in the accumulation of osmolytes, spikelet fertility and yield under drought stress. A similar phenotype was observed for the overexpression lines as well. Taken together these results suggest that an optimal expression is required for basal level drought tolerance (Du et al., 2011).

6.2 LEA proteins

Late embryogenesis abundant proteins are a conserved group of hydrophilic proteins implicated to play a role in drought stress as osmoprotectants or cryoprotectants (Todaka et al., 2015). The overexpression of *OsLEA3-1* in rice constitutively and under control of a drought-inducible promoter showed increased drought tolerance, improved spikelet fertility and increased yield under drought stress in comparison to the non-transformed rice plants (Xiao et al., 2007).

6.3 Osmoprotectant synthesis

Drought-stress-induced free radicals are scavenged by antioxidant enzymes and osmolytes, for example, proline, that forms the first line of defence against stress-induced damage. Ornithine δ-aminotransferase (*OsOAT*) is a pyridoxal-5-phosphate-dependent enzyme involved in proline and arginine metabolism. *OsOAT* is induced by multiple stresses and phytohormone treatments (You et al., 2012). The overexpression of *OsOAT* in rice enhanced drought and oxidative stress tolerance and increased seed-setting rate, a major component trait of GY in the transgenic rice plants exposed to drought stress. *OsOAT* was identified as a direct target of *SNAC2*, and tolerance to stresses in rice plants overexpressing *OsOAT* was a result of ROS scavenging mediated by an increased accumulation of antioxidant enzymes and proline (You et al., 2012). Ascorbate peroxidases (APX) are antioxidant enzymes that help in catalysing the dismutation of H_2O_2 into H_2O and O_2 and thereby maintaining the ROS homeostasis in plants. Overexpression of *OsAPX2* in rice plants increased spikelet fertility in comparison to the wild-type plants under drought stress (Zhang et al., 2013).

6.4 Other genes

Farnesyl transferases comprising two subunits play a role in post-translational modification of the protein by adding a farnesyl group to the –SH of the cysteine. Squalene synthase (*SQS*) located in the endoplasmic reticulum is a key enzyme in isoprenoid biosynthesis catalysing the conversion of two farnesyl pyrophosphate molecules into squalene that is then directed to sterol biosynthesis. RNAi-mediated disruption of rice *SQS* by maize *SQS* leads to enhanced drought tolerance and increased yield under drought than the wild-type plants. The enhanced drought tolerance might be due to reduced stomatal conductance and water loss (Manavalan et al., 2012). Lipid transfer proteins (LTPs) are small ubiquitous family of proteins catalysing the transfer of phospholipids between membranes and are known to play a variety of roles like plant defence and embryogenesis (Kader, 1996). Rice lines overexpressing the LTP *OsDIL* gene exhibit drought resistance at the vegetative stage, with a reduction in defective anther sacs and tapetum under drought at the reproductive stage. *OsDIL* downregulates *ABAOX3*, an ABA degradation enzyme, increasing ABA biosynthesis and stabilizing the expression of genes related to anther biosynthesis (Guo et al., 2013). *SKIP* proteins identified by yeast two-hybrid screening using oncogene v-Ski as bait are characterized as spliceosome components and coregulators of transcription. *OsSKIPa*, a rice homolog of human *SKIP* protein, was found inducible to various abiotic stresses and phytohormone treatments. Transgenic rice overexpressing *OsSKIPa* showed drought resistance at seedling stage, with increased spikelet fertility and yield at the reproductive stage. Drought stress resistance mediated by

OsSKIPa in transgenic rice plants was achieved by efficient ROS scavenging machinery and diverse regulation of stress-related genes (Hou et al., 2009). Stress-associated proteins, SAP, containing an A20/AN1 domain are known to play an important role in abiotic stress tolerance. The overexpression of *OsiSAP8* in rice conferred drought tolerance at the anthesis stage with no yield penalty in comparison to the unstressed transgenic plants (Kanneganti and Gupta, 2008). Posttranscriptional regulation of RNA metabolism is known to play a key role in drought stress response of plants. To investigate the role of RNA-binding proteins (RBP) in stress tolerance mechanisms, transgenic rice plants were generated overexpressing *AtGRP2* or *AtGRP7* encoding glycine-rich RBP that have been reported to have RNA chaperone activity. Transgenic rice plants expressing the *AtGRP* genes showed an increased yield under drought caused by improved grain filling and modulation of various stress responsive genes like *OSE2*, *Dip1* and *PBZ1* (Yang et al., 2014). Drought tolerance of seven stress-inducible candidate genes: *CBF3/DREB1A* an AP2/ERF transcription factor, *SOS2* a serine/threonine protein kinase, *NCED2* and *LOS5* ABA biosynthesis genes, *NPK1* a mitogen-activated protein kinase, *ZAT10* a zinc finger transcription factor and NHX1 a vacuolar Na^+/H^+ antiporter were tested by overexpression in rice under the control of constitutive/stress-inducible promoters. Transgenic rice plants showed higher relative yield and spikelet fertility under drought stress both in the field and testing in PVC tubes. In a comparison of the seven genes, transgenic rice plants overexpressing *LOS5* and *ZAT10* showed better effect both in the field and in PVC tubes than the other genes (Xiao et al., 2009).

7 Plant hormones affecting drought tolerance

Plant hormones act through a complex interacting network and integrate signalling components to regulate plant growth and development as well as environmental stress response (Kohli et al., 2013). A central role for major phytohormones such as ABA, cytokinin (CK), gibberellic acid (GA), auxin and ethylene in drought stress response is becoming increasingly evident (Wilkinson et al., 2012). The mechanisms by which these hormones translate signals for plant adaptations at vegetative stage such as root and shoot growth, plant architecture, plant-water relations and gas exchange have been intensively studied (Peleg and Blumwald, 2011). However, more direct roles of phytohormones in regulating yield-related traits are very limited. Studies involving hormonal homeostasis and hormone interactions under reproductive drought, which mainly affects GY have started emerging recently. Drought at pre-anthesis stage can affect flowering mainly causing pollen sterility and abortion and at post-anthesis stage it can affect grain filling by modifying source-sink relationships (Xing and Zhang, 2010).

7.1 Abscisic acid

Plant adaptation to drought stress through ABA-mediated regulation of stomatal closure and water loss at the vegetative stage has been well documented (Tardieu and Davies, 1992; Schroeder et al., 2001). Increase in ABA under drought can rapidly limit transpiration through stomatal closure, which becomes a problem under reproductive drought. Improving GY of rice through ABA remains a great challenge partly because of the potentially inadvertent reductions in carbon gain upon stomatal closure and ABA-induced pollen sterility and abortion (Ji et al., 2011). It has been very well shown in wheat by studying the ABA content and homeostasis in the anthers at reproductive drought. A drought-tolerant wheat variety was shown to synthesize ABA catabolic enzyme earlier, thereby decreasing the accumulation of ABA in the anthers as an early drought response (Ji et al., 2011). In addition, increased ABA during reproductive drought can suppress invertases needed to maintain sink strength resulting in pollen abortion. Conversely, the role of ABA in processes other than stomatal regulation can be exploited to improve GY under reproductive drought. For example, ABA can also increase hydraulic conductivity by improved aquaporin functioning and osmotic regulation potentially leading to sustained growth even under reproductive drought (Tardieu et al., 2010; Travaglia et al., 2010). In addition, ABA has been shown to play a positive role during post-anthesis stage. A study has shown that drought stress at the late grain-filling stage substantially increased ABA content and maximum grain filling, and partitioning of fixed ^{14}C was highly correlated with ABA content. The exogenous application of ABA showed similar results as drought stress, suggesting that increased ABA content during grain filling enhanced remobilization of stored carbon from source tissue and accelerated grain filling rate (Yang et al., 2001a, 2001b, 2002, 2003, 2004). ABA in leaves, stems and roots was positively correlated with the remobilization of pre-stored carbon under drought (Yang et al., 2002; Yang and Zhang, 2010).

Several studies aimed at enhancing drought tolerance through ABA homeostasis by modifying ABA biosynthetic and catabolic pathways have failed to show good performance of plants under terminal drought, which mainly impacts GY. However, there are ABA-signalling genes, which have improved yield under reproductive drought. For example, senescence-associated and ABA-dependent genes such as *OsNAP* and *OsNAC5* improved yield under drought at the flowering stage (Sperotto et al., 2009; Chen et al., 2014; Liang et al., 2014). In *OsNAP*-overexpressing plants, the expression of ABA-signalling genes such as *OsPP2C06/OsABI2*, *OsPP2C09* and *OsPP2C68* was upregulated. These findings suggest that ABA-signalling genes can be used to improve the source-sink relationship during grain filling. Another gene, *DSM2* encodes a putative β-carotene hydroxylase that is involved in the synthesis of zeaxanthin, a precursor of ABA biosynthesis. T-DNA mutants showed hypersensitivity to drought stress with reduced photosynthesis and yield under drought stress. The overexpression of *DSM2* leads to enhanced drought resistance mediated by an increase in xanthophylls and non-photochemical quenching, with increased spikelet fertility and seed setting rate (Du et al., 2010).

7.2 Cytokinin

Cytokinin (CK) regulates cell proliferation and differentiation in plants, in addition to delaying senescence. Under water-limited conditions CK delays premature leaf senescence and death. This trait is very useful during reproductive drought and grain-filling period. CK content has been shown to transiently increase at an early grain-filling stage with rapid induction at the late grain-filling stage, suggesting the role of CK in regulating grain development in rice (Yang et al., 2001b). Spraying 6-benzylaminopurine, a synthetic CK, at the heading stage improved GY and quality of the two super hybrid rice genotypes (Pan et al., 2013). Expression of *isopentenyltransferase* (*IPT*) gene, which is involved in CK biosynthesis, has improved stress tolerance of plants (Rivero et al., 2007; Qin et al., 2011; Kuppu et al., 2013). In addition, *IPT* expression under a stress- and maturation-induced promoter in rice increased GY under pre- and post-anthesis drought by delaying stress response (Peleg et al., 2011). Transgenic plants showed a higher expression of brassinosteroid-related genes and the decreased expression of jasmonate-related genes, indicating that CK represses negative effects of JA resulting in improved yield.

7.3 Gibberellic acid

Gibberellic acid (GA) plays a significant role in regulating plant growth under drought stress (Colebrook et al., 2014), with DELLA proteins (transcription factors containing the amino acid motif DELLA near the N-terminal) emerging as central modulators of GA signalling (Davière and Achard, 2013). The content of GA has been shown to be higher at the early grain-filling stage than at the late grain-filling stage. The rapid increase in GA content at early grain filling was associated with a rapid enlargement of the embryo, suggesting that GA plays a crucial role in embryo development (Yang et al., 2001b).

However, water stress substantially reduces the accumulation of GA. The maximal grain-filling rate and partitioning of ^{14}C into grains was negatively correlated with GA content in the grains, indicating a negative role of GA in grain filling (Yang et al., 2001b). Testing the association between single nucleotide polymorphism (SNP) markers and GY under reproductive-stage drought using a recombinant inbred line (RIL) population (IR64 × Cabacu) showed a region close to the *sd1* locus (functional *SD1* gene gibberellin 20-oxidase) affecting GA and GY (Trijatmiko et al., 2014). The region of *qDTY1.1* (QTL for GY under drought) in repulsion to the *sd1* allele was broken by recombination, generating semi-dwarf 'green-revolution' ideotype *sd1* genotypes with the drought tolerance allele (Vikram et al., 2015).

7.4 Auxin

Auxin is well known for its role in plant morphogenesis and antagonistic interaction with CK in determining root and shoot cell differentiation (Zhao, 2010; Su et al., 2011). During early grain-filling stage, the content of auxin increases sharply, and low auxin has been shown to be associated with poor grain filling in rice, suggesting a role of auxin in regulating grain filling at the early grain-filling stage (Yang et al., 1999, 2001b; Tamaki et al., 2015). Under drought stress, there was a substantial downregulation of auxin biosynthetic pathway genes with concomitant reduction in auxin content in rice (Du et al., 2013). Though several genes in the auxin signalling, biosynthesis and transport have been implicated under drought stress in rice, their effect on GY has not been studied (Peleg and Blumwald, 2011; Zhang et al., 2012). Recently, the *DEEPER ROOTING 1* (*DRO1*) gene, identified by positional cloning from the *DRO1* QTL region controlling root growth angle was shown to be negatively regulated by auxin. Higher expression of *DRO1* increased the root growth angle and hence reaching deeper soil layer. Introducing *DRO1* into a shallow-rooting rice cultivar resulted in drought avoidance and high yield under drought (Uga et al., 2013).

7.5 Ethylene

Ethylene is well known for its role in stress-induced leaf senescence, abscission and flooding response (Munn et al., 2004; Fukao et al., 2006; Perata and Voesenek, 2007). However, the increased content of ethylene under drought stress has been shown to inhibit root growth and development, shoot/leaf expansion and photosynthesis (Rajala and Peltonen-Sainio, 2001; Sharp, 2002). Ethylene production under stress can also directly affect yield by increasing embryo and grain abortion, and reducing the grain-filling rate (Wilkinson et al., 2012; Tamaki et al., 2015). Under drought, the concentration of ethylene was shown to increase in the rice grain at the early grain-filling stage and decrease during grain development (Yang et al., 2004). The ethylene concentration has been shown to negatively correlate with the grain-filling rate. It has been proposed that ethylene and ABA work antagonistically and their effect on grain filling depends not only on their concentration but also on their balance with high ABA to ethylene ratio enhancing grain-filling rate (Yang and Zhang, 2006; Zhu et al., 2011).

7.6 Jasmonic acid

Jasmonic acid (JA) is mainly involved in plant development and defence response (Turner et al., 2002). However, the role of JA in response to abiotic stresses is relatively less known. The endogenous level of JA has been shown to markedly increase under drought stress (Creelman and Mullet, 1995; Du et al., 2013). Under reproductive drought, methyl jasmonate (MeJA) levels were increased by 19-fold in rice panicles, resulting in reduced GY in rice (Kim et al., 2009a; Kim et al., 2009b). This phenotype has been reproduced by increasing the levels of MeJA in young panicles through the constitutive expression of *JA carboxyl methyltransferase* gene (*AtJMT*) from *Arabidopsis*. GY was significantly reduced in *Ubi1:AtJMT* plants due to a lower numbers of spikelets and lower filling rates. Transgenic plants also had altered numbers of spikelet organs, including the lemma/palea, lodicule, anther and pistil. All these phenotypes were mimicked with exogenous MeJA suggesting that increased levels of MeJA in panicles under drought will inhibit spikelet development.

8 Future trends in research

Improvement of drought tolerance in rice is a challenging task that requires an exhaustive understanding of the various morphological, biochemical, physiological and molecular characters governing yield under drought. Physiological parameters like maintenance of water potential using deeper rooting system, synchronized modulation of metabolic processes and the level of osmolytes and antioxidants are key factors in the selection for drought tolerance. The endogenous levels of phytohormones play an important role in drought resistance as there is a crosstalk between different signalling pathways mediated by the phytohormones. Hence, stabilized yield and increased WUE should be the essential focus of all research aimed at improving drought tolerance.

Selection of rice genotypes using advanced phenotyping platforms for drought resistance traits will help in the identification of genomic regions and QTLs contributing to drought resistance and hence to yield under drought. Yield under drought is a complex phenomenon arising out of the cumulative effect of several traits and also it needs to be established as to which trait is the primary contributor to yield under drought in the target region. Nonetheless, yield under drought-based selection approach will be the practical way of selecting promising rice genotypes in breeding programmes and lead to the development of high-yielding drought-resistant cultivars. In the light of molecular techniques, several QTLs with consistent effect on GY under drought have been mapped and their effects have been tested under different varietal backgrounds and environments. In addition to molecular breeding approaches, genetic engineering of regulatory and functional genes in rice has improved crop productivity under drought. Though extensive basic research is under progress, there is still an extreme necessity for understanding the response and mechanisms controlling yield under drought in farmers' fields. Thus, it is essential to integrate crop physiology, molecular genetics and breeding approaches to dissect the complex trait of yield under drought and develop crops with high productivity under normal and drought stress environmental conditions.

9 Where to look for further information

A list of all genetic loci characterized in rice for drought tolerance and grain yield under drought, along with references to the publications and further information, is provided in Table 1. In addition, a number of genes have been shown to confer drought stress tolerance in rice by transformation experiments that describe and support mechanisms of drought response and tolerance. A comprehensive list of genes shown to be involved in drought stress tolerance in rice is listed in Table 2, describing the growth trait for improved drought stress, along with references for access of detailed information on the genes and mechanisms in rice.

10 Acknowledgements

This work was supported by the National Science Foundation award numbers DBI–0922747 and ABI1062472.

11 References

Ambavaram, M. M., Basu, S., Krishnan, A., Ramegowda, V., Batlang, U. and Pereira, A. (2014), Coordinated regulation of photosynthesis in rice increases yield and tolerance to environmental stress. *Nat. Commun.* 5: 5302.

Babu, R. C., Shashidhar, H. E., Lilley, J. M., Thanh, N. D., Ray, J. D., Sadasivam, S., Sarkarung, S., O'Toole, J. C. and Nguyen, H. T. (2001), Variation in root penetration ability, osmotic adjustment and dehydration tolerance among accessions of rice adapted to rainfed lowland and upland ecosystems. *Plant Breed.* 120: 233–8.

Babu, R. C., Nguyen, B. D., Chamarerk, V., Shanmugasundaram, P., Chezhian, P., Jeyaprakash, P., Ganesh, S. K., Palchamy, A., Sadasivam, S., Sarkarung, S., Wade, L. J. and Nguyen, H. T. (2003), Genetic analysis of drought resistance in rice by molecular markers: association between secondary traits and field performance. *Crop Sci.* 43: 1457–69.

Bengough, A. G., McKenzie, B. M., Hallett, P. D. and Valentine, T. A. (2011), Root elongation, water stress, and mechanical impedance: a review of limiting stresses and beneficial root tip traits. *J. Exp. Bot.* 62: 59–68.

Bernier, J., Kumar, A., Ramaiah, V., Spaner, D. and Atlin, G. (2006), A large-effect QTL for grain yield under reproductive-stage drought stress in upland rice. *Crop Sci.* 47: 507–16.

Blum, A., Shpiler, L., Golan, G. and Mayer, J. (1989), Yield stability and canopy temperature of wheat genotypes under drought stress. *Field Crop Res.* 22: 289–96.

Blum, A. (2005), Drought resistance, water-use efficiency, and YP: are they compatible, dissonant, or mutually exclusive. *Aust. J. Agric. Res.* 56: 1159–68.

Chen, X., Wang, Y., Lv, B., Li, J., Luo, L., Lu, S., Zhang, X., Ma, H. and Ming, F. (2014), The NAC family transcription factor *OsNAP* confers abiotic stress response through the ABA pathway. *Plant Cell Physiol.* 55: 604–19.

Colebrook, E. H., Thomas, S. G., Phillips, A. L. and Hedden, P. (2014), The role of gibberellin signalling in plant responses to abiotic stress. *J. Exp. Biol.* 217(1): 67–75.

Creelman, R. A. and Mullet, J. E. (1995), Jasmonic acid distribution and action in plants: regulation during development and response to biotic and abiotic stress. *Proc. Natl. Acad. Sci. USA* 92(10): 4114–19.

Datta, K., Baisakh, N., Ganguly, M., Krishnan, S., Yamaguchi, S. and Datta, S. K. (2012), Overexpression of Arabidopsis and rice stress genes inducible transcription factor confers drought and salinity tolerance to rice. *Plant Biotechnol. J.* 10: 579–86.

Davière, J. and Achard, P. (2013), Gibberellin signaling in plants. *Development* 140(6): 1147–51.

Delphine, F., Jefferies, S., Kuchel, H. and Langridge, P. (2010), Genetic and genomic tools to improve drought tolerance in wheat. *J. Exp. Bot.* 61(12): 3211–22.

Dixit, S., Swamy, B. and Vikram, P. (2012), Increased drought tolerance and wider adaptability of qDTY12.1 conferred by its interaction with qDTY2.3 and qDTY3.2. *Mol. Breed.* 30: 1767–79.

Dixit, S., Singh, A. and Kumar, A. (2014a), Rice breeding for high grain yield under drought: a strategic solution to a complex problem. *Int. J. Agron.* 2014: 1–15.

Dixit, S., Singh, A., Sta Cruz, M. T., Maturan, P. T., Amante, M. and Kumar, A. (2014b), Multiple major QTL lead to stable yield performance of rice cultivars across varying drought intensities. *BMC Genet.* 15: 16.

Du, H., Wang, N., Cui, F., Li, X., Xiao, J. and Xiong, L. (2010), Characterization of the beta-carotene hydroxylase gene *DSM2* conferring drought and oxidative stress resistance by increasing xanthophylls and abscisic acid synthesis in rice. *Plant Physiol.* 154: 1304–18.

Du, H., Liu, L., You, L., Yang, M., He, Y., Li, X. and Xiong, L. (2011), Characterization of an inositol 1,3,4-trisphosphate 5/6-kinase gene that is essential for drought and salt stress responses in rice. *Plant Mol. Biol.* 77: 547–63.

Fukao, T., Xu, K., Ronald, P. C. and Bailey-Serres, J. (2006), A variable cluster of ethylene response factor–like genes regulates metabolic and developmental acclimation responses to submergence in rice. *Plant Cell* 18(8): 2021–34.

Fukao, T. and Xiong, L. (2013), Genetic mechanisms conferring adaptation to submergence and drought in rice: simple or complex? *Curr. Opin. Plant Biol.* 16: 196–204.

Garris, A. J., Tai, T. H., Coburn, J., Kresovich, S. and McCouch, S. (2005), Genetic structure and diversity in *Oryza sativa* L. *Genetics* 169: 1631–8.

Gu, J., Yin, X., Stomph, T.-J., Wang, H. and Struik, P. C. (2012), Physiological basis of genetic variation in leaf photosynthesis among rice (*Oryza sativa* L.) introgression lines under drought and well-watered conditions. *J. Exp. Bot.* 63: 5137–53.

Guan, Y. S., Serraj, R., Liu, S. H., Xu, J. L., Ali, J., Wang, W. S., Venus, E., Zhu, L. H. and Li, Z. K. (2010), Simultaneously improving yield under drought stress and non-stress conditions: a case study of rice (*Oryza sativa* L.). *J. Exp. Bot.* 61: 4145–56.

Guo, C., Ge, X. and Ma, H. (2013), The rice *OsDIL* gene plays a role in drought tolerance at vegetative and reproductive stages. *Plant Mol. Biol.* 82: 239–53.

Hemamalini, G. S., Shashidhar, H. E. and Hittalmani, S. (2000), Molecular marker assisted tagging of morphological and physiological traits under two contrasting moisture regimes at peak vegetative stage in rice (*Oryza sativa* L.). *Euphytica* 112: 69–78.

Hou, X., Xie, K., Yao, J., Qi, Z. and Xiong, L. (2009), A homolog of human ski-interacting protein in rice positively regulates cell viability and stress tolerance. *Proc. Natl. Acad. Sci. USA* 106: 6410–15.

Hsiao, T. C., O'Toole, J. C., Yambao, E. B. and Turner, N. C. (1984), Influence of osmotic adjustment on leaf rolling and tissue death in rice (*Oryza sativa* L.). *Plant Physiol.* 75: 338.

Hu, H., Dai, M., Yao, J., Xiao, B., Li, X., Zhang, Q. and Xiong, L. (2006), Overexpressing a NAM, ATAF, and CUC (NAC) transcription factor enhances drought resistance and salt tolerance in rice. *Proc. Natl. Acad. Sci. USA* 103: 12987–92.

Ishizaki, T., Maruyama, K., Obara, M., Fukutani, A., Yamaguchi-Shinozaki, K., Ito, Y. and Kumashiro, T. (2013), Expression of Arabidopsis *DREB1C* improves survival, growth, and yield of upland New Rice for Africa (NERICA) under drought. *Mol. Breed.* 31: 255–64.

Jeong, J. S., Kim, Y. S., Baek, K. H., Jung, H., Ha, S. H., Do Choi, Y., Kim, M., Reuzeau, C. and Kim, J. K. (2010), Root-specific expression of *OsNAC10* improves drought tolerance and grain yield in rice under field drought conditions. *Plant Physiol.* 153: 185–97.

Jeong, J. S., Kim, Y. S., Redillas, M. C., Jang, G., Jung, H., Bang, S. W., Choi, Y. D., Ha, S. H., Reuzeau, C. and Kim, J. K. (2013), *OsNAC5* overexpression enlarges root diameter in rice plants leading to enhanced drought tolerance and increased grain yield in the field. *Plant Biotechnol. J.* 11: 101–14.

Ji, X., Dong, B., Shiran, B., Talbot, M. J., Edlington, J. E., Hughes, T., White, R. G., Gubler, F. and Dolferus, R. (2011), Control of abscisic acid catabolism and abscisic acid homeostasis is important for reproductive stage stress tolerance in cereals. *Plant Physiol.* 156(2): 647–62.

Kader, J. C. (1996), Lipid-transfer proteins in plants. *Annu. Rev. Plant Physiol. Plant Mol. Biol.* 47: 627–54.

Kamoshita, A., Rodriguez, R., Yamauchi, A. and Wade, I. J. (2004), Genotypic variation in response of rainfed lowland rice to prolonged drought and rewatering. *Plant Prod. Sci.* 7: 406–20.

Kamoshita, A., Babu, R., Boopathi, N. M. and Fukai, S. (2008), Phenotypic and genotypic analysis of drought-resistance traits for development of rice cultivars adapted to rainfed environments. *Field Crops Res.* 109: 1–23.

Kanneganti, V. and Gupta, A. K. (2008), Overexpression of *OsiSAP8*, a member of stress associated protein (SAP) gene family of rice confers tolerance to salt, drought and cold stress in transgenic tobacco and rice. *Plant Mol. Biol.* 66: 445–62.

Kato, Y., Abe, J., Kamoshita, A. and Yamagishi, J. (2006), Genotypic variation in root growth angle in rice (*Oryza sativa* L.) and its association with deep root development in upland fields with different water regimes. *Plant and Soil* 287: 117–29.

Kato, Y., Kamoshita, A., Yamagishi, J., Imoto, H. and Abe, J. (2007), Growth of rice (Oryza sativa L.) cultivars under upland conditions with different levels of water supply. 3. Root system development, soil moisture change and plant water status. *Plant Prod. Sci.* 10: 3–13.

Khush, G. S. (2001), Green revolution: the way forward. *Nat. Rev.* 2: 815–22.

Kim, E. H., Kim, Y. S., Park, S. H., Koo, Y. J., Choi, Y. D., Chung, Y. Y., Lee, I. J. and Kim, J. K. (2009a), Methyl jasmonate reduces grain yield by mediating stress signals to alter spikelet development in rice. *Plant Physiol.* 149: 1751–60.

Kim, E. H., Park, S. H. and Kim, J. K. (2009b), Methyl jasmonate triggers loss of grain yield under drought stress. *Plant Signal Behav.* 4: 348–9.

Kohli, A., Sreenivasulu, N., Lakshmanan, P. and Kumar, P. P. (2013), The phytohormone crosstalk paradigm takes center stage in understanding how plants respond to abiotic stresses. *Plant Cell Rep.* 32(7): 945–57.

Kumar, A., Bernier, J., Verulkar, S., Lafitte, H. R. and Atlin, G. N. (2008), Breeding for drought tolerance: direct selection for yield, response to selection and use of drought-tolerant donors in upland and lowland-adapted populations. *Field Crops Res.* 107(3): 221–31.

Kumar, A., Dixit, S., Ram, T., Yadaw, R. B., Mishra, K. K. and Mandal, N. P. (2014), Breeding high-yielding drought-tolerant rice: genetic variations and conventional and molecular approaches. *J. Exp. Bot.* 65(21): 6265–78.

Kumar, R., Venuprasad, V. and Atlin, G. N. (2007), Genetic analysis of rainfed lowland rice drought tolerance under naturally-occurring stress in eastern India: heritability and QTL effects. *Field Crops Res.* 103: 42–52.

Lafitte, H. R., Price, A. H. and Courtois, B. (2004), Yield response to water deficit in an upland rice mapping population: associations among traits and genetic markers. *Theor. Appl. Genet.* 109: 1237–46.

Lanceras, J. C., Pantuwan, G., Jongdee, B. and Toojinda, T. (2004), Quantitative trait loci associated with drought tolerance at reproductive stage in rice. *Plant Physiol.* 135: 384–99.

Lauteri, M., Haworth, M., Serraj, R., Monteverdi, M. C. and Centritto, M. (2014), Photosynthetic diffusional constraints affect yield in drought stressed rice cultivars during flowering. *PLoS ONE* 9: e109054.

Li, Z. K. and Xu, J. L. (2007), Breeding for drought and salt tolerant rice (*Oryza sativa* L.): progress and perspective. In M. A. Jenks, P. M. Hasegawa and S. M. Jain (eds), *Advances in Molecular Breeding Towards Salinity and Drought Tolerance*. Springer, Netherland, pp. 531–64.

Liang, C., Wang, Y., Zhu, Y., Tang, J., Hu, B., Liu, L., Ou, S., Wu, H., Sun, X., Chu, J. and Chu, C. (2014), *OsNAP* connects abscisic acid and leaf senescence by fine-tuning abscisic acid biosynthesis and directly targeting senescence-associated genes in rice. *Proc. Natl. Acad. Sci. USA* 111(27): 10013–18.

Liang, J., Lu, Y., Xiao, P., Sun, M., Corke, H. and Bao, J. (2010), Genetic diversity and population structure of a diverse set of rice germplasm for association mapping. *Theor. Appl. Genet.* 121: 475–87.

Liu, C., Mao, B., Ou, S., Wang, W., Liu, L., Wu, Y., Chu, C. and Wang, X. (2014), *OsbZIP71*, a *bZIP* transcription factor, confers salinity and drought tolerance in rice. *Plant Mol. Biol.* 84: 19–36.

Liu, G. L., Mei, H. W., Yu, X. Q., Zou, G. H., Liu, H. Y., Hu, S. P., Li, M. S., Wu, J. H., Chen, L. and Luo, L. J. (2008), QTL analysis of panicle neck diameter, a trait highly correlated with panicle size, under well-watered and drought conditions in rice (*Oryza sativa* L.). *Plant Sci.* 174: 71–7.

Mae (1997), Physiological nitrogen efficiency in rice: Nitrogen utilization, photosynthesis, and yield potential. *Plant Soil* 196: 201–10.

Manavalan, L. P., Chen, X., Clarke, J., Salmeron, J. and Nguyen, H. T. (2012), RNAi-mediated disruption of squalene synthase improves drought tolerance and yield in rice. *J Exp. Bot.* 63: 163–75.

Maruyama, K., Todaka, D., Mizoi, J., Yoshida, T., Kidokoro, S., Matsukura, S., Takasaki, H., Sakurai, T., Yamamoto, Y. Y., Yoshiwara, K., Kojima, M., Sakakibara, H., Shinozaki, K. and Yamaguchi-Shinozaki, K. (2012), Identification of cis-acting promoter elements in cold- and dehydration-induced transcriptional pathways in arabidopsis, rice, and soybean. *DNA Res.* 19: 37–49.

Maruyama, K., Urano, K., Yoshiwara, K., Morishita, Y., Sakurai, N., Suzuki, H., Kojima, M., Sakakibara, H., Shibata, D., Saito, K., Shinozaki, K. and Yamaguchi-Shinozaki, K. (2014), Integrated analysis of the effects of cold and dehydration on rice metabolites, phytohormones, and gene transcripts. *Plant Physiol.* 164: 1759–71.

Mishra, K. K., Vikram, P., Yadaw, R. B., Swamy, B. P., Dixit, S., Cruz, M. T., Maturan, P., Marker, S. and Kumar, A. (2013), qDTY12.1: a locus with a consistent effect on grain yield under drought in rice. *BMC Genet.* 14: 12.

Mitra, J. (2001), Genetics and genetic improvement of drought resistance in crop plants. *Curr. Sci.* 80: 758–62.

Munn, B. S. and Alegre, L. (2004), Die and let live: leaf senescence contributes to plant survival under drought stress. *Funct. Plant Biol.* 31(3): 203–16.

Ning, J., Li, X., Hicks, L. M. and Xiong, L. (2010), A Raf-like MAPKKK gene *DSM1* mediates drought resistance through reactive oxygen species scavenging in rice. *Plant Physiol.* 152: 876–90.

Nguyen, H. T., Babu, R. C. and Blum, A. (1997), Breeding for drought tolerance in rice: physiology and molecular genetics considerations. *Crop Sci.* 37: 1426–34.

Nguyen, T. T., Klueva, N., Chamareck, V., Aarti, A., Magpantay, G., Millena, A. C., Pathan, M. S. and Nguyen, H. T. (2004), Saturation mapping of QTL regions and identification of putative candidate genes for drought tolerance in rice. *Mol. Genet. Genomics* 272: 35–46.

Nuruzzaman, M., Manimekalai, R., Sharoni, A. M., Satoh, K., Kondoh, H., Ooka, H. and Kikuchi, S. (2010), Genome-wide analysis of NAC transcription factor family in rice. *Gene* 465: 30–44.

Oh, S. J., Kim, Y. S., Kwon, C. W., Park, H. K., Jeong, J. S. and Kim, J. K. (2009), Overexpression of the transcription factor AP37 in rice improves grain yield under drought conditions. *Plant Physiol.* 150: 1368–79.

Olsen, A. N., Ernst, H. A., Leggio, L. L. and Skriver, K. (2005), NAC transcription factors: structurally distinct, functionally diverse. *Trends Plant Sci.* 10: 79–87.

O'Toole, J. C. (1982), Adaptation of rice to drought-prone environments. In *Drought Resistance in Crops, with Emphasis on Rice*. International Rice Research Institute, Manila, Philippines.

Pan, S., Rasul, F., Li, W., Tian, H., Mo, Z., Duan, M. and Tang, X. (2013), Roles of plant growth regulators on yield, grain qualities and antioxidant enzyme activities in super hybrid rice (*Oryza sativa* L.). *Rice* 6(1): 9.

Pandey, G. K., Kanwar, P., Singh, A., Steinhorst, L., Pandey, A., Yadav, A. K., Tokas, I., Sanyal, S. K., Kim, B.-G., Lee, S.-C., Cheong, Y.-H., Kudla, J. and Luan, S. (2015), Calcineurin B-like protein-interacting protein kinase CIPK21 regulates osmotic and salt stress responses in Arabidopsis. *Plant Physiol.* 169: 780–92.

Peleg, Z. and Blumwald, E. (2011), Hormone balance and abiotic stress tolerance in crop plants. *Curr. Opin. Plant Biol.* 14(3): 290–5.

Peleg, Z., Reguera, M., Tumimbang, E., Walia, H. and Blumwald, E. (2011), Cytokinin-mediated source/sink modifications improve drought tolerance and increase grain yield in rice under water-stress. *Plant Biotechnol. J.* 9: 747–58.

Perata, P. and Voesenek, L. A. (2007), Submergence tolerance in rice requires Sub1A, an ethylene-response-factor-like gene. *Trends Plant Sci.* 12: 43–6.

Price, A., Steele, K., Moore, B. and Jones, R. (2002), Upland rice grown in soil-filled chambers and exposed to contrasting water-deficit regimes: II. Mapping quantitative trait loci for root morphology and distribution. *Field Crop. Res.* 76: 25–43.

Prince, S. J., Beena, R., Gomez, S. M., Senthivel, S. and Babu, R. C. (2015), Mapping consistent rice (*Oryza sativa* L.) yield QTLs under drought stress in target rainfed environments. *Rice (N Y)* 8: 53.

Raineri, J., Wang, S., Peleg, Z., Blumwald, E. and Chan, R. L. (2015), The rice transcription factor *OsWRKY47* is a positive regulator of the response to water deficit stress. *Plant Mol. Biol.* 88: 401–13.

Rajala, A. and Peltonen-Sainio, P. (2001), Plant growth regulator effects on spring cereal root and shoot growth. *Agron. J.* 93: 936–43.

Raman, A., Verulkar, S., Mandal, N. and Variar, M. (2012), Drought yield index to select high yielding rice lines under different drought stress severities, 31.

Ramegowda, V., Basu, S., Krishnan, A. and Pereira, A. (2014), Rice GROWTH UNDER DROUGHT KINASE is required for drought tolerance and grain yield under normal and drought stress conditions. *Plant Physiol.* 166: 1634–45.

Ravikumar, G., Manimaran, P., Voleti, S. R., Subrahmanyam, D., Sundaram, R. M., Bansal, K. C., Viraktamath, B. C. and Balachandran, S. M. (2014), Stress-inducible expression of *AtDREB1A* transcription factor greatly improves drought stress tolerance in transgenic indica rice. *Transgenic Res.* 23: 421–39.

Redillas, M. C., Jeong, J. S., Kim, Y. S., Jung, H., Bang, S. W., Choi, Y. D., Ha, S. H., Reuzeau, C. and Kim, J. K. (2012), The overexpression of *OsNAC9* alters the root architecture of rice plants enhancing drought resistance and grain yield under field conditions. *Plant Biotechnol. J.* 10: 792–805.

Rivero, R. M., Kojima, M., Gepstein, A., Sakakibara, H., Mittler, R., Gepstein, S. and Blumwald, E. (2007), Delayed leaf senescence induces extreme drought tolerance in a flowering plant. *Proc. Natl. Acad. Sci. USA* 104: 19631–6.

Sangodele, E. A., Hanchinal, R. R, Hanamaratti, N. G., Shenoy, V. and Kumar, V. M. (2014), Analysis of drought tolerant QTL linked to physiological and productivity component traits under water-stress and non-stress in rice (*Oryza sativa* L.). *Int. J. Curr. Res. Acad. Rev.* 2(5): 108–13.

Schroeder, J. I., Allen, G. J., Hugouvieux, V., Kwak, J. M. and Waner, D. (2001), Guard cell signal transduction. *Annu. Rev. Plant Physiol. Plant Mol. Biol.* 52(1): 627–58.

Sellamuthu, R., Chandrababu, R. and Serraj, R. (2015), Mapping QTLs for reproductive stage drought resistance traits using an advanced backcross population in upland rice. *Crop Sci.* 55: 1–12.

Sharp, R. E. (2002), Interaction with ethylene: changing views on the role of abscisic acid in root and shoot growth responses to water stress. *Plant Cell Environ.* 25(2): 211–22.

Shi, J., Kim, K.-N., Ritz, O., Albrecht, V., Gupta, R., Harter, K., Luan, S. and Kudla, J. (1999), Novel protein kinases associated with Calcineurin B–like calcium sensors in Arabidopsis. *Plant Cell* 11: 2393–405.

Sinha, A. K., Jaggi, M., Raghuram, B. and Tuteja, N. (2011), Mitogen-activated protein kinase signaling in plants under abiotic stress. *Plant Signal Behav.* 6: 196–203.

Singh, S., Pradhan, S., Singh, A. and Singh, O. (2012), Marker validation in recombinant inbred lines and random varieties of rice for drought tolerance. *Australian J. of Crop Sci.* 6: 606–12.

Sperotto, R. A, Ricachenevsky, F. K., Duarte, G., Boff, T., Lopes, K., Sperb, E. R., Grusak, M. A. and Fett, J. P. (2009), Identification of up- regulated genes in flag leaves during rice grain filling and characterization of *OsNAC5*, a new ABA-dependent transcription factor. *Planta* 230(5): 985–1002.

Srividhya, A., Vemireddy, L. R., Sridhar, S., Jayaprada, M., Ramanarao, P. V., Hariprasad, A. S., Reddy, H. K., Anuradha, G. and Siddiq, E. (2011), Molecular mapping of QTLs for yield and its components under two water supply conditions in rice (*Oryza sativa* L.). *J. Crop Sci. Biotech* 14(1): 45–56.

Su, Y. H., Liu, Y. B. and Zhang, X. S. (2011), Auxin-cytokinin interaction regulates meristem development. *Mol. Plant* 4: 616–25.

Suji, K. K., Biji, K. R., Poornima, R., Prince, K. S. J., Amudha, K., Kavitha, S., Mankar, S. and Babu, R. C. (2012), Mapping QTLs for plant phenology and production traits using indica rice (*Oryza sativa* L.) lines adapted to rainfed environment. *Mol. Biotechnol.* 52: 151–60.

Sullivan, C. Y. and Ross, W. M. (1979), *Stress Physiology in Crop Plants.* Wiley, New York, 263–81.

Swamy, B. P. M, Ahmed, H. U., Henry, A., Mauleon, R., Dixit, S., Vikram, P., Tilatto, R., Verulkar, S, B., Perraju, P., Mandal, N. P., Variar, M., Robin, S., Chandrababu, R., Singh, O. N., Dwivedi, J. L., Das, S. P., Mishra, K. K., Yadaw, R. B., Aditya, T. L., Karmakar, B., Satoh, K., Moumeni, A., Kikuchi, S., Leung, H. and Kumar, A. (2013), Genetic, physiological, and gene expression analyses reveal that multiple QTL enhance yield of rice mega-variety IR64 under drought. *PLos ONE* 8(5): e62795.

Tamaki, H., Reguera, M., Abdel, T., Yasser, M., Takebayashi, Y., Kasahara, H. and Blumwald, E. (2015), Targeting hormone-related pathways to improve grain yield in rice: A chemical approach. *PLoS ONE* 10(6): e0131213.

Tang, N., Zhang, H., Li, X., Xiao, J. and Xiong, L. (2012), Constitutive activation of transcription factor *OsbZIP46* improves drought tolerance in rice. *Plant Physiol.* 158: 1755–68.

Tardieu, F. and Davies, W. J. (1992), Stomatal response to abscisic acid is a function of current plant water status. *Plant Physiol.* 98(2): 540–5.

Tardieu, F., Parent, B. and Simonneau, T. (2010), Control of leaf growth by abscisic acid hydraulic or non hydraulic processes? *Plant Cell Environ.* 33(4): 636–47.

Todaka, D., Shinozaki, K. and Yamaguchi-Shinozaki, K. (2015), Recent advances in the dissection of drought stress regulatory networks and strategies for development of drought-tolerant transgenic rice plants. *Front Plant Sci.* 6: 84.

Travaglia, C., Reinoso, H., Cohen, A., Luna, C., Tommasino, E., Castillo, C. and Bottini, R. (2010), Exogenous ABA increases yield in field-grown wheat with moderate water restriction. *J. Plant Growth Regul.* 29(3): 366–74.

Trijatmiko, K. R., Supriyanta, P. J., Thomson, M. J., Vera Cruz, C. M., Moeljopawiro, S. and Pereira, A. (2014), Meta-analysis of quantitative trait loci for grain yield and component traits under reproductive-stage drought stress in an upland rice population. *Mol. Breed.* 34(2): 283–95.

Tuberosa, R. and Salvi, S. (2006), Genomics-based approaches to improve drought tolerance of crops. *Trends Plant Sci.* 11: 405–12.

Turner N. C. (1979) Drought resistance and adaptation to water deficits in crop plants. In H. Mussell and C. R. Staples (eds), *Stress Physiology in Crop Plants.* John Wiley & Sons, New York, pp. 343–72.

Turner John, G., Ellis, C. and Devoto, A. (2002), The Jasmonate Signal Pathway. *Plant Cell* 14 (suppl 1): S153–64.

Uga, Y., Okuno, K. and Yano, M. (2011), *Dro1*, a major QTL involved in deep rooting of rice under upland field conditions. *J. Exp. Bot.* 62: 2485–94.

Uga, Y., Sugimoto, K., Ogawa, S., Rane, J., Ishitani, M., Hara, N., Kitomi, Y., Inukai, Y., Ono, K., Kanno, N., Inoue, H., Takehisa, H., Motoyama, R., Nagamura, Y., Wu, J., Matsumoto, T., Takai, T., Okuno, K. and Yano, M. (2013), Control of root system architecture by *DEEPER ROOTING 1* increases rice yield under drought conditions. *Nat. Genet.* 45: 1097–102.

Venuprasad, R., Cruz, M. T., Amante, M., Magbanua, R., Kumar, A. and Atlin, G. N. (2008), Response to two cycles of divergent selection for grain yield under drought stress in four rice breeding populations. *Field Crop Res.* 107(3): 232–44.

Verma, S. K., Saxena, R. R., Saxena, R. R., Xalxo, M. S. and Verulkar, S. B. (2014), QTL for grain yield under water stress and non-stress conditions over years in rice (*Oryza sativa* L.). *Aust. J. Crop Sci.* 8(6): 916–26.

Vikram, P., Swamy, B. P., Dixit, S., Ahmed, H. U., Teresa Sta Cruz, M. and Kumar, A. (2011), *qDTY1.1*, a major QTL for rice grain yield under reproductive-stage drought stress with a consistent effect in multiple elite genetic backgrounds. *BMC Genet.* 12: 89.

Vikram, P., Swamy, B. P. M., Dixit, S., Ahmed, H., Sta Cruz, M. T., Singh, A. K., Ye, G. and Kumar, A. (2012), Bulk segregant analysis: 'An effective approach for mapping consistent-effect drought grain yield QTLs in rice'. *Field Crops Res.* 134: 185–92.

Vikram, P., Swamy, B. P., Dixit, S., Singh, R., Singh, B. P., Miro, B., Kohli, A., Henry, A., Singh, N. K. and Kumar, A. (2015), Drought susceptibility of modern rice varieties: an effect of linkage of drought tolerance with undesirable traits. *Sci. Rep.* doi: 10.1038/srep14799.

Wang, H., Inukai, Y. and Yamauchi, A. (2006), Root development and nutrient uptake. *Crit. Rev. Plant Sci.* 25: 279–301.

Wang, Y., Zang, J., Sun, Y., Ali, J., Xu, J. L. and Li, Z. (2013), Background-independent quantitative trait loci for drought tolerance identified using advanced backcross introgression lines in rice. *Crop Sci.* 53: 430–41.

Wang, Y., Zhang, Q., Zheng, T., Cui, Y., Zhang, W., Xu, J. and Zhikang, L. (2014), Drought-tolerance QTLs commonly detected in two sets of reciprocal introgression lines in rice. *Crop Pasture Sci.* 65: 171–84.

Wilkinson, S., Kudoyarova, G. R., Veselov, D. S., Arkhipova, T. N. and Davies, W. J. (2012), Plant hormone interactions: innovative targets for crop breeding and management. *J. Exp. Bot.* 63: 3499–509.

Xiang, Y., Tang, N., Du, H., Ye, H. and Xiong, L. (2008), Characterization of *OsbZIP23* as a key player of the basic leucine zipper transcription factor family for conferring abscisic acid sensitivity and salinity and drought tolerance in rice. *Plant Physiol.* 148: 1938–52.

Xiao, B., Huang, Y., Tang, N. and Xiong, L. (2007), Over-expression of a LEA gene in rice improves drought resistance under the field conditions. *Theor. Appl. Genet.* 115: 35–46.

Xiao, B. Z., Chen, X., Xiang, C. B., Tang, N., Zhang, Q. F. and Xiong, L. Z. (2009), Evaluation of seven function-known candidate genes for their effects on improving drought resistance of transgenic rice under field conditions. *Mol. Plant* 2: 73–83.

Xing, Y. and Zhang, Q. (2010), Genetic and molecular bases of rice yield. *Ann. Rev. Plant Biol.* 61(1): 421–42.

Xing, W., Zhao, H. and Zou, D. (2014), Detection of main-effect and epistatic QTL for yield-related traits in rice under drought stress and normal conditions. *Canadian J. Plant Sci.* 94: 633–41.

Xu, K., Chen, S., Li, T., Ma, X., Liang, X., Ding, X., Liu, H. and Luo, L. (2015), *OsGRAS23*, a rice *GRAS* transcription factor gene, is involved in drought stress response through regulating expression of stress-responsive genes. *BMC Plant Biol.* 15: 141.

Yang, D. H., Kwak, K. J., Kim, M. K., Park, S. J., Yang, K. Y. and Kang, H. (2014), Expression of Arabidopsis glycine-rich RNA-binding protein *AtGRP2* or *AtGRP7* improves grain yield of rice (*Oryza sativa*) under drought stress conditions. *Plant Sci.* 214: 106–12.

Yang, J., Wang, Z., Zhu, Q. and Lang, Y. (1999), Regulation of ABA and GA to rice grain filling. *Acta Agron. Sin.* 25: 341–8.

Yang, J., Zhang, J., Wang, Z., Zhu, Q. and Liu, L. (2001a), Water deficit–induced senescence and its relationship to the remobilization of pre-stored carbon in wheat during grain filling. *Agronomy J.* 93: 196–206.

Yang, J., Zhang, J., Wang, Z., Zhu, Q. and Wang, W. (2001b), Hormonal changes in the grains of rice subjected to water stress during grain filling. *Plant Physiol.* 127(1): 315–23.

Yang, J., Zhang, J., Wang, Z., Zhu, Q. and Liu, L. (2002), Abscisic acid and cytokinins in the root exudates and leaves and their relationship to senescence and remobilization of carbon reserves in rice subjected to water stress during grain filling. *Planta* 215(4): 645–52.

Yang, J., Zhang, J., Wang, Z., Zhu, Q. and Liu, L. (2003), Involvement of abscisic acid and cytokinins in the senescence and remobilization of carbon reserves in wheat subjected to water stress during grain filling. *Plant Cell Environ.* 26(10): 1621–31.

Yang, J., Zhang, J., Wang, Z., Xu, G. and Zhu, Q. (2004), Activities of key enzymes in sucrose-to-starch conversion in wheat grains subjected to water deficit during grain filling. *Plant Physiol.* 135: 1621–9.

Yang, J. and Zhang, J. (2006), Grain filling of cereals under soil drying. *New Phytol.* 169(2): 223–36.

Yang, J. and Zhang, J. (2010), Crop management techniques to enhance harvest index in rice. *J. Exp. Bot.* 61: 3177–89.

Yang, W., Kong, Z., Omo-Ikerodah, E., Xu, W., Li, Q. and Xue, Y. (2008), Calcineurin B-like interacting protein kinase *OsCIPK23* functions in pollination and drought stress responses in rice (*Oryza sativa* L.). *J. Genet. Genom.* 35: 531–43.

You, J., Hu, H. and Xiong, L. (2012), An ornithine δ-aminotransferase gene *OsOAT* confers drought and oxidative stress tolerance in rice. *Plant Sci.* 197: 59–69.

You, J., Zong, W., Li, X., Ning, J., Hu, H., Xiao, J. and Xiong, L. (2013), The *SNAC1*-targeted gene *OsSRO1c* modulates stomatal closure and oxidative stress tolerance by regulating hydrogen peroxide in rice. *J. Exp. Bot.* 64: 569–83.

Yu, L., Chen, X., Wang, Z., Wang, S., Wang, Y., Zhu, Q., Li, S. and Xiang, C. (2013), Arabidopsis enhanced drought tolerance1/HOMEODOMAIN GLABROUS11 confers drought tolerance in transgenic rice without yield penalty. *Plant Physiol.* 162: 1378–91.

Zhang, Q., Li, J., Zhang, W., Yan, S., Wang, R., Zhao, J., Li, Y., Qi, Z., Sun, Z. and Zhu, Z. (2012), The putative auxin efflux carrier *OsPIN3t* is involved in the drought stress response and drought tolerance. *Plant J.* 72(5): 805–16.

Zhang, Z., Zhang, Q., Wu, J., Zheng, X., Zheng, S., Sun, X., Qiu, Q. and Lu, T. (2013), Gene knockout study reveals that cytosolic ascorbate peroxidase 2 (*OsAPX2*) plays a critical role in growth and reproduction in rice under drought, salt and cold stresses. *PLoS ONE* 8: e57472.

Zhao, K., Wright, M., Kimball, J., Eizenga, G., McClung, A., Kovach, M., Tyagi, W., Ali, M. L., Tung, C. W., Reynolds, A., Bustamante, C. D. and McCouch, S. R. (2010), Genomic diversity and introgression in *Oryza sativa* reveal the impact of domestication and breeding on the rice genome. *PLoS ONE* 5(5): e10780.

Zhao, X. Q., Xu, J. L., Zhao, M., Lafitte, R., Zhu, L. H., Fu, B.-Y., Gao, Y.-M. and Li, Z.-K. (2008), QTLs affecting morph-physiological traits related to drought tolerance detected in overlapping introgression lines of rice (*Oryza sativa* L.). *Plant Sci.* 174: 618–25.

Zhao, Y. (2010), Auxin biosynthesis and its role in plant development. *Ann. Rev. Plant Biol.* 61(1): 49–64.

Zhu, G., Ye, N., Yang, J., Peng, X. and Zhang, J. (2011), Regulation of expression of starch synthesis genes by ethylene and ABA in relation to the development of rice inferior and superior spikelets. *J. Exp. Bot.* 62(11): 3907–16.

Zou, G., Mei, H., Liu, H., Liu, G., Hu, S., Yu, X., Li, M., Wu, J. and Luo, L. (2005), Grain yield responses to moisture regimes in a rice population: association among traits and genetic markers. *Theor. Appl. Genet.* 112: 106–13.

Part 2

Rice nutritional and processing quality

Chapter 7

Advances in understanding the role of rice in nutrition

Melissa Fitzgerald, University of Queensland, Australia; Adoracion Resurreccion, International Rice Research Institute, The Philippines; and Julie Pua Ferraz, Diabetes Foundation Marikina and Calamba Doctors' Hospital, The Philippines

1 Introduction

Rice is a major global source of nutrients for many reasons. It is the staple food for many people, supplying the majority of carbohydrates and protein consumed each day. In many countries, there is now widespread understanding that rice with the bran layer intact provides a host of nutritional benefits because of the wealth of minerals and phytochemicals present inside the bran layer (Dipti et al., 2012). As a whole grain, rice is an attractive option for food processors because it does not contain the widely allergenic properties of wheat or soy, although some people do have an allergic response to one of the proteins. An additional nutritional opportunity rice offers is to address a host of chronic diseases associated with malnutrition. Interestingly, and perhaps counter-intuitively, malnutrition includes both under- and over-nutrition. Rice can contribute to solutions for both of these conditions. In some areas significant progress has already been made, whilst, in other areas, more needs to be done. These opportunities for rice in contributing solutions to the problem of malnutrition will form the focus of this chapter.

Malnutrition is a term used to describe a diet that is poor in nutrients, leading to loss of health and wellness. The term is generally associated with starvation, stunting and food unavailability as consequences of poverty (Tanumihardjo et al., 2007). However, malnutrition can be viewed from a different perspective where poor dietary choices and over-consumption of specific food types combine to provide calories of low dietary quality (Lopez et al., 2006). In combination with other factors, the impact of a low-quality diet can then result in obesity and the onset of chronic disease (Lopez et al., 2006). In developing countries, the rates of obesity and its consequences for chronic disease are at epidemic

http://dx.doi.org/10.19103/AS.2016.0003.30

levels, placing a significant economic burden upon the countries affected (Popkin, Adair and Ng, 2012). For many rice-consuming countries, perhaps paradoxically, both forms of malnutrition exist. It is important therefore, to invest in developing rice varieties that improve dietary quality through the provision of micronutrients as well as particular forms of starch and other compounds that could begin to reverse both types of malnutrition.

2 Rice and under-nutrition

Under-nutrition refers to dietary deficiency of key nutrients, including calories and micronutrients. Micronutrient deficiency is described as hidden hunger because it does not manifest as hunger, but as negative impact on growth and development. Micronutrients include vitamins, minerals and a host of other compounds that are essential to health and development. Micronutrient deficiency leads to poor development, stunting, blindness, cognitive issues and poor immune function (Bouis and Hunt, 1999). Most staple foods contain very few micronutrients in the way that they are consumed, for example, as white grains of rice or white flour.

Many studies have shown that unpolished rice contains high levels of minerals, many essential vitamins and a host of other phytochemicals (Dipti et al., 2012) but, once the bran is removed and the grains are white, the mineral and vitamin content is very low (Dipti et al., 2012; Lu et al., 2013; Hansen et al., 2012). Therefore, in countries with the triumvirate of polished rice as the staple food, high poverty and low dietary diversity, micronutrient deficiency is a significant component of malnutrition.

Fortification and supplementation programmes are mechanisms to deliver micronutrients to people. One successful example is iodised salt, which ensures that sufficient amounts of iodine are consumed. However, fortification and supplementation do not work in places where people do not purchase the fortified food, or where countries cannot afford to implement a fortification policy (White and Broadley, 2005). The strategy to biofortify staple crops with essential micronutrients was first developed in the early 1990s (Graham, Welch and Bouis, 2001), and has continued until now, with programmes actively working to develop staple crops biofortified to deliver key nutrients. The biofortification programme developed aspirational targets for increasing the amount of iron, zinc and β-carotene in six staple foods, including rice, that would reach the malnourished in rural areas and improve the nutritional status of those who consumed them (Bouis et al., 2011). Deficiencies in these three micronutrients accounted for about 6% of disability-adjusted life years (DALYs) (Black, 2003).

2.1 Iron

Anaemia due to iron deficiency affects more than one billion people worldwide, but the majority of the burden is carried by developing countries (Shaw and Friedman, 2011). As an example, a recent study has shown that haemoglobin concentrations are lowest and anaemia is highest in parts of Asia and Africa (Stevens et al., 2013). These findings can be easily understood in rice-consuming countries both because the iron content in polished rice is very low, in the order of 2–3 ppm (Lu et al., 2013; Hansen et al., 2012; Gregorio et al., 2000; Doesthale et al., 1979; Bryant et al., 2005), and because the bioavailability of that iron is low (Shaw and Friedman, 2011). The concentration of iron in polished rice that is considered sufficient to address and combat iron deficiency is 13–15 ppm (Graham, Welch and Bouis, 2001).

In the early stages of the biofortification projects in rice, significant efforts were made to identify germplasm which accumulates high iron in grains (White and Broadley, 2005; Gregorio et al., 2000; Gregorio, 2002). Germplasm was soon found with significant variability for iron content in unpolished grains, up to 20 ppm, but once polished to white rice, the amount of iron plummeted to less than 5 ppm, and all variability was lost (Dipti et al., 2012). A huge amount of effort has been undertaken to reach the target of 15 ppm of iron in polished grains using conventional breeding techniques (White and Broadley, 2005; Graham, Welch and Bouis, 2001; Gregorio et al., 2000; Gregorio, 2002; Jiang et al., 2007; Liang et al., 2007; Liang et al., 2008; Ma et al., 2005; Tucker, 2003; Virk and Barry, 2007), but variation for the iron content of polished grains has not been found. This suggests either that there is no biological reason for significant amounts of iron to accumulate in the endosperm, or some physiological or physical barrier prevents its transport to the endosperm. Rice improvement programmes, faced with the failure of reaching the targets for iron with conventional breeding, turned to transgenic solutions in order to determine whether iron could in fact accumulate in the rice endosperm.

Transgenic studies using the ferritin gene from soy (Goto et al., 1999; Vasconcelos et al., 2003), bean (Qu et al., 2005), or the phytase gene from *Aspergillus* (Lucca, Hurrell and Potrykus, 2002) were all able to show a significant increase in the iron concentration of grains, well above the HarvestPlus target of 15 ppm. These studies clearly indicate that no physical barrier prevents iron from reaching grains, but rather a sink must be activated in the grains to bring iron into the endosperm. The transgenic grains showed that the ferritin in the grains was able to reverse the symptoms of anaemia in rats (Murray-Kolb et al., 2002). These early studies paved the way for harnessing biotechnology to optimise the iron content in polished rice, and provided promise to malnourished rice consumers that nutritionally enhanced rice was on the horizon.

Nicotianamine (NA) is a non-protein amino acid found in all higher plants (Noma and Noguchi, 1976). Its role is to chelate divalent metal cations such as Fe^{2+} and Zn^{2+}, and redistribute these metals at the cellular level to maintain metal homeostasis (Scholz et al., 1992). NA is the product of the NA synthase (*NAS*) gene. There are three *NAS* genes in rice, each involved in the long distance transport of iron, and they are differentially regulated by iron (Inoue et al., 2003). *OsNAS1* and *OsNAS2* are expressed in root cells and respond to iron deficiency, and *OsNAS3* is expressed in companion cells of roots and leaves and is not iron sensitive (Inoue et al., 2003). Scientists have used this information to manipulate the expression of these genes to increase the iron content of rice grains so that the plant supplies the grain with iron.

Johnson et al. (2011) transformed Nipponbare, a temperate japonica variety of rice, to construct populations over-expressing *OsNAS1*, *OsNAS2* or *OsNAS3*. In the unpolished grains, iron was threefold higher in the transformed lines than in the control, and zinc was twofold higher. Grains were polished in selected transformed lines, and the highest iron concentration was achieved from *OsNAS2* lines, at 19 ppm (Johnson et al., 2011). Recently the works on ferritin and *OsNAS* have been combined and a number of promoters were tested to find the best combination of genes that affect iron using the variety IR64, a very popular variety in Asia. This work has shown that selection of the right promoter and combining a soy ferritin gene with *OsNAS2* provides reproducibly high iron in the polished rice, at and above the target of 15 ppm (Trijatmiko et al., 2016). They further show that the iron is bioavailable. Since NA transports divalent ions, Zn is also elevated in the grains (Trijatmiko et al., 2016). While work still must be done in testing different fields and postharvest conditions, there is enormous promise that rice biofortified with iron and zinc will soon be ready for delivery to target populations.

2.2 Zinc

Zinc deficiency is one of the major factors contributing to the burden of disease in developing countries (WHO, 2002). Zinc contributes to neurological development and function and therefore impacts on child mortality and foetal growth (Tamura and Goldenberg, 1996). For these reasons, zinc has been listed as a top priority for the biofortification strategy for staple crops (Graham, Welch and Bouis, 2001; Gregorio et al., 2000).

Zinc plays a key role in the biology of plants. Plant proteins typically bind zinc more than any other metal (Gladeshev et al., 2004). It is therefore intuitive that plants will be sensitive to the zinc status of the soil. However, the chemical complexities of growing rice in a waterlogged environment impact on the availability of zinc for uptake (Impa and Johnson-Beebou, 2010). In addition, the genotype of the plant determines whether zinc is sequestered in the maternal or filial parts of the plant (Wissuwa, Ismail and Graham, 2008). However, research shows that genotypes able to take up and use zinc do respond to zinc status in soil (Wissuwa, Ismail and Graham, 2008), and that the zinc reaches the grain either by uptake and distribution during grain filling, or by remobilisation of sequestered zinc from the roots or shoots (Impa et al., 2013). This genetic diversity has provided a basis for approaching the HarvestPlus target for zinc of 28 ppm (Welch and Graham, 2004) using conventional breeding. A variety has now been released in Bangladesh with higher levels of zinc in the grain (www.irri.org).

In order to reach the most malnourished with biofortified rice, there is substantial value in ensuring that one variety delivers both iron and zinc. Breeders have long found that elevated iron in the grain is usually accompanied by elevated zinc. This has been confirmed in all the varieties transgenically modified to accumulate iron in the grain (Vasconcelos et al., 2003; Murray-Kolb et al., 2002; Trijatmiko et al., 2016; Lee et al., 2009; Qu et al., 2005). The finding is unsurprising since the biochemistry for transporting divalent cations around the plant is likely to transport both Zn^{2+} and Fe^{2+}, as seen with NA. Therefore, once the high iron varieties are released into the rice community, it is highly likely that both iron and zinc will be elevated in the same grain.

2.3 Vitamin A

Vitamin A is a fat-soluble molecule key to development of retinal receptors, cochlear function and to more generic functions of cognitive function, differentiation of epithelial cells and immune functions (Sommer, 2001). The first clinical manifestations of vitamin A deficiency impact on iron utilisation and anaemia, immune incompetence and growth retardation. More severe levels of deficiency impact vision and lead to the condition of night blindness (xerophthalmia). This condition is rapidly reversed by systemic vitamin A. However, if deficiency continues, significant ocular degeneration and blindness occurs (Sommer, 2001).

Vitamin A is derived from food such as eggs and dairy products or as provitamin A carotenoids, such as β-carotene, from plant products such as green leafy vegetables, yellow vegetables and orange-coloured fruit (Morris et al., 2015). Rice contains no β-carotene (Dipti et al., 2012; Fitzgerald, 2007), and in low socio-economic groups where dietary diversity is low, vitamin A deficiency is high. For this reason, the biofortification strategy prioritised β-carotene as one of its nutritional targets, leading to the development of Golden Rice.

The history of Golden Rice is well documented from the first event in the 1990s, with the transfer of the β-carotene pathway from daffodils to rice, followed by the second event utilising genes from maize (Beyer et al., 2002; Grusak, 2005; Lucca, Poletti and Sautter, 2006). In brief, Golden Rice has been developed to synthesise and accumulate β-carotene in the polished grain at a level that will make an impact on people with vitamin A deficiency. The pathway to place Golden Rice in the hands of those who need it has been littered with obstacles and regulatory politics (Tang et al., 2015; Dubock, 2014; Moghissi, Pei and Liu, 2016; Tang et al., 2012; Yang, Xu and Rodriguez, 2014; Krawinkel, 2007); these have ensured that Golden Rice remains in experimental fields.

The bioavailability of β-carotene is known to be about sixfold higher from a lipid-based matrix than a fruit-based matrix and fourteen-fold higher than β-carotene from green leafy vegetables (Zhu et al., 2016). Studies on the bioavailability of β-carotene from Golden Rice indicate that it is between an oil-based and fruit-based matrix (Tang et al., 2012; Tang et al., 2009), and using that value, modelling shows that the impact of replacing white rice for Golden Rice would decrease vitamin A deficiency significantly in women and children in Bangladesh, Indonesia and the Philippines (De Moura et al., 2016).

Over the past two decades, investment into strategic biofortification of rice for iron, zinc and β-carotene has brought together research teams with a wide range of expertise. These teams have combed the genus for existing diversity and innovated to produce germplasm with grains that reproducibly reach the target levels of zinc, iron and β-carotene. The bioavailability of these compounds has been shown in a small number of studies, and the data has been extrapolated, using modelling, to illustrate the nutritional impact on the malnourished. With the breeding and selection technologies that have developed alongside the biofortification process, these nutritional traits can be introgressed into other germplasm using conventional breeding and newer selection techniques. This enables the nutritional traits to be widely disseminated in varieties preferred by different consumers. The same strategy was used to disseminate the *Sub1* gene (Septiningsih et al., 2009), and its use would provide a convenient and rapid way to reach populations where dietary diversity is low, hidden hunger is high, and supplementation and conventional fortification has failed. However, the increased nutritional value has been mainly achieved using transgenic techniques and, before the plants can be grown and consumed, the germplasm must pass through regulatory processes, which have already proven to be protracted and complex (Dubock, 2014).

3 Rice and over-nutrition

3.1 The problem of over-nutrition

Malnutrition usually conjures images of under-nourished people, but paradoxically, the definition of poor nutrition also includes over-nutrition (Tanumihardjo et al., 2007). Over-nutrition is the condition where adequate calories are available, but their dietary quality is poor, and this, in combination with other lifestyle factors, can lead to obesity and a host of other chronic diseases. In recent years, we have, for the first time, witnessed the deadly impacts of chronic disease such as obesity and type II diabetes outstripping those of infectious disease (Gersh et al., 2010; Ghaffar, Reddy and Singhi, 2004; Hossain, Kawar and El Nahas, 2007; James, 2008; Kopelman, 2000; Misra, Singhal and Khurana, 2010; Nugent, 2008).

Obesity and type II diabetes are considered to be the two most serious chronic diseases (Vos, et al., 2015). Obesity is often a precursor to type II diabetes because (i) excessive weight places strain on the metabolic organs of the body, and (ii) the dietary and lifestyle choices that initially lead to uncontrolled weight management are usually those that increase susceptibility to type II diabetes (James, 2008, Kopelman, 2000). Type II diabetes is a condition that is characterised by high blood sugar, insulin resistance and low secretion of insulin (Misra, Singhal and Khurana, 2010). People with type II diabetes are also at much higher risk of cardiovascular disease than others (Popkin, Adair and Ng, 2012; Pedersen, et al., 2016) and cancer (Gallagher and LeRoith, 2015). These chronic diseases of obesity, type II diabetes and cardiovascular disease are not just found in western countries but are occurring in many countries of Asia at pandemic rates (Popkin, Adair and Ng, 2012; Misra, Singhal and Khurana, 2010; Nugent, 2008; Yoon et al., 2006; Shaw, Sicree and Zimmet, 2010). For many of these countries, rice is the staple food (Dipti et al., 2012). The burden of all chronic diseases in low-income countries measured in disability-adjusted life years (DALYs) is 50% and 12% for obesity and type II diabetes, respectively (Abegunde et al., 2008).

3.2 Rice consumption and the prevention of obesity and and/or chronic disease

Factors such as the type of carbohydrate, the glycaemic index (GI) of the food and the digestibility of the starch are all implicated in the prevention and management of chronic diseases (Hodge et al., 2004; Munter et al., 2007; Nisak et al., 2010). There is significant diversity in the GI of rice (Fitzgerald et al., 2011), but a number of studies suggest that rice consumption is a risk factor for obesity and type II diabetes (Hu et al., 2012; Malik, Willett and Hu, 2013; Sun et al., 2010). Furthermore, other studies demonstrate that the nutritional value of a diet rich in carbohydrates is key to protection against chronic disease (Gallagher and LeRoith, 2015). Clearly, there is confusion about the role and opportunity for rice in contributing to the prevention of obesity and chronic diseases. In spite of the epidemic nature of over-nutrition in rice-consuming countries, there has not been the same level of ambition to biofortify rice to improve nutritional properties for people with chronic diseases resulting from poor dietary quality.

Carbohydrate quality

Ingested carbohydrates are either digested to provide energy to fuel daily activity or fermented in the large intestine (Topping, 2007). Both types are required for health and nutrition. Most of the carbohydrate in a rice grain is starch. Starch is synthesised by a coordinated set of enzymes that includes six isoforms of starch synthase, and three isoforms of the branching enzyme (SBE) and debranching enzymes (DBE) (Cuevas et al., 2010). Based on haplotypes defined by single-nucleotide polymorphisms, there are five haplotypes of the waxy (*Wx*) gene (Chen et al., 2008), leading to different expression levels of granule-bound starch synthase 1 (GBSS1) (Mikami et al., 2008), which associates with different proportions of the linear starch, amylose. Variation is also known at the locus of *SSIIa*, with four haplotypes leading to different proportions of chain lengths in each amylopectin cluster (Cuevas et al., 2010). The third locus where variation is known is *SBEIIb* and mutations at this locus lead to significant alteration of amylopectin structure (Liu et al., 2009; Wang, Kozlowski and Delgado, 2001; Yang et al., 2006; Yano et al., 1985; Yao, Paez and White, 2009; Yeh, Garwood and Shannon, 1981), which lowers the GI (Fitzgerald et al., 2011) and increases the proportion of resistant starch (RS) in the grain (Butardo et al., 2011). One such mutant in rice is one of the varieties used in the case study described below.

Resistant starch (RS)

RS is a form of dietary fibre. RS escapes digestion in the small intestine and moves into the large intestine, where it is fermented. Five categories of RS have been identified (Wallis, 2004). The categories relevant to rice are as follows: RS1, which is starch protected from digestion, for example, by the bran layer in whole grains; RS2, which is the naturally occurring RS; and RS3, which is retrograded starch.

Fermentation of RS by anaerobic bacteria in the large intestine leads to the production of short-chain fatty acids (SCFA) such as butyrate, propionate and acetate (Chang et al., 2014). These SCFA mediate a plethora of health benefits. SCFA lower the pH in the colon, which inhibits the growth of pathogenic bacteria, and strengthen the colon so that pathogens such as *Salmonella* cannot penetrate and infect. Butyrate has anti-inflammatory properties, and propionate is able to reduce cholesterol concentration in the blood (Haenen et al., 2013). Populations with high dietary intake of RS are often characterised by low incidence of chronic disease (Topping, Fukushima and Bird, 2003; Topping and Clifton, 2001). The gut is colonised by a diverse microflora, and different microbes have been associated with different dietary factors (Fukuda et al., 2011; O'Keefe et al., 2015; Vipperla and O'Keefe, 2012).

The development of genotyping technology has overcome the problems of working with bacteria that are highly anaerobic, and enabled nutritionists to probe the microbial environment of our digestive tract (Bäckhed et al., 2004). This research has unlocked information about the role of the microbiome in our health, and the importance of diet in maintaining homeostasis in the relationship between the human host and their microbiome (O'Keefe et al., 2015; Vipperla and O'Keefe, 2012). It has been found that a diet high in RS leads to microbiota dominated by the genus *Prevotella,* which degrades starch, ferments carbohydrates and produces butyrate (Ou et al., 2013). By contrast, when a diet is high in animal protein and saturated fat, the dominant microbes are from the genus *Bacteroides,* which produce secondary bile acids (Ou et al., 2013); these are inflammatory and carcinogenic (Nagengast, Grubben and Van Munster, 1995).

A recent study investigated the microbiome, diet and colonic inflammation status of African Americans consuming a diet rich in meat, and rural South Africans who consume a diet rich in RS (O'Keefe et al., 2015). The study demonstrated high levels of inflammatory biomarkers in the African Americans compared with the rural Africans, and different microbial compositions in each group. The diets of the two groups were switched for two weeks. After just two weeks, it was found that the microbial groups required for butyrogenesis were in greater abundance, as were their metabolites, and the inflammatory biomarkers were much lowered in the African Americans switched to the high-fibre diet, and the reverse for the rural Africans (O'Keefe et al., 2015). Findings such as these, which not only illustrate the benefits of a diet high in RS but also demonstrate how rapidly the microbial alliance can be brought into symbiosis, suggest that RS is a very important component of our diet.

The RS of cooked rice ranges from 1 to 3%. Elevating the RS in rice has been done in two ways. The first has been by identifying varieties that carry a mutation in *BEIIb*. The absence of this enzyme in starch synthesis leads to a significant alternation in the structure of the starch which impacts its digestibility in maize (Kasemsuwan et al., 1995) and in rice (Cuevas et al., 2010). This mutation has been found or created a few times in rice, and acts in the same way in terms of starch structure and digestibility, independently of genetic background (Yano et al., 1985; Butardo et al., 2012; Choi et al., 2006; Choi et al., 2005; 2006; Lee et al., 2006; Juliano et al., 1990; Regina et al., 2005). There are two major issues with all these mutants. First, the impact on starch leads to a penalty to yield potential of the plant, and secondly, the texture of the cooked rice has been described as being similar to wet concrete (Juliano et al., 1990; Shu et al., 2007). In order to combine the phenotype of high RS with better sensory quality, transgenic techniques have been used to elevate RS (Butardo et al., 2011; Wei et al., 2010). The grains look the same as the mutant grains, so it is likely that sensory quality may not be improved. While these varieties high in RS might not make it to the table as steamed rice, it is highly possible that food-processing solutions could be considered for their use.

Glycaemic index (GI)

The GI of a food is defined as the increase in blood glucose (over the fasting level) that is observed in the 2 h following ingestion of a fixed amount of carbohydrate in an individual food. This value is then compared with the response to a reference food (glucose or white bread) containing an equivalent amount of carbohydrate (Tsihlias et al., 2000). Carbohydrates of high GI (>70) are rapidly digested, causing a rapid rise in blood glucose and in most cases a quick rise in insulin. Conversely, carbohydrates that have low GI (~55 and below) are more slowly digested and subsequently cause relatively small increases in blood sugar and insulin (Psaltopoulou, Ilias and Alevizaki, 2010). High GI foods lead to rapid rises in postprandial plasma glucose and insulin excursions that are associated with an increased risk of developing diabetes (Halton et al., 2008; Barclay et al., 2008). As such, interventions to reduce postprandial plasma glucose and insulin peaks are one of the essential therapeutic objectives for treating diabetic patients to decrease the risk of patients developing cardiovascular complications (Salwa, 2004). Reducing postprandial plasma glucose can be achieved either through drugs or through diet. For families with low incomes, such as those in developing countries, the drugs are not affordable; therefore, dietary intervention becomes the only treatment available to clinicians. Several studies have shown that consumption of white rice is a risk factor for type II diabetes, and therefore other cancers (Hu et al., 2012; Malik, Willett and Hu, 2013; Sun et al., 2010), despite variation in GI of rice (Fitzgerald et al., 2011). In an attempt to bring clarity to the apparently conflicting results for the role of GI and diet in modulating blood sugar status, a case study is presented where rice varieties differing in GI were fed to people with and without type II diabetes in a study in the Philippines.

4 Case study: the impact of the glycaemic index (GI) of rice on blood sugar levels

For most of the Asian population, white rice is consumed as a staple food, and is eaten several times a day in large quantities. Different varieties of rice have been shown to differ in predicted GI (Fitzgerald et al., 2011). Three varieties were selected: IR24, with a predicted GI of 90; IR64, with a predicted GI of 64; and IR36ae, with a mutation in SBEIIb and a predicted GI of 48; and these were fed to volunteers to determine the impact of these on their blood sugar status.

The experimental procedure was approved by the Institute of Science in Diabetes Foundation Inc. (ISDFI) ethics committee. Ten healthy volunteers and eighteen diabetic volunteers were recruited to participate in the feeding trial. All volunteers consented to the research protocol. The study required two months for completion. Participants were financially compensated at the end of the feeding. Volunteers were screened by the clinical investigator (JPF) through medical chart review, medical interview, physical examination and laboratory testing for fasting blood sugar and glycosylated haemoglobin (HbA1c). Healthy volunteers were used to determine the difference in glycaemic response to rice over different time points, and diabetic volunteers were screened at fasting and two hours postprandial to determine the effect of each variety on their blood sugar status.

Diabetic volunteers who were at least 18 years old were diagnosed for diabetes mellitus type II by a physician or verified during chart review. The subjects had an HbA1c of 6–10%, were on a stable dose of treatment, were on diet for 2 months prior to the experiment and did not use any form of insulin. These subjects were also instructed on the level of physical activity a day prior to the feeding experiment. Subjects under age 18 or above 80 and those with a body mass index of below 22 or above 45 were excluded. Also excluded were subjects with type 1 diabetes, secondary diabetes and subjects with symptomatic heart failure or who had a myocardial infarct, an unstable angina or a coronary artery surgery the previous year. Pregnant or lactating women, those with a major malignancy within the previous 5 years, or those who had a mental status that would not permit an understanding of the scope, nature and consequences of the study were also excluded. Other exclusion criteria were liver disease, thyroid disease, uncontrolled hypertension, peripheral oedema, infections and history of a metabolic acidosis including ketoacidosis, hypoglycaemia or concurrent use of glucocorticoids. Current use of disallowed therapy such as glucosidase inhibitors was also excluded. The volunteers reported on the allocated day of testing and blood samples were withdrawn at fasting (FB). Cooked rice was presented (50 g available carbohydrate) and at two hours postprandial blood (PPB) was again withdrawn. Each volunteer attended three sessions.

The healthy volunteers were trainee priests residing in a seminary. These subjects had all followed the same diet and exercise routines for the previous fortnight at least. Blood samples were withdrawn at time zero (fasting blood sugar) and the volunteers were given

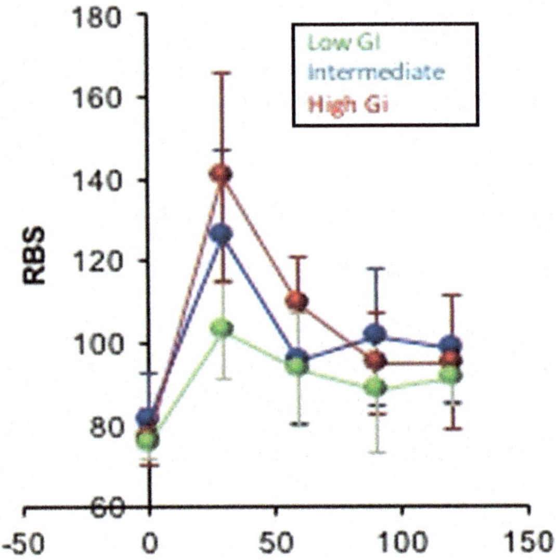

Figure 1 Random blood sugar test at 30 minute intervals in healthy subjects.

samples of the cooked rice (50 g available carbohydrate) and a bottle of water to consume. All volunteers consumed the three varieties of rice over three sessions. Blood samples were withdrawn every 30 min, and blood sugar concentration was measured.

The random blood sugar (RBS) test of the healthy volunteers (Fig. 1) clearly showed that rice varieties with different predicted GI followed the same trend when the healthy volunteers consumed them: an increase at 30 min and a decrease thereafter, with the largest spike seen for consumers of IR24. Most notable are the small deviations over time for consumers of IR36ae. These data clearly illustrate that consuming rice of different predicted GI will follow a trend for blood sugar status in the first hour after consumption. Using Tukey's analysis of *p* value and ignoring the fact that the data are not independent, a significant difference between the RBS for each variety after 30 minutes was observed, and after 60 minutes there was a significant difference in RBS between those who consumed IR24 and IR36ae. There was no significant difference in RBS that could be attributed to any variety later than 60 minutes after the meal.

Table 1 shows the baseline data for the diabetic volunteers. Most of the volunteers were female.

The blood sugar status of the diabetic volunteers was taken at fasting and two hours after eating the rice.

Table 2 shows the difference in the PPB glucose (PPBG) concentration of the blood in the diabetic patients 2 hours after consuming the rice. While the trend of decreasing blood glucose concentration is similar to that seen for the healthy patients, the variability between these patients is high and the data are therefore not significantly different. The data was analysed in more detail to determine if the absence of statistical difference on PPBG among the consumers was influenced by the HbA1c status of the patients at the commencement of the test. The patients were grouped according to their baseline HbA1c levels and their mean PPBG are compared across the three rice varieties (Table 3).

Table 1 Baseline characteristics of diabetic patients (n = 17)

Variable	Average	95% confidence interval
Age, years	56 ± 1.4	52.58, 58.72
Female %	82 ± 5	71.5, 93.2
Systolic blood pressure, mmHG	126.3 ± 4.3	117.18, 135.53
Diastolic blood pressure, mmHg	75.9 ± 3.1	69.31, 82.45
HBa1c	7.9 ± 0.3	7.32, 8.54
Waist circumference, cm	88.7 ± 1.8	84.80, 92.49
Hip circumference, cm	90.7 ± 1.5	87.58, 93.95
Waist–hip ratio	0.9 ± 0	0.95, 1.00

Table 2 Blood glucose concentration of 17 diabetics 2h post consumption

Rice variety	Average blood glucose	P value
IR24	163.83 ± 41.45	0.5422
IR64	158.38 ± 39.26	
IR36ae	148.43 ± 41.92	

Table 3 Mean and standard deviation of averaged 2-Hour PPBG by FBS level

FBS Group	N	IR24	IR64	AE36
≥ 130	8	191.18 ± 42.49	181.05 ± 39.35	176.85 ± 41.64
< 130	9	138.92 ± 23.40	138.84 ± 25.57	123.17 ± 21.72

When the 2 h postprandial values are expressed as a function of the baseline HbA1c group, it can be clearly seen that after two hours, the blood glucose concentration of the group with HbA1c below 7.3 shows decreasing PPBG with decreasing GI of the rice; there is a significant difference between the blood glucose concentration after eating IR64 and IR36ae. By contrast, for patients with higher HbA1c at fasting, there was no significant difference between blood sugar status after consuming the different rice varieties.

The data in this case study illustrate three major points that could be used to inform further studies on the digestibility profiles of rice, their impact on blood sugar status and the utility of GI as a tool to manage type II diabetes. The first point is that Fig. 1 shows that the largest and most significant difference in blood sugar status is seen just 30 minutes after consuming the rice, and again there is a significant difference between the blood sugar status of participants who consumed the high and low GI rice. At 90 and 120 minutes, there is no significant difference between the blood sugar status of any of the volunteers. This indicates that the major difference is seen at 30 minutes after consumption, and by sampling volunteers at 2 hours after consumption, any differences are likely to have been missed.

The second point raised by the data is that the subsampling of the same person over time limits the statistical power of the assay. The power would be far greater if independent data could be obtained for each of the time points. However, this would involve a much bigger and more costly experiment.

The third interesting point illustrated by these data is that diabetic patients of different HbA1c status differ in the glycaemic response due to the rice varieties. When blood sugar status is managed, their HbA1c status is optimally below 7 (Harris, 2000). Only five participants fell into that category, but they showed a significant response to eating the different rice varieties. The remaining participants, those whose HbA1c status was elevated, did not show a significant response to consuming the different rice varieties, and the data was highly variable (Table 3). This suggests that GI could be the most useful tool for diabetic patients who are able to manage their blood sugar status.

5 Conclusions

Malnutrition includes varying types of inadequate nutrition. Over the past two decades, significant financial investment has been made to biofortify rice to carry iron, zinc and β-carotene to those who need it. Rice scientists have harnessed transgenic technology to develop new rice varieties that carry levels of those micronutrients in the grains that the medical community predicts will have positive health outcomes. These varieties are not yet ready to be placed in the hands of consumers, as they are caught in the political quagmire of international biosafety approval due to the necessity of using transgenic techniques.

Obesity and chronic disease are another aspect of malnutrition. Obesity and its consequences are significant issues in rice-consuming countries and elsewhere. In Australia, the economic cost of obesity and its consequences was $8.6 billion in 2011/12, and is growing each year. The global burden of chronic diseases is significant in developing countries and, measured by DALYs, is greater than or equivalent to hidden hunger. Even though it is clear that there is diversity in the GI of rice variety, and that low GI rice leads to a smaller glycaemic response, very little political or financial investment has been made for developing varieties of rice that assist with managing blood sugar status. This could be because images of under-nutrition, showing malnourished people blind from micronutrient deficiency, elicit a stronger emotional response in donors response than images of obesity.

If we are to reduce the burden and costs of chronic disease to individuals, families and societies, and advance our understanding of the role of rice in all relevant aspects of malnutrition, there must be investment in understanding the traits of rice that can play a role in preventing or managing the onset of chronic disease, and then harnessing the science to enable development of varieties that carry those traits.

6 References

Abegunde, D.O., et al., 2008. The burden and costs of chronic diseases in low-income and middle-income countries. *The Lancet* 370 (9603): pp. 1929–1938.

Bäckhed, F., et al., 2004. The gut microbiota as an environmental factor that regulates fat storage. *Proceedings of the National Academy of Sciences of the USA* 101(44): pp. 15718–15723.

Barclay, A.W., et al., 2008. Glycemic index, glycemic load, and chronic disease risk–a meta-analysis of observational studies. *American Journal of Clinical Nutrition* 87(3): pp. 627–637.

Beyer, P., et al., 2002. Golden Rice: Introducing the b-carotene biosynthesis pathway into rice endosperm by genetic engineering to defeat Vitamin A deficiency. *Journal of Nutrition* 132(3): pp. 506S–510S.

Black, R., 2003. Micronutrient deficiency: an underlying cause of morbidity and mortality. *Bulletin of the World Health Organization* 81: pp. 79.

Bouis, H., and J. Hunt, 1999. Linking food and nutrition security: past lessons and future opportunities. *Asian Development Review* 17: pp. 168–213.

Bouis, H.E., et al., 2011. Biofortification: a new tool to reduce micronutrient malnutrition. *Food and Nutrition Bulletin* 32(Suppl 1): pp. S31–S40.

Bryant, R.J., et al., 2005. Phosphorus and mineral concentrations in whole grain and milled low phytic acid (lpa) 1-1 rice. *Cereal Chemistry* 82(5): pp. 517–522.

Butardo, V., et al., 2011. Impact of down-regulation of starch branching Enzyme IIb in rice by artificial microRNA and Hairpin-RNA-Mediated RNA silencing. *Journal of Experimental Botany* 62(14): pp. 4927–4941.

Butardo, V.M., et al., 2012. Biomolecular analyses of starch and starch granule proteins in the High-Amylose rice mutant Goami 2. *Journal of Agricultural and Food Chemistry* 60(46): pp. 11576–11585.

Chang, P.V., et al., 2014. The microbial metabolite butyrate regulates intestinal macrophage function via histone deacetylase inhibition. *Proceedings of the National Academy of Sciences* 111(6): pp. 2247–2252.

Chen, M.-H., et al., 2008. Waxy gene haplotypes: Associations with apparent amylose content and the effect by the environment in an international rice germplasm collection. *Journal of Cereal Science* 47(3): pp. 536–545.

Choi, H.C., et al., 2005. A new non dietary starch and high amylose rice cultivar, Goamy 2. *Korean Journal of Breeding* 37(2): pp. 131–132.

Choi, H.-C., et al., 2006. New mutants of specialty rice induced from Ilpumbyeo, a high-quality rice cultivar, by MNU(N-methyl-N-nitrosourea) treatment on fertilized egg cells. *Korean Journal of Breeding* 38(3): pp. 154–160.

Choi, I.D., et al., 2006. Indigestible carbohydrate contents and physical properties of Goami2 harvested at the maximized milling quality. *Food Science and Biotechnology* 15(2): pp. 254–259.

Cuevas, R., et al., 2010. Melting the secrets of gelatinisation temperature. *Functional Plant Biology* 37: pp. 439–447.

Cuevas, R.P., et al., 2010. Using chain-length distributions to diagnose genetic diversity in starch biosynthesis. *Carbohydrate Polymers* 81: pp. 120–127.

De Moura, F.F., et al., 2016. Biofortified β-carotene rice improves vitamin A intake and reduces the prevalence of inadequacy among women and young children in a simulated analysis in Bangladesh, Indonesia, and the Philippines. *The American Journal of Clinical Nutrition* 104(3): pp. 769–775.

Dipti, S.S., et al., 2012. The potential of rice to offer solutions for malnutrition and chronic diseases. *Rice* 5: p. 19.

Doesthale, Y.G., et al., 1979. Effect of milling on mineral and trace element composition of raw and parboiled rice. *Journal of the Science of Food and Agriculture* 30(1): pp. 40–46.

Dubock, A., 2014. The politics of Golden Rice. *GM Crops & Food* 5(3): pp. 210–222.

Fitzgerald, M., 2007. *Screening the International Rice Genebank Collection for Variation in Carotenoid Content 2007*, Harvestplus and IRRI, p. 5.

Fitzgerald, M., et al., 2011. Identification of a major genetic determinant of glycaemic index in rice. *Rice* 4: pp. 66–74.

Fukuda, S., et al., 2011. Bifidobacteria can protect from enteropathogenic infection through production of acetate. *Nature* 469(7331): pp. 543–547.

Gallagher, E.J. and D. LeRoith, 2015. Obesity and diabetes: the increased risk of cancer and cancer-related mortality. *Physiological Reviews* 95(3): pp. 727–748.

Gersh, B., et al., 2010. The epidemic of cardiovascular disease in the developing world: global implications. *European Heart Journal* 31: pp. 642–648.

Ghaffar, A., K. Reddy and M. Singhi, 2004. Burden of non-communicable diseases in South Asia. *British Medical Journal* 328: pp. 807–810.

Gladeshev VN, et al., 2004. Identification of trace element-containing proteins in genomic databases. *Annual Review of Nutrition* 24: pp. 579–596.

Goto, F., et al., 1999. Iron fortification of rice seed by the soybean ferritin gene. *Nature Biotechnology* 17(3): pp. 282–286.

Graham, R., R. Welch and H. Bouis, 2001. Addressing micronutrient malnutrition through the nutritional quality of staple foods: principles, perspectives, and knowledge gaps. *Advances in Agronomy* (70): pp. 77–142.

Gregorio, G.B., 2002. Progress in breeding for trace minerals in staple crops. *Journal of Nutrition* 132(3): pp. 500S–502S.

Gregorio, G.B., et al., 2000. Breeding for trace mineral density in rice. *Food and Nutrition Bulletin* 21: pp. 382–386.

Grusak, M.A., 2005. Golden Rice gets a boost from maize. *Nature Biotechnology* 23(4): pp. 429–430.

Haenen, D., et al., 2013. A diet high in resistant starch modulates microbiota composition, SCFA concentrations, and gene expression in pig intestine. *Journal of Nutrition* 143(3): pp. 274–283.

Halton, T.L., et al., 2008. Low-carbohydrate-diet score and risk of type 2 diabetes in women. *American Journal of Clinical Nutrition* 87(2): pp. 339–346.

Hansen, T.H., et al., 2012. Losses of essential mineral nutrients by polishing of rice differ among genotypes due to contrasting grain hardness and mineral distribution. *Journal of Cereal Science* 56(2): pp. 307–315.

Hansen, T.H., et al., 2012. Losses of essential mineral nutrients by polishing of rice differ among genotypes due to contrasting grain hardness and mineral distribution. *Cerea Chemistry* 56: pp. 307–15.

Harris, M.I., 2000. Health care and health status and outcomes for patients with type 2 diabetes. *Diabetes Care* 23(6): pp. 754–758.

Hodge, A.M., et al., 2004. Glycemic Index and dietary fiber and the risk of type 2 diabetes. *Diabetes Care* 27(11): pp. 2701–2706.

Hossain, P., B. Kawar and M. El Nahas, 2007. Obesity and diabetes in the developing world – a growing challenge. *New England Journal of Medicine* 356(3): pp. 213–215.

Hu, E.A., et al., 2012. White rice consumption and risk of type 2 diabetes: meta-analysis and systematic review. *BMJ* 344.

Impa, S., and T.S. Johnson-Beebou, 2010. Mitigating zinc deficiency and achieving high grain Zn in rice through integration of soil chemistry and plant physiology research. *Plant and Soil* 361: pp. 3–41.

Impa, S.M., et al., 2013. Zn uptake, translocation, and grain Zn loading in rice (Oryza sativa L.) genotypes selected for Zn deficiency tolerance and high grain Zn. *Journal of Experimental Botany* 64(10): pp. 2739–2751.

Inoue, H., et al., 2003. Three rice nicotianamine synthase genes, OsNAS1, OsNAS2 and OsNAS3, are expressed in cells involved in long-distance transport of iron and differentially regulated by iron. *Plant Journal* 36: pp. 366–381.

James, W.P.T., 2008. The fundamental drivers of the obesity epidemic. *Obesity Reviews* 9(s1): pp. 6–13.

Jiang, S.L., et al., 2007. Correlation analysis of mineral element contents and quality traits in milled rice (Oryza sativa L.). *Journal of Agricultural and Food Chemistry* 55(23): pp. 9608–13.

Johnson, A.A.T., et al., 2011. Constitutive Overexpression of the OsNAS Gene Family Reveals Single-Gene Strategies for Effective Iron- and Zinc-Biofortification of Rice Endosperm.

Juliano, B.O., et al., 1990. Some grain properties of IR36-based starch mutants. *Starch – Stärke* 42: pp. 256–260.

Kasemsuwan, T., et al., 1995. Characterization of the dominant mutant amylose-extender (Ae1–5180) maize starch. *Cereal Chemistry* 72(5): pp. 457–464.

Kopelman, P.G., 2000. Obesity as a medical problem. *Nature* 404(6778): pp. 635–643.
Krawinkel, M.B., 2007. What we know and don't know about Golden Rice. *Nature Biotechnology* 25(6): p. 623.
Lee, K.W., et al., 2006. The effects of Goami No. 2 rice, a natural fiber-rich rice, on body weight and lipid metabolism. *Obesity* 14(3): pp. 423–430.
Lee, S., et al., 2009. Iron fortification of rice seeds through activation of the nicotianamine synthase gene. *Proceedings of the National Academy of Sciences* 106(51): pp. 22014–22019.
Liang, J., et al., 2008. Milling characteristics and distribution of phytic acid and zinc in long-, medium- and short-grain rice. *Journal of Cereal Science* 48(1): pp. 83–91.
Liang, J., F, et al., 2007. Iron, zinc and phytic acid content of selected rice varieties from China. *Journal of the Science of Food and Agriculture* 87: pp. 504–10.
Liu, F., et al., 2009. The amylose extender mutant of maize conditions novel protein-protein interactions between starch biosynthetic enzymes in amyloplasts. *Journal of Experimental Botany* 60(15): pp. 4423–40.
Lopez AD, et al., 2006. *Global Burden of Disease And Risk Factors*. New York, Oxford University Press.
Lu, L., et al., 2013. Analysis of Metal Element Distributions in Rice (Oryza sativa L.) Seeds and Relocation during Germination Based on X-Ray Fluorescence Imaging of Zn, Fe, K, Ca, and Mn. *PLoS ONE* 8(2): p. e57360.
Lucca, P., R. Hurrell and I. Potrykus, 2002. Fighting iron deficiency anemia with iron-rich rice. *Journal of the American College of Nutrition* 21(3): pp. 184S–190S.
Lucca, P., S. Poletti and C. Sautter, 2006. Genetic engineering approaches to enrich rice with iron and vitamin A. *Physiologia Plantarum* 126(3): pp. 291–303.
Ma, G., et al., 2005. Phytate, calcium, iron, and zinc contents and their molar ratios in foods commonly consumed in China. *Journal of Agricultural and Food Chemistry* 53(26): pp. 10285–90.
Malik, V.S., W.C. Willett and F.B. Hu, 2013. Global obesity: trends, risk factors and policy implications. *Nature Reviews Endocrinology* 9(1): pp. 13–27.
Mikami, I., et al., 2008. Allelic diversification at the wx locus in landraces of Asian rice. *Theoretical and Applied Genetics* 116(7): pp. 979–989.
Misra, A., N. Singhal and L. Khurana, 2010. Obesity, the metabolic syndrome, and type 2 diabetes in developing countries: Role of dietary fats and oils. *Journal of the American College of Nutrition* 29(Suppl. 1): pp. 289S–2301.
Moghissi, A.A., S. Pei and Y. Liu, 2016.Golden rice: scientific, regulatory and public information processes of a genetically modified organism. *Critical Reviews in Biotechnology* 36(3): pp. 535–541.
Morris, M., et al., 2015. Relations to cognitive change with age of micronutrients found in green leafy vegetables. *The FASEB Journal* 29: pp. 260–263.
Munter, d.J.S., et al., 2007. Whole grain, bran, and germ Intake and risk of type 2 diabetes: A prospective cohort study and systematic review. *PLoS Medicine* 4(8): p. 261.
Murray-Kolb, L.E., et al., 2002. Transgenic rice is a source of iron for iron-depleted rats. *Journal of Nutrition* 132(5): pp. 957–960.
Nagengast, F., M. Grubben and I. Van Munster, 1995. Role of bile acids in colorectal carcinogenesis. *European Journal of Cancer* 31(7): pp. 1067–1070.
Nisak, M.Y.B., et al., 2010. Improvement of dietary quality with the aid of a low glycemic index diet in Asian patients with type 2 diabetes mellitus. *Journal of the American College of Nutrition* 29(3): pp. 161–170.
Noma, M., and M. Noguchi, 1976. Occurrence of nicotianamine in higher-plants. *Phytochemistry* 15: pp. 1701–1702.
Nugent, R., 2008. Chronic Diseases in Developing Countries. *Annals of the New York Academy of Sciences* 1136(1): pp. 70–79.
O'Keefe, S.J.D., et al., 2015. Fat, fibre and cancer risk in African Americans and rural Africans. *Nature Communications* 6.
Ou, J., et al., 2013. Diet, microbiota, and microbial metabolites in colon cancer risk in rural Africans and African Americans. *The American Journal of Clinical Nutrition* 98(1): pp. 111–120.

Pedersen, H.K., et al., 2016. Human gut microbes impact host serum metabolome and insulin sensitivity. *Nature* 535(7612): pp. 376–381.

Popkin, B.M., L.S. Adair and S.W. Ng, 2012. Global nutrition transition and the pandemic of obesity in developing countries. *Nutrition Reviews* 70(1): pp. 3–21.

Psaltopoulou, T., I. Ilias and M. Alevizaki, 2010. The role of diet and lifestyle in primary, secondary, and tertiary diabetes prevention: a review of meta-analyses. *Reviews of Diabetic Studies* 7: pp. 26–35.

Qu, L.Q., et al., 2005. Iron accumulation does not parallel the high expression level of ferritin in transgenic rice seeds. *Planta* 222(2): pp. 225–233.

Regina, A., et al., 2005. Starch branching enzyme IIb in wheat is expressed at low levels in the endosperm compared to other cereals and encoded at a non-syntenic locus. *Planta (Berlin)* 222(5): pp. 899–909.

Salwa, R., 2004. Improved plasma glucose control, whole-body Glucose utilization, and lipid profile on a Low-glycemic index diet in type 2 diabetic men. *Diabetes Care* 27: pp. 1866–1872.

Scholz, G., et al., 1992. Nicotianamine – a common constituent of strategies I and II of iron acquisition by plants: a review. *Journal of Plant Nutrition* 15: pp. 1647–1665.

Septiningsih, E.M., et al., 2009. Development of submergence-tolerant rice cultivars: the Sub1 locus and beyond. *Annals of Botany* 103(2): pp. 151–160.

Shaw, J.E., R. Sicree and P. Zimmet, 2010. Global estimates of the prevalence of diabetes for 2010 and 2030. *Diabetes Research and Clinical Practice* 87: pp. 4–14.

Shaw, J.G., and J.F. Friedman, 2011. iron deficiency anemia: focus on infectious diseases in lesser developed countries. *Anemia, Review Article* 2011: p. 10. doi:10.1155/2011/260380.

Shu, X.-l., et al., 2007. The influences of chain length of amylopectin on resistant starch in rice (*Oryza sativa* L.). *Starch – Stärke* 59(10): pp. 504–509.

Sommer, A., 2001. *Vitamin A Deficiency*. John Wiley & Sons Ltd.

Stevens, G.A., et al., 2013. Global, regional, and national trends in haemoglobin concentration and prevalence of total and severe anaemia in children and pregnant and non-pregnant women for 1995–2011: a systematic analysis of population-representative data. *Lancet Glob Health* 1(1): pp. e16–e25.

Sun, Q., et al., 2010. White rice, brown rice, and risk of type 2 diabetes in US men and women. *Archives of Internal Medicine* 170(11): pp. 961–969.

Tamura, T., and R.L. Goldenberg, 1996. Zinc nurture and pregnancy outcome. *Nutrition Research* 16: pp. 139–181.

Tang G, Y. Hu, S.-A. Yin, Y. Wang, G.E. Dallal, M.A. Grusak and R.M. Russell, 2012. β-Carotene in Golden Rice is as good as β-carotene in oil at providing vitamin A to children. *The American Journal of Clinical Nutrition* 96: pp. 658–64. Retraction in *Am. J. Clin. Nutr.* 102(3): p. 715.

Tang, G., et al., 2009. Golden Rice is an effective source of vitamin A. *American Journal of Clinical Nutrition* 89: pp. 1776–1783.

Tang, G., et al., 2012. β-Carotene in Golden Rice is as good as β-carotene in oil at providing vitamin A to children. *The American Journal of Clinical Nutrition* 96(3): pp. 658–664.

Tanumihardjo, S., et al., 2007. Poverty, obesity, and malnutrition: An international perspective recognizing the paradox. *Journal of the American Dietetic Association* 107(11): pp. 1966–1971.

Topping, D., M. Fukushima and A.R. Bird, 2003. Resistant starch as a prebiotic and synbiotic: state of the art. *Proceedings of the Nutrition Society* 62(1): pp. 171–176.

Topping, D.L. and P.M. Clifton, 2001. Short-chain fatty acids and human colonic function: Roles of resistant starch and nonstarch polysaccharides. *Physiological Reviews* 81(3): pp. 1031–1064.

Topping, D.L., 2007. Cereal complex carbohydrates and their contribution to human health. *Journal of Cereal Science* 46(3): pp. 220–229.

Trijatmiko, K.R., et al., 2016. Biofortified indica rice attains iron and zinc nutrition dietary targets in the field. *Scientific Reports* 6: p. 19792.

Tsihlias E, et al., 2000. Comparison of high and low-glycemic-index breakfast cereals with monounsaturated fat in the long term dietary management of type 2 diabetes. *American Journal of Clinical Nutrition* 72: pp. 439–449.

Tucker, G., 2003. Nutritional enhancement of plants. *Current Opinion in Biotechnology* 14(2): pp. 221–225.

Vasconcelos, M., et al., 2003. Enhanced iron and zinc accumulation in transgenic rice with the ferritin gene. *Plant Science* 164(3): pp. 371–378.

Vipperla, K., and S.J. O'Keefe, 2012. The Microbiota and its metabolites in colonic mucosal health and cancer risk. *Nutrition in Clinical Practice* 27(5): pp. 624–635.

Virk, P. and G. Barry, 2007. *Biofortified Rice – Towards Combating Human Micronutrient Deficiencies.* International Rice Research Institute, DAPO Box 7777, Metro Manila, Philippines.

Vos, T., et al., 2015. Global, regional, and national incidence, prevalence, and years lived with disability for 301 acute and chronic diseases and injuries in 188 countries, 1990–2013: a systematic analysis for the Global Burden of Disease Study 2013. *The Lancet* 386(9995): pp. 743–800.

Wallis, K., 2004. Resistant starch – a natural functional food ingredient. *Food Australia* 56(6): p. 242.

Wang, Y.-J., R. Kozlowski, and G.A. Delgado, 2001. Enzyme resistant dextrins from high amylose corn mutant starches. *Starch – Stärke* 53(1): pp. 21–26.

Wei, C., et al., 2010. Microstructure and ultrastructure of high-amylose rice resistant starch granules modified by antisense RNA inhibition of starch branching enzyme. *Journal of Agricultural and Food Chemistry* 58(2): pp. 1224–1232.

Welch, R.M., and R.D. Graham, 2004. Breeding for micronutrients in staple food crops from a human nutrition perspective. *Journal of Experimental Botany* 55(396): pp. 353–64.

White, P.J., and M.R. Broadley, 2005. Biofortifying crops with essential mineral elements. *Trends in Plant Science* 10(12): pp. 586–93.

WHO, 2002. Health situation in the South East Asia Region 1998–2000, Regional Office for South East Asia, New Delhi.

Wissuwa, M., A.M. Ismail and R.D. Graham, 2008. Rice grain zinc concentrations as affected by genotype, native soil-zinc availability, and zinc fertilization. *Plant Soil* 306: pp. 37–48.

Yang, C.Z., et al., 2006. Starch properties of mutant rice high in resistant starch. *Journal of Agricultural and Food Chemistry* 54: pp. 523–528.

Yang, J., K. Xu and L. Rodriguez, 2014. The rejection of science frames in the news coverage of the golden rice experiment in Hunan, China. *Health, Risk & Society* 16(4): pp. 339–354.

Yano, M., et al., 1985. High amylose mutants of rice, *Oryza sativa* L. *Theoretical & Applied Genetics* 69(3): pp. 253–257.

Yao, N., A.V. Paez and P.J. White, 2009. Structure and function of starch and resistant starch from corn with different doses of mutant amylose-extender and floury-1 alleles. *Journal of Agricultural and Food Chemistry* 57(5): pp. 2040–2048.

Yeh, J.Y., D.L. Garwood and J.C. Shannon, 1981. Characterization of starch from maize endosperm mutants. *Starch – Stärke* 33(7): pp. 222–230.

Yoon, K., et al., 2006. Epidemic obesity and type 2 diabetes in Asia. *The Lancet* 368(9548): pp. 1681–1688.

Zhu, C.H., et al., 2016. Consumption of canned citrus fruit meals increases human plasma β-cryptoxanthin concentration, whereas lycopene and β-carotene concentrations did not change in healthy adults. *Nutrition Research* 36(7): pp. 679–688.

Chapter 8

The nutraceutical properties of rice

Lu Yu, University of Maryland, USA; and Margaret Slavin and Mengyi Dong, George Mason University, USA

1 Introduction

Rice is the third largest crop globally and provides about 20% of the energy supply to the world's population (IRRI, 2015). It is an essential part of the diet for many countries around the world, especially in Asia and Africa. This chapter reviews the following topics relating to the nutritional quality of rice:

- Rice varieties and variations in nutritional content
- Bio-fortification
- Antioxidants in rice
- The impact of breeding, cultivation and processing on nutritional quality

The principal components of rice grain are:

- Husk (or hull): the outer covering
- Bran: a hard outer layer of rice grain, which can be separated after milling
- Endosperm: a reserve tissue inside the seeds of rice grain
- Germ (embryo): reproductive part of rice grain that germinates to grow into a plant

These are shown in Fig. 1, which summarizes their nutritional function and value.

http://dx.doi.org/10.19103/AS.2016.0011.31

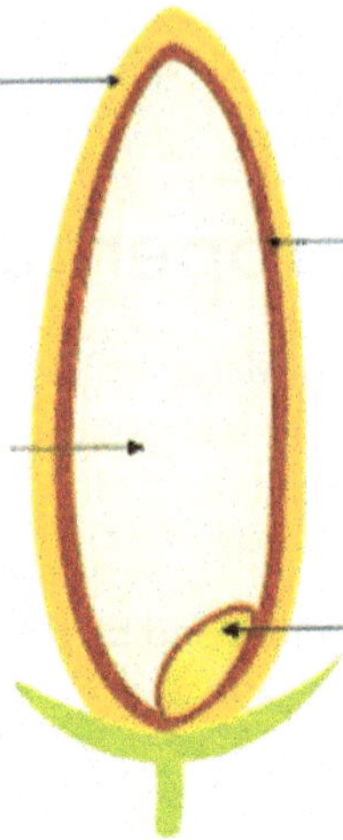

Figure 1 Rice grain components.

2 Rice varieties, nutritional content and bio-fortification

There are an estimated 120000 different rice varieties in the world. Rice can be classified into two major cultivated species and twenty-three wild species (IRRI.org). The two major cultivated species are (Khush, 1997) *Oryza sativa* (Asian rice), which is grown worldwide, and *Oryza glaberrima* (African rice), which is grown in West Africa.

Asia contains 60% of the world's population and grows more than 90% of the world's rice (Londo et al., 2006). Asian rice (*O. sativa*) feeds more than one-half of the world's population (Chen et al., 2004). Based on geographical analysis, two major subspecies for *O. sativa* are *O. sativa indica* (domesticated in regions such as eastern India, Myanmar and Thailand) and *O. sativa japonica* (domesticated in southern China).

A large array of intermediates between these two groups also exists (Second, 1982; Garris et al., 2005). Rice can also be classified into groups according to colour: white rice, brown rice, black rice, purple rice and red rice (Oka, 1988).

The genetic differences between cultivated varieties of rice contribute greatly to variations in nutritional content (Tables 1 and 2). An example is carbohydrate content such as starch: *indica* rice, for example, generally contains more amylose than *japonica* rice (Houston and Kohler, 1970). In one study of brown rice using eight different sets of genotypes (n = 1138), mineral concentrations of iron and zinc varied by up to four-fold, from 6.3 to 24.4 μg/g for iron and 13.5 to 58.4 μg/g for zinc, respectively (Gregorio, 2002). In addition, across a number of rice varieties (Forster et al., 2013), the following variations are observed:

- Total lipid content varies from 5 to 16%, with individual fatty acids ranging from 1.4- to 1.9-fold differences.
- Vitamin E isoforms (α-, γ-, δ-tocotrienols and tocopherols) showed 1.3- to 15.2-fold differences.
- γ-oryzanol and total phenolics differ between 100–275 ng/mg and 57–146 ng gallic acid equivalent/mg, respectively.

Table 1 Selected antioxidant profile of rice varieties

Rice varieties	Bran colour	Subpopulation	Total phenolics (ng/mg)	γ-Oryzanol (ng/mg)	Vitamin E isoforms (ng/mg)
Jasmine 85	Brown	*Indica*	4.50 ± 0.10	247 ± 11	339 ± 2
IAC 600	Purple	*Indica*	6.35 ± 0.93	197 ± 15	323 ± 2
Red wells	Red	*Tropical japonica*	13.56 ± 0.06	274 ± 11	350 ± 1
IL 121-1-1	Red	*Tropical japonica*	13.64 ± 0.05	100 ± 19	322 ± 2
Cypress	Brown	*Tropical japonica*	5.40 ± 0.06	195 ± 14	367 ± 2.48
Wells	Brown	*Tropical japonica*	~3.99 ± 0.11	252 ± 10	330 ± 4
Shu 121	Brown	*Indica*	3.99 ± 0.11	110 ± 6	281 ± 1

Source: Forster et al., 2013, p. 1546.

Table 2 Rice bran fatty acid profiles of selected varieties (% of total lipid reported as % of rice bran)

Rice varieties	Jasmine 85	IAC 600	Red wells	IL 121-1-1	Cypress	Wells	Shu 121
Total lipid (%)	5.02	9.8	9.96	12.1	12.3	12.6	16.2
Myristic acid (14:0)	0.31	0.36	0.31	0.34	0.26	0.26	0.35
Palmitic acid (16:0)	18.95	16.63	13.73	14.21	13.25	14.76	19.7
Palmitoleic acid (16:1)	0.22	0.22	0.16	0.14	0.15	0.17	0.19
Stearic acid (18:0)	2.22	3.06	2.32	2.84	1.91	1.63	1.81
Oleic acid (18:1n9)	44.23	38.89	43.32	43.5	44.3	42.36	40.78
Vaccenic acid (18:1n7)	0.98	1.44	0.99	0.95	0.93	1.02	1.74
Linoleic acid (18:2n6)	29.15	34.82	35.22	33.8	35.5	36.27	31.08
α-Linolenic acid (18:3n3)	1.2	1.09	1.09	0.97	1.22	1.17	1.27
Arachidic acid (20:0)	1.08	1.27	1	1.37	0.85	0.75	1.46
Gadoleic acid (20:1)	0.61	0.55	0.69	0.67	0.72	0.7	0.66
Behenic acid (22:0)	0.41	0.48	0.4	0.44	0.32	0.32	0.4
Lignoceric acid (24:0)	0.64	0.81	0.78	0.68	0.58	0.6	0.57

Source: Forster et al., 2013, p. 1550.

Brown, red and purple/black rice usually contain greater amounts of phenolic acids compared with white rice (Tian et al., 2004; Zhou et al., 2004). Purple/black rice has been reported to contain bioactive compounds, such as anthocyanins, not seen in any other type of rice (Hou et al., 2013).

Rice represents the world's most important cereal crop with regard to the supply of calories (Londo et al., 2006). It is responsible for more than one-fifth of the calories consumption worldwide (Smith, 1998). Generally, 100 g of rice contains 10 g of water, 7 g of rice protein, less than 3 g of fat, about 80 g of carbohydrate, 3 g of fibre and small amounts of vitamins B, E and K, as well as micronutrients (Nutrient Data Laboratory, USDA, 2015). Each gram of rice supplies about 3.6 kcal of energy. Previous research has also reported many types of chemical compounds from rice, including phenolic acids, phytosterols, unsaturated fatty acids and so on (Tian et al., 2004; Dunford and King, 2000; Oluremi et al., 2013). Oil extracted from rice bran contains high levels of unsaturated fatty acids, including oleic acid and linoleic acid, which is beneficial to human health as it lowers the levels of total cholesterol and low-density lipoprotein (LDL) in the blood.

From a nutritional stand point, no single food is perfectly balanced. Despite being a great source of carbohydrate fuel for human beings, rice does not provide sufficiency when it comes to some vitamins, minerals and proteins. Rice, like other cereals, is essentially devoid of vitamins A and D (Houston and Kohler, 1970). Although brown rice has adequate amount of vitamins of the B complex, (milled) white rice has an inadequate amount of thiamine (vitamin B1), riboflavin (vitamin B2) and niacin (vitamin B3) (Houston and Kohler, 1970). Iron and calcium content is low in brown rice and even lower in white rice (Houston and Kohler, 1970). It is estimated that, based on the daily dietary allowances recommended by the National Research Council (National Research Council, 1989), an adequate intake of thiamine and riboflavin is necessary and a daily intake of about 2 kg or 6 kg of white rice, respectively, would meet this requirement (Houston and Kohler, 1970).

The lack of certain nutritional aspects in rice can be overcome by rice enrichment and fortification. Rice enrichment and fortification have been in commercial use for years in the United States and other countries of the world (Dexter, 1998). Under the regulation of the US Food and Drug Administration (2015), enriched rice can provide 2–4 mg of thiamine, 1.2–2.4 mg of riboflavin, 16–32 mg of niacin, 0.7–1.4 mg of folic acid and 13–26 mg of iron in each pound of rice. Besides, each pound of enriched rice may contain 250–1000 U.S.P. units of vitamin D and 500–1000 mg of calcium. A remarkable benefit of consuming fortified rice is the huge reduction of beriberi incidence (by 89%), as evidenced in a large-scale field trial of rice enriched with 2 mg of thiamine, 13 mg of iron and 16 mg of nicotinic acid in the Philippines (Salcedo et al., 1950). Another example is the fortification of rice with iron, which showed a significant increase in the amount of iron absorption after consuming iron-fortified rice. In addition, a clinical trial showed a greater increase in haemoglobin in subjects who received iron-fortified rice than in those who did not (Florentino et al., 1998).

One type of rice enrichment technology involves spraying a liquid mix (e.g. containing select micronutrients such as thiamine) to the rice in several layers to form a rice premix with a waxy layer, which prevents the micronutrients from rinsing off (Dexter, 1998). Chemical modification is another effective strategy to improve the bioactive property of rice. Evidence from the research of Wang et al. indicates that sulphated defatted rice bran polysaccharides prepared by chlorosulphonic acid–pyridine (CSA–Pyr) method exhibited relatively stronger anti-tumour activity in vitro (Wang et al., 2009).

3 Antioxidants in rice

Bioactive compounds in rice, especially from brans and hulls, are reported to have numerous health-promoting effects, including anti-inflammation, anti-cancer, cholesterol-lowering and anti-allergic activities in cells, animals and humans (Friedman, 2013; Akihisa et al., 2000). Rice contains a number of phenolic components, which contribute to the antioxidant activity of rice. There is increasing evidence that intake of natural antioxidants is closely associated with lowered risks of many types of chronic diseases (Iqbal et al., 2005). Dietary rice has bioactive phytochemicals with health-promoting effects, including the inhibition of inflammation, prevention of cancers, blood pressure control and cholesterol-lowering properties (Roschek et al., 2009; Henderson et al., 2012; Kim et al., 2013; Ha et al., 2005).

Compared to pearled rice (without husk and bran), which is most commonly consumed for a high content of starch, the whole grain with husk and bran layer contains more bioactive compounds (Butsat and Siriamornpun, 2010). The major health-promoting components in rice include sterols, oryzanols, tocopherols, tocotrienols and phenolic compounds such as flavonoids (Aguilar-Garcia et al., 2007; Chotimarkorn et al., 2008). A number of components promote antioxidant activity to rice, and particularly in rice bran, namely:

- γ-oryzanol;
- phenolic acids and flavonoids;
- the vitamin E family tocopherols and tocotrienols; and
- anthocyanins and proanthocyanidins, in the case of pigmented rices.

In addition, rice bran comprises some special phenolic acids that are not present in significant quantities in fruit and vegetables, such as ferulic acid, *p*-coumaric acid and diferulate (Lin and Lai, 2011).

Gamma (γ) oryzanol is a major lipophilic antioxidant present in rice, with a higher concentration than the vitamin E family and with a higher antioxidative activity (Chen and Bergman, 2005; Xu et al., 2001). Gamma oryzanol refers to a mixture of a phytosterol phenolates, consisting of a phenolic acid bound to a phytosterol or triterpene alcohol by an ester bond. *Trans*-ferulic acid is joined to one of four phytosterols (to a phytosterol ferulate): predominately cycloartenol, β-sitosterol, 24-methylenecycloartenol or campesterol; though other combinations are found as minor components (Lerma-García et al., 2009). Because of this chemical composition, γ-oryzanol is of interest for both its anti-hypercholesterolaemic and its antioxidative activities (Lerma-García et al., 2009).

Rice bran contains the vast majority of γ-oryzanol content in the grain, to the extent that white rice contains less than 1% of the per weight concentration of bran, at 26.12 and 3 296.50 mg/kg, respectively (Tuncel and Yılmaz, 2011). Levels of γ-oryzanol in the grain are known to vary by genotype and due to environmental factors. In one study, the amount of γ-oryzanol varied from 0.180 to 0.627 mg/g in brown rice (Lerma-García et al., 2009).

The antioxidant activities of γ-oryzanol and its individual components have been demonstrated in a variety of assays, ranging from purely chemical to fatty acid and cholesterol oxidation models (Nyström et al., 2005; Xu et al., 2001), to in vitro tissue assays. Interestingly, the lipophilic antioxidant activity of synthetic phytosterol ferulates was found to have a higher antioxidant activity than ferulic acid alone, as assessed by the oxygen radical absorbance capacity assay and their ability to prevent thiobarbituric acid reactive substances accumulation in a cooked pork product (Tan and Shahidi, 2011). Their use as antioxidants has thus been suggested in a lipophilic, albeit not bulk oil, environment (Tan and Shahidi, 2014).

Phenolics exist in rice bran in two subgroups: phenolic acids and flavonoids. Phenolic acids are primarily found in the bran, mostly bound to the fibrous cellulose matrix, followed by soluble conjugates and finally the least amount as a soluble-free form (Shao and

Bao, 2015). The complete analysis of phenolic acids requires sequential extraction and hydrolysis steps; moreover, the bioavailability and health impact of the different fractions may vary.

Phenolic acids are a series of chemical components that have potent antioxidant properties and free radical scavenging capabilities. Phenolic acids are reported to have been identified and isolated from rice bran or whole rice samples, including protocatechuic, hydroxy-benzoic, vanillic, syringic, ferulic, *p*-coumaric, caffeic and sinapinic acid (Tian et al., 2004; Oluremi et al., 2013). The main rice phenolic acids are ferulic, *p*-coumaric and sinapic acids (Shao and Bao, 2015). These phenolic acids have been recognized as one of the major bioactive compounds in rice.

Of the flavonoids, anthocyanins are present in black rice (sometimes called purple), with the vast majority in the bran fraction, whereas red rice contains only minimal amounts (Shao and Bao, 2015). Cyanidin-3-glucoside (Cy-3-glc) is the primary anthocyanin in pigmented rice, followed by peonidin-3-glucoside (Deng et al., 2013; Escribano-Bailón et al., 2004). Anthocyanin content, even in pigmented varieties of the same colour, varies widely. Shao and Bao report a range of 0.05–2.3 mg Cy-3-glc/g dry weight of whole grain black rice across numerous studies, with the extraction using acidified methanol (Shao and Bao, 2015). Also, proanthocyanidins appear in red rice and in some black rice varieties. In one study of red rice, the total proanthocyanidin content ranged from 191 to 2273 μg/g of red whole grain rice by dry weight (Shao and Bao, 2015).

The vitamin E family is also present primarily in the rice bran fraction. Both tocopherols and tocotrienols are present in significant quantities, whereas great variation in ratios exists across varieties, growing conditions and extraction methods (Lerma-García et al., 2009). The predominant tocol in rice, rice bran and rice bran oil varies between α and γ isomers of tocopherols and tocotrienols. Total tocols in a brown rice sample was 43–236 μg/g of brown rice grain, whereas concentrations in rice bran oil were considerably higher, exceeding 1000 μg/g in some samples (Lerma-García et al., 2009) (Table 3).

Table 3 Selected antioxidant content from rice fractions (TEAC represents Trolox equivalent antioxidant capacity. DPPH represents 2,2-diphenyl-1-picrylhydrazyl as free radicals)

Rice fractions antioxidant content	Raw seed	Bran	Brown rice (cargo)	Unpolished rice	White rice	Chalky rice (milky piece)
γ-Oryzanol (mg/kg)	454.60 ± 25.81ab	3 296.50 ± 83.12a	408.90 ± 64.46ab	152.69 ± 7.73bc	26.12 ± 0.85cd	12.19 ± 0.13d
Total free phenol (mg ferulic acid equivalents/g)	0.91 ± 0.02ab	4.12 ± 0.13a	0.75 ± 0.01bc	0.43 ± 0.01cd	0.30 ± 0.01de	0.14 ± 0.01e
Total free phenolic acids (μg/g)	2.96	3.48	1.85	1.23	0.82	0.91
TEAC (mM trolox/g)	15.12 ± 0.48ab	58.65 ± 0.89a	9.82 ± 0.29bc	6.58 ± 0.20cd	4.93 ± 0.16de	2.76 ± 0.04e
DPPH scavenging rate (%)	70.07 ± 0.77ab	92.01 ± 0.38a	64.42 ± 0.70bc	61.71 ± 0.31cd	60.39 ± 0.23de	36.74 ± 0.46e

Source: Tuncel and Yılmaz, 2011, p. 582.

Free radicals are a group of molecules with unpaired electrons that interact with various tissue components, resulting in damage to cell structures and DNA and causing protein dysfunction. (Kehrer, 1993; Valko et al., 2007). Free-radical-induced oxidative stress is closely associated with chronic diseases (Rao et al., 2010). The anti-inflammatory and anti-cancer effects of rice bran might be attributed to the free radical scavenging capacity of bioactive components through the blocking of chronic inflammatory responses or cancer cell proliferation (Henderson et al., 2012; Ravishankar et al., 2013).

In 2013, isoferulic, syringic and chlorogenic acids were observed in red rice for the first time. Correlated with the contents of newly discovered and other antioxidant components, including *p*-coumaric, syringic, vanillic, ferulic and isoferulic acids, significant anti-inflammatory potential was first reported in red rice (Niu et al., 2013). An in vitro study of extract of defatted rice bran from rice species *Oryza sativa L.* has reported that defatted rice bran may have a significant anti-inflammatory effect through reducing the expression of pro-inflammatory factors and lipopolysaccharide (cell wall component from Gram-negative bacteria) induced signalling pathway (Kim et al., 2013).

The chemical profiles and anti-cancer activities of rice vary among rice varieties (Forster et al., 2013). The differential chemopreventive bioactivities of rice against cancer development are considered to be positively correlated with the phytochemical contents such as phenols and flavonoids in rice, especially in the bran and hull from pigmented rice (e.g. black and red rice) (Baek et al., 2015). Bran and hull extracts from pigmented rice showed significantly ($p < 0.05$) higher contents of phenolics and flavonoids and greater inhibitive effect on growth of human breast, melanoma and oral cancer cells than extracts from un-pigmented rice (e.g. light brown rice) (Baek et al., 2015). In addition, γ-oryzanol, which contains sterols and ferulic acids called cycloartanyl ferulate or triterpene alcohol ferulate, is unique to the rice plant and reported of anti-cancer properties in many animal studies (Srinivasan et al., 2007; Kim et al., 2013).

4 Breeding and cultivation techniques to improve the nutritional value of rice

Conventional breeding methods have been used to improve the level of minerals and vitamins in crops. The Harvest-Plus has invested $14 million annually in conventional breeding of 12 crops to improve the vitamin A, iron and zinc content (Pfeiffer and McClaffery, 2007). The first zinc-fortified rice was bred and released in Bangladesh in 2013, which contains 20 parts per million (ppm) zinc, with the aim of increasing the zinc concentration from 16 to 28 ppm (Bouis et al., 2011). However, breeding methods are

Table 4 Nutritionally enhanced cultivars from conventional breeding

Rice varieties	Enhanced nutrients	Reference
Indica	Provitamin A	Sautter et al., 2006
Japonica	Iron	Lucca et al., 2002
Indica	Iron, zinc	Graham et al., 1999

limited by genetic variation in the gene pool. Distant relatives are used in breeding, requiring many years to breed commercial cultivars (Hirschi, 2009) (Table 4).

Transgenic method is used when genetic variation of nutrient content is not distinct (Hirschi, 2009; Zhu et al., 2007). Genetic technologies, including whole-genome sequencing, high-throughput physical mapping, global gene expression analysis and metabolite profiling, can quickly identify and characterize the function of DNA fragments, and this information can be used to modify plants (Christou and Twyman, 2004; Galili et al., 2002). For example, Lucca et al. used ferritin gene from *Phaseolus vulgaris* (common bean) to successfully double the iron content of rice grain (Lucca et al., 2002). Table 5 shows some examples of the use of transgenic technology to increase or add certain nutrient content in rice. Transgenic methods are currently limited by consumer concerns about safety and strict regulatory requirements (Hirschi, 2009; Powell, 2007).

Growing conditions and cultivation techniques have also been shown to have an impact on nutritional content. An example is climate change. During the past 100 years, the CO_2 concentration in the atmosphere has increased from 290 to 375 μmol/mL and is expected to reach 550 μmol/mL by 2050. The increase of CO_2 concentration in the atmosphere will have an impact on crop growth, yield and quality (Shao and Bao, 2015). High CO_2 levels have been linked to reduced phenolic and total flavonoid content of brown rice, white rice and rice bran (Goufo et al., 2014).

Rice germination has been shown to affect the nutritional value and antioxidant content of grains (Andarwulan et al., 1999; Yang, 2001). Germination has been reported to increase vitamin C, vitamin E, ferulic acid and total phenolic acid contents in wheat and *Pangium edule* Reinw (Andarwulan et al., 1999; Yang, 2001). Germinated brown rice was reported to have higher nutrition value, better flavour and better digestion and absorption properties than white rice and brown rice (Zhang et al., 2005). The germinating process has been reported to increase the nutritional value, reduce the phytic acid and oligosaccharides, increase protein digestibility and improve the sensory attributes of rice seeds (Eyzaguirre et al., 2006; Khattak et al., 2007; Lintschinger et al., 1997; Rubio et al., 2002; Schulze et al., 1997). Tian et al. (2004) studied the phenolic antioxidants in white rice, brown rice and germinated brown rice. When comparing white rice with brown rice and germinated brown rice, almost all phenolic compounds are more abundant in brown rice and germinated brown rice. During the germinating process, ferulic acid (0.48 mg/100 g flour), sinapinic

Table 5 Cases of rice nutritionally fortified using transgenic methods

Nutrients	Content	References
Flavonoids	Increased flavonoid content	Shin et al., 2006
Glycinin	Increased protein content	Katsube et al., 1999
Iron	Increased to twofold of the original Increased bioavailability	Lucca et al., 2002
Starch	Increased amylose Decreased amylopectin Increased phosphorus	Schwall et al., 2000
Provitamin A	Add β-carotene biosynthetic pathway	Ye et al., 2000
α-Linolenic acid	Increased omega-3 fatty acid content	Anai et al., 2003

acid (0.21 mg/100 g flour) and insoluble phenolic compounds (24.78 mg/100 g flour) significantly increased (Tian et al., 2004).

Germination of rice significantly improved the levels of γ-oryzanol, α-tocopherol and phenolic contents by two, three and five times, respectively, compared with that of untreated rice bran (Sungsopha et al., 2009). During germination, a significant increase of free phenolic acid was observed, and the ferulic acid content of brown rice was increased by 1.5 times, making it the most abundant phenolic compound in germinated brown rice. The total content of insoluble phenolic compounds in brown rice was increased by 1.34 times after germination (Tian et al., 2004).

The insoluble phenolic compounds (particularly hydroxycinnamates) are ester linked to polysaccharides in the cell wall, which contains insoluble fibre, polysaccharides and lignin components (Faulds and Williamson, 1999; Shibuya, 1984). The reason for the increase in insoluble phenolics can be that, during germination, the cell wall is dismantled and the free forms of phenolic compounds are released by alkaline hydrolysis (Tian et al., 2004). It has been reported that germination can reduce the phytic acid level by 67–87% in sorghum and up to 60% in rice (Liang et al., 2008; Mahgoub and Elhag, 1998). The decrease of phytic acid leads to an increase of bioavailability of minerals. For example, the HCl-extractability of calcium, iron and zinc in pearl millet was reported to increase as the germination time increased (Badau et al., 2005).

Gamma-aminobutyric acid (GABA) is a non-protein amino acid and an important functional compound found in brown rice. GABA content in brown rice can be significantly increased during the germinating process, and the accumulation of GABA will reach the highest level at 72 h of germination under hypoxic conditions, ten times greater than milled rice (Ding et al., 2011; Kayahara et al., 2001; Moongngarm and Saetung, 2010; Zhang et al., 2014). Brown rice is also reported to have significantly higher dietary fibre, ferulic acid, tocotrienols, magnesium, potassium, zinc, γ-oryzanol and prolylendopeptidas inhibitor content after germinating (Kayahara et al., 2001).

Fertilizer has also been used during planting to enhance the mineral content of rice. The effectiveness of fertilizer depends on factors such as application methods, soil composition, geographical locations and the mobility of metal ions in soil and in plants (Hirschi, 2009; Zhu et al., 2007). As iodine and selenium have high mobility in soil and plants, they are successfully used as fertilizers for vegetable and cereal, respectively (Dai et al., 2004; Hartikainen, 2005). Iron fertilizer has not been successfully used due to its low mobility in soil (Grusak and DellaPenna, 1999). However, micronutrient fertilizers can damage the environment, and a large quantity of metal fertilizer in soil can be harmful to the growth of plant as well as other organisms in soil (Grusak and Cakmak, 2004).

5 The effects of processing and cooking on the nutritional value of rice

Bioactive components in rice are affected by different preparation/processing techniques (Deng et al., 2013; Tian et al., 2004). Dehulling and milling are crucial steps during rice processing for better cooking characteristics, cooking texture and quality. However, dehulling has been shown to result in a loss about 20% and 60% of the total antioxidants (such as tocols, γ-oryzanols and polyphenols) of the original raw white and red rice, respectively (Dors et al., 2011; Shao and Bao, 2015). It has been shown that milling process caused a loss of 40% antioxidants and bioactive components in red rice, while undermilling of rice slightly reduced the antioxidant loss in red rice by 2% as compared to milling (Finocchiaro et al., 2007). Thus, the consumption of whole rice or partially milled rice would be suggested for customers to preserve antioxidants and bioactive components.

Fermentation is widely used in grain processing and is associated with reductions in phytic acid, which affects mineral availability (Leenhardt et al., 2005; Mahgoub and Elhag, 1998). Animal studies have shown that the bioavailability of zinc and iron in wheat flour increased by more than 30% after fermentation (Lopez et al., 2003). As reported by Liang et al. (2008), fermentation can reduce the phytic acid by up to 95% in brown rice, thus increasing the availability of mineral nutrients. In addition, the stale rice, typically considered a discarded waste part, exhibited greater level of nutrients and superoxide dismutase activity after solid-state fermentation with *Cordyceps sinensis* (Zhang et al., 2008). The application of enzymatic fermentation treatment switches the low-valued stale rice to a good resource for health-beneficial bioactive material. The observations indicate that appropriate germination and enzymatic treatment of rice might be suggested as methods to improve the availability of bioactive components and health-related benefits from rice.

The habit of eating rice varies in different countries. Individual preference and traditional culture differs with the cooking methods of rice dishes in different regions. Rice is usually eaten plain in East Asian countries (China, Japan and Korea). In West Asia and the Middle East, rice is flavoured with oil and spices. And in Europe, 'risotto' and 'paella' are the most famous rice dishes (Keiko et al., 1997). Cooking results in a dramatic decrease of nutrients. Total antioxidant capacity of dehulled red rice is shown to have lost almost completely when rice is boiled (Finocchiaro et al., 2007). One study group investigating the influence of water cooking methods using rice cooker, pressure cooker or a gas range on anthocyanins in black rice has shown that all cooking methods caused significant ($P < 0.001$) degradation of cyanidin-3-glucoside and an increased production of protocatechuic acid (Hiemori et al., 2009). Among all of the cooking methods, pressure cooking led to the greatest loss of anthocyanins (79.8%), followed by the rice cooker (74.2%) and gas range (65.4%) (Hiemori et al., 2009). As compared to other traditional water cooking methods, which cause consistent loss of nutrients, Italian procedure (risotto cooking) has shown to have significantly reduced the nutrients loss in red rice and to have not affected the total antioxidant capacity of white rice before and after water cooking (Finocchiaro et al., 2007). Water cooking is a predominant cooking method for the majority of rice consumers, but as compared among all of the cooking methods, risotto cooking results in the better availability of nutritional properties of rice.

Extraction technologies such as solvent extraction (methanol), supercritical fluid extraction (supercritical carbon dioxide) and microwave-assisted (microwave with isopropanol and hexane) extraction have been used to extract polyphenols from rice bran (Shen et al., 1997; Chotimarkorn et al., 2008; Zigoneanu et al., 2008). However, these extraction methods use a large amount of chemicals and require a long extraction period (Tabaraki and Nateghi, 2011).

Ultrasound technology is recognized as an environment-friendly technique and has gained much attention in recent years. Extraction using low frequency ultrasound (20–100 KHz) can provide high reproducibility in shorter time; simplify manipulation; and reduce solvent consumption, temperature and energy input (Chemat et al., 2008). Ultrasound may increase the extent of hydroxylation, which releases antioxidants from the fibre. It has been applied to herb and fruit studies and has succeeded in improving the availability of antioxidants (Liu et al., 2005; Carrera et al., 2012; d'Alessandro et al., 2012). An optimal extraction condition was reported by Tabaraki and Nateghi (2011) to extract antioxidants from rice bran, which is 65–67% ethanol using ultrasound at 51–54 °C for 40–45 min.

6 Future trends

Rice has been recognized as an excellent source of natural components with health-promoting functions. Significant levels of antioxidant activities and phenolic components have been detected in rice. Current research interest has been placed more on the pigmented rice and rice bran/hull for higher nutritional benefits, though polished rice is more preferred by customers. And using these approaches that help with the enhancement of available nutraceutical values has drawn great attentions from researchers.

Despite the numerous studies of chemical composition, antioxidant capacity and health-beneficial effects of rice, most of the current researches are focusing on in vitro analyses, resulting in limited information about the in vivo observation, which is more physiologically related to human health. With the nationwide growth of interest in health, more focus will be placed on the biofunctions of rice, especially for the disease prevention and health promotion purpose. In vivo or clinical data collected will be used more frequently for the functionality study of rice. Second, the discovery of novel chemical components (such as γ-oryzanol) with newly discovered functional properties would be another trend for researchers. Chemical or biological modifications of natural components from rice with novel biofunctions are likely to generate future research interest. Third, as nutrient loss and degradation during rice processing is still problematic, study of novel methods for the inhibition of the loss of nutraceuticals and improvement of health-beneficial effects of rice would be of importance for health concern and market interest.

7 Where to look for further information

A concise overview of the nutraceutical properties of rice is provided by Bao (2012). A broader overview of the subject can be found in Watson et al. (2014). The use of bio-fortification is reviewed in Brooks (2010). Developments in breeding to improve nutritional properties are discussed in Novartis (2001).

8 References

Aguilar-Garcia, C., Gavino, G., Baragaño-Mosqueda, M., Hevia, P. and Gavino, V. (2007). Correlation of tocopherol, tocotrienol, c-oryzanol and total polyphenol content in rice bran with different antioxidant capacity assays. *Food Chem.*, 102, 1228–32.

Akihisa, T., Yasukawa, K., Yamaura, M., Ukiya, M., Kunura, Y., Shimizu, N. and Arai, K. (2000). Triterpene alcohol and sterol ferulates from rice bran and their anti-inflammatory effects. *J. Agric. Food Chem.*, 48(6), 2313–9.

Anai, T., Koga, M., Tanaka, H., Kinoshita, T., Rahman, S. and Takagi, Y. (2003). Improvement of rice (*Oryza sativa L.*) seed oil quality through introduction of a soybean microsomal omega-3 fatty acid desaturase gene. *Plant Cell Rep.*, 21(10), 988–92.

Andarwulan, N., Fardiaz, D., Wattimena, G. A. and Shetty, K. (1999). Antioxidant activity associated with lipid and phenolic mobilization during seed germination of Pangium edule Reinw. *J. Agric. Food Chem.*, 47(8), 3158–63.

Badau, M. H., Nkama, I. and Jideani, I. A. (2005). Phytic acid content and hydrochloric acid extractability of minerals in pearl millet as affected by germination time and cultivar. *Food Chem.*, 92(3), 425–35.

Baek, J. A., Chung, N. J., Choi, K. C., Hwang, J. M. and Lee, J. C. (2015). Hull extracts from pigmented rice exert antioxidant effects associated with total flavonoid contents and induce apoptosis in human cancer cells. *Food Sci. Biotechnol.*, 24 (1), 241–7.

Bao, J. (2012). Nutraceutical properties and health benefits of rice. In L. Yu, R. Tsao and F. Shahidi (eds), *Cereals and Pulses: Nutraceutical Properties and Health Benefits.* Wiley-Blackwell, New York, USA. 37–64.

Bouis, H. E., Hotz, C., McClafferty, B., Meenakshi, J. V. and Pfeiffer, W. H. (2011). Biofortification: a new tool to reduce micronutrient malnutrition. *Food Nutr. Bull.*, 32(1), 31S–40S.

Brooks, S. (2010). Rice bio-fortification: lessons for global science and development. Routledge, London, UK, Earthscan.

Butsat, S. and Siriamornpun, S. (2010). Antioxidant capacities and phenolic compounds of the husk, bran and endosperm of Thai rice. *Food Chem.*, 119, 606–13.

Carrera, C., Ruiz-Rodríguez, A., Palma, M. and Barroso, C. G. (2012). Ultrasound assisted extraction of phenolic compounds from grapes. *Anal. Chim. Acta.*, 732, 100–4.

Chemat, F., Tomao, V. and Virot, M. (2008). Ultrasound-assisted extraction in food analysis. In Handbook of Food Analysis Instruments. CRC press, USA, 85–103.

Chen, L. J., Lee, D. S., Song, Z. P., Suh, H. S. and Lu, B. R. (2004). Gene flow from cultivated rice (*Oryza sativa*) to its weedy and wild relatives. *Ann. Bot.*, 93(1), 67–73.

Chen, M.-H. and Bergman, C. J. (2005). A rapid procedure for analysing rice bran tocopherol, tocotrienol and γ-oryzanol contents. *J. Food Comp. Anal.*, 18(2–3), 139–51. http://doi.org/10.1016/j.jfca.2003.09.004.

Chotimarkorn, C., Benjakul, S. and Silalai, N. (2008). Antioxidant components and properties of five long-grained rice bran extracts from commercial available cultivars in Thailand. *Food Chem.*, 111(3), 636–41.

Christou, P. and Twyman, R. M. (2004). The potential of genetically enhanced plants to address food insecurity. *Nutr. Res. Rev.*, 17(1), 23–42.

Dai, J.-L., Zhu, Y.-G., Zhang, M. and Huang, Y.-Z. (2004). Selecting iodine-enriched vegetables and the residual effect of iodate application to soil. *Biol. Trace Elem. Res.*, 101(3), 265–76.

D'Alessandro, L. G., Kriaa, K., Nikov, I. and Dimitrov, K. (2012). Ultrasound assisted extraction of polyphenols from black chokeberry. *Sep. Purif. Technol.*, 93, 42–7.

Deng, G.-F., Xu, X.-R., Zhang, Y., Li, D., Gan, R.-Y. and Li, H.-B. (2013). Phenolic compounds and bioactivities of pigmented rice. *Crit. Rev. Food Sci. Nutr.*, 53(3), 296–306. http://doi.org/10.1080/10408398.2010.529624.

Dexter, P. B. (1998). Rice fortification for developing countries. Department of Food Science, University of Arkansas-Fayetteville. OMNI/USAID publication, p. 12.

Ding, J., Liu, Z., Zhao, S. and Xiong, S. (2011). Changes of endogenous enzyme activities and major components in brown rice during germination. *Food Sci.*, 11, 8.

Dors, G. C., Primel, E. G., Fagundes, C. A. A., Mariot, C. H. P. and Badiale-Furlong, E. (2011). Distribution of pesticide residues in rice grain and in its coproducts. *J. Braz. Chem. Soc.*, 22(10), 1921–30. doi: 10.1590/S0103-50532011001000013.

Dunford, N. T. and King, J. W. (2000). Phytosterol enrichment of rice bran oil by a supercritical carbon dioxide fractionation technique. *J. Food Sci.*, 65, 1395–9.

Escribano-Bailón, M. T., Santos-Buelga, C. and Rivas-Gonzalo, J. C. (2004). Anthocyanins in cereals. *J. Chromatogr. A*, 1054(1–2), 129–41. http://doi.org/10.1016/j.chroma.2004.08.152.

Eyzaguirre, R. Z., Nienaltowska, K., De Jong, L. E., Hasenack, B. B. and Nout, M. J. (2006). Effect of food processing of pearl millet (Pennisetum glaucum) IKMP-5 on the level of phenolics, phytate, iron and zinc. *J. Sci. Food Agric.*, 86(9), 1391–8.

Faostat. Faostat.fao.org. Retrieved October 2015.

Faulds, C. B. and Williamson, G. (1999). The role of hydroxycinnamates in the plant cell wall. *J. Sci. Food Agric.*, 79(3), 393–5.

Finocchiaro, F., Ferrari, B., Gianinetti, A., Asta, C.D., Galaverna, G., Scazzina, F. and Pellegrini, N. (2007). Characterization of antioxidant compounds of red and white rice and changes in total antioxidant capacity during processing. *Mol. Nutr. Food Res.*, 51(8), 1006–19.

Florentino, R. F., Pedro, M. and Regina, A. (1998). Update on rice fortification in the Philippines. *Food Nutr. Bull.*, 19(2), 149–53.

Forster, G. M., Raina, K., Kumar, A., Kumar, S., Agarwal, R., Chen, M. H., Bauer, J. E., McClung, A. M. and Ryan, E. P. (2013). Rice varietal differences in bioactive bran components for inhibition of colorectal cancer cell growth. *Food Chem.*, 141(2), 1545–52.

Friedman, M. (2013). Rice brans, rice bran oils, and rice hulls: composition, food and industrial uses, and bioactivities in humans, animals, and cells. *J. Agric. Food Chem.*, 61, 10626–41.

Galili, G., Galili, S., Lewinsohn, E. and Tadmor, Y. (2002). Genetic, molecular, and genomic approaches to improve the value of plant foods and feeds. *Crit. Rev. Plant Sci.*, 21(3), 167–204.

Garris, A. J., Tai, T. H., Coburn, J., Kresovich, S. and McCouch, S. (2005). Genetic structure and diversity in *Oryza sativa L. Genetics*, 169(3), 1631–8.

Goufo, P., Pereira, J., Figueiredo, N., Oliveira, M. B. P., Carranca, C., Rosa, E. A. and Trindade, H. (2014). Effect of elevated carbon dioxide (CO_2) on phenolic acids, flavonoids, tocopherols, tocotrienols, γ-oryzanol and antioxidant capacities of rice (*Oryza sativa L.*). *J. Cereal Sci.*, 59(1), 15–24.

Graham,R., Senadhira, D., Beebe, S., Iglesias, C. and Monasterio, I. (1999). Breeding for micronutrient density in edible portions of staple food crops: conventional approaches. *Field Crops Res.*, 60, 57–80.

Gregorio, G. B. (2002). Progress in breeding for trace minerals in staple crops. *ASNS*, 132, 500S–502S.

Grusak, M. A. and Cakmak, I. (2004). Methods to improve the crop-delivery of minerals to humans and livestock. *Plant Nutr. Genomics*, 22(13), 265–86.

Grusak, M. A. and DellaPenna, D. (1999). Improving the nutrient composition of plants to enhance human nutrition and health. *Annu. Rev. Plant Biol.*, 50(1), 133–61.

Ha, T. Y., Han, S., Kim, S. R., Kim, I. H., Lee, H. Y. and Kim, H. K. (2005). Bioactive components in rice bran oil improve lipid profiles in rats fed a high-cholesterol diet. *Nutr. Res.*, 25(6), 597–606.

Hartikainen, H. (2005). Biogeochemistry of selenium and its impact on food chain quality and human health. *J. Trace Elem. Med. Biol.*, 18(4), 309–18.

Henderson, A. J., Ollila, C. A., Kumar, A., Borresen, E. C., Raina, K., Agarwal, R. and Ryan, E. P. (2012). Chemopreventive properties of dietary rice bran: current status and future prospects. *Adv. Nutr.*, 3, 643–53.

Hiemori, M., Koh, E. and Mitchell, A. E. (2009). Influence of cooking on anthocyanins in black rice (*Oryza sativa L.* japonica var. SBR). *J. Agric. Food Chem.*, 57 (5), 1908–14.

Hirschi, K. D. (2009). Nutrient Biofortification of Food Crops. *Annu. Rev. Nutr.*, 29(1), 401–21. doi: 10.1146/annurev-nutr-080508-141143.

Hou, Z., Qin, P., Zhang, Y., Cui, S. and Ren, G. (2013). Identification of anthocyanins isolated from black rice (*Oryza sativa L.*) and their degradation kinetics. *Food Res. Int.*, 50, 691–7.

Houston, D. F. and Kohler, G. O. (1970). Nutritional properties of rice. National Academy of Science, Print, Washington, DC.

Iqbal, S., Bhanger, M. I. and Anwar, F. (2005). Antioxidant properties and components of some commercially available varieties of rice bran in Pakistan. *Food Chemistry*, 93, 265–72.

IRRI (2015). http://irri.org/rice-today/trends-in-global-rice-consumption. Retrieved October 2015.

Katsube, T., Kurisaka, N., Ogawa, M., Maruyama, N., Ohtsuka, R., Utsumi, S. and Takaiwa, F. (1999). Accumulation of soybean glycinin and its assembly with the glutelins in rice. *Plant Physiol.*, 120(4), 1063–74.

Kehrer, J. P. (1993). Free radicals as mediators of tissue injury and disease. *Crit. Rev. Toxicol.*, 23(1), 21–48.

Keiko, H., Mitsuko, K., Yoko, T., Atsuko, S., Keiko, N., Yin, Z. R., Maria, I. and Amalia, M. (1997). The habit of eating rice in four countries: Japan, China, Spain and France. *J. Consumer Stud. Home Econ.*, 21, 349–61.

Khattak, A. B., Zeb, A., Bibi, N., Khalil, S. A. and Khattak, M. S. (2007). Influence of germination techniques on phytic acid and polyphenols content of chickpea (Cicer arietinum L.) sprouts. *Food Chem.*, 104(3), 1074–9. doi: 10.1016/j.foodchem.2007.01.022.

Khush, G. S. (1997). Origin, dispersal, cultivation and variation of rice. *Plant Mol. Biol.*, 35, 25–34.

Kim, M. H., Park, J. S. and Shin, T. Y. (2013). Anti-Inflammatory effects of deffated rice bran ethanol extract in RAW 264.7 macrophage cells. *JCIT*, 8(13), 324–31.

Leenhardt, F., Levrat-Verny, M.-A., Chanliaud, E. and Rémésy, C. (2005). Moderate decrease of pH by sourdough fermentation is sufficient to reduce phytate content of whole wheat flour through endogenous phytase activity. *J. Agric. Food Chem.*, 53(1), 98–102.

Lerma-García, M. J., Herrero-Martínez, J. M., Simó-Alfonso, E. F., Mendonça, C. R. B. and Ramis-Ramos, G. (2009). Composition, industrial processing and applications of rice bran γ-oryzanol. *Food Chem.*, 115(2), 389–404. http://doi.org/10.1016/j.foodchem.2009.01.063.

Liang, J., Han, B.-Z., Nout, M. J. R. and Hamer, R. J. (2008). Effects of soaking, germination and fermentation on phytic acid, total and in vitro soluble zinc in brown rice. *Food Chem.*, 110(4), 821–8. doi: 10.1016/j.foodchem.2008.02.064.

Lin, P. Y. and Lai, H. M. (2011). Bioactive compounds in rice during grain development. *Food Chem.*, 127, 86–93.

Lintschinger, J., Fuchs, N., Moser, H., Jäger, R., Hlebeina, T., Markolin, G. and Gössler, W. (1997). Uptake of various trace elements during germination of wheat, buckwheat and quinoa. *Plant Food Hum. Nutr.*, 50(3), 223–37.

Liu, C.-T., Wu, C.-Y., Weng, Y.-M. and Tseng, C.-Y. (2005). Ultrasound-assisted extraction methodology as a tool to improve the antioxidant properties of herbal drug Xiao-chia-hu-tang. *J. Ethnopharmacol.*, 99(2), 293–300. doi: 10.1016/j.jep.2005.02.018.

Londo, J. P., Chiang, Y. C., Hung, K. H., Chiang, T. Y. and Schaal, B. A. (2006). Phylogeography of Asian wild rice, Oryza rufipogon, reveals multiple independent domestications of cultivated rice, *Oryza sativa*. Current Issue, 103(25), 9578–83.

Lopez, H. W., Duclos, V., Coudray, C., Krespine, V., Feillet-Coudray, C., Messager, A., Demigné, C. and Rémésy, C. (2003). Making bread with sourdough improves mineral bioavailability from reconstituted whole wheat flour in rats. *Nutrition*, 19(6), 524–30.

Lucca, P., Hurrell, R. and Potrykus, I. (2002). Fighting iron deficiency anemia with iron-rich rice. *J. Am. Coll. Nutr.*, 21(sup 3), 184S–190S.

Mahgoub, S. E. and Elhag, S. A. (1998). Effect of milling, soaking, malting, heat-treatment and fermentation on phytate level of four Sudanese sorghum cultivars. *Food Chem.*, 61(1), 77–80.

Moongngarm, A. and Saetung, N. (2010). Comparison of chemical compositions and bioactive compounds of germinated rough rice and brown rice. *Food Chem.*, 122(3), 782–8. doi: 10.1016/j.foodchem.2010.03.053.

National Research Council (1989). Recommended dietary allowances, 10th edition, National Academy Press.

Niu, Y. G., Gao, B. Y., Slavin, M., Zhang, X. W., Yang, F., Bao, J. S., Shi, H. M., Xie, Z. H. and Yu, L. L. (2013). Phytochemical compositions, and antioxidant and anti-inflammatory properties of twenty-two red rice samples grown in Zhejiang. *Food Sci. Technol.*, 54(2), 521–7.

Novartis (2001). *Rice Biotechnology: Improving Yield, Stress Tolerance and Grain Quality*. Wiley-Blackwell, New York, USA.

Nyström, L., Mäkinen, M., Lampi, A.-M. and Piironen, V. (2005). Antioxidant activity of steryl ferulate extracts from rye and wheat bran. *J. Agric. Food Chem.*, 53(7), 2503–10. http://doi.org/10.1021/jf048051t.

Oka, H. I. (1988). *Origin of Cultivated Rice*. Tokyo, Japan Scientific Societies Press, Print.

Oluremi, O., Solomon, A. O. and Saheed, A. A. (2013). Fatty acids, metal composition and physic-chemcial parameters of Igbemo Ekiti rice bran oil. *J. Environ. Chem. Ecotoxicol.*, 5, 39–46.

Pfeiffer, W. and McClaffery, B. (2007). HarvestPlus: breeding crops for better nutrition. *Crop Sci.*, 47, S88–S105.

Powell, K. (2007). Functional foods from biotech – an unappetizing prospect?. *Nat. Biotechnol.*, 25(5), 525–31. doi: 10.1038/nbt0507-525.

Rao, A. S., Reddy, S. G., Babu, P. P. and Reddy, A. R. (2010). *BMC Complementary and Alternative Medicine*, 10(4), 1–9.

Ravishankar, D., Rajora, A. K., Greco, F. and Osborn, H. M. I. (2013). Flavonoids as prospective compounds for anti-cancer therapy. *Int. J. Biochem. Cell Biol.*, 45 (12), 2821–31.

Roschek, B., Fink, R. C., Li, D., McMichael, M., Tower, C. M., Smith, R. D. and Alberte, R. S. (2009). Inflammatory enzymes, cyclooxygenase 1, cyclooxygenase 2, and 5-Lipooxygenase, inhibited by stabilized rice bran extracts. *J. Med. Food*, 12(3), 615–23.

Rubio, L. A., Muzquiz, M., Burbano, C., Cuadrado, C. and Pedrosa, M. M. (2002). High apparent ileal digestibility of amino acids in raw and germinated faba bean (Vicia faba)-and chickpea (Cicer arietinum)-based diets for rats. *J. Sci. Food Agric.*, 82(14), 1710–17.

Salcedo, J., Jr., Bamba, M. D., Carrasco, E. O., Chan, G. S., Concepcion, I., Jose, F. R. and DeLeon, J. F. (1950). Artificial enrichment of white rice as a solution to endemic beriberi. Report of field trials in Bataan, Philippines. *J. Nutr.*, 42(4), 501–23.

Sautter, C., Poletti, S., Zhang, P. and Gruissem, W. (2006). Biofortification of essential nutritional compounds and trace elements in rice and cassava. *Proc. Nutr. Soc.*, 65, 153–9.

Schulze, H., Savelkoul, F. H., Verstegen, M. W., Van der Poel, A. F., Tamminga, S. and Groot Nibbelink, S. (1997). Nutritional evaluation of biologically treated white kidney beans (*Phaseolus vulgaris L.*) in pigs: ileal and amino acid digestibility. *J. Anim. Sci.*, 75(12), 3187–94.

Schwall, G. P., Safford, R., Westcott, R. J., Jeffcoat, R., Tayal, A., Shi, Y.-C., Gidley, M. J. and Jobling, S. A. (2000). Production of very-high-amylose potato starch by inhibition of SBE A and B. *Nat. Biotechnol.*, 18(5), 551–4.

Second, G. (1982). Origin of the genic diversity of cultivated rice (*Oryza* spp.): study of the polymorphism scored at 40 isozyme loci. *Jpn. J. Genet.*, 57, 25–57.

Shao, Y. and Bao, J. (2015). Polyphenols in whole rice grain: Genetic diversity and health benefits. *Food Chem.*, 180, 86–97. http://doi.org/10.1016/j.foodchem.2015.02.027.

Shen, Z., Palmer, M. V., Ting, S. S. and Fairclough, R. J. (1997). Pilot scale extraction and fractionation of rice bran oil using supercritical carbon dioxide. *J. Agric. Food Chem.*, 45(12), 4540–4.

Shibuya, N. (1984). Phenolic acids and their carbohydrate esters in rice endosperm cell walls. *Phytochemistry*, 23(10), 2233–7.

Shin, Y.-M., Park, H.-J., Yim, S.-D., Baek, N.-I., Lee, C.-H., An, G. and Woo, Y.-M. (2006). Transgenic rice lines expressing maize C1 and R-S regulatory genes produce various flavonoids in the endosperm. *Plant Biotech. J.*, 4(3), 303–15.

Smith, B. D. (1998). *The Emergence of Agriculture*. Scientific American Library, A Division of HPHLP, New York, ISBN 0-7167-6030-4.

Spanier, A. M., Shahidi, F., Parliment, T. H., Mussinan, C., Ho, C. T., Contis, E. T., Kayahara, H., Tsukahara, K., Tatai, T., (2001). Flavor, health and nutritional quality of pre-germinated brown rice. In *Food Flavors and Chemistry: Advances of the New Millennium*. Proceedings of the 10th International Flavor Conference, Paros, Greece, 4–7 July 2000. Royal Society of Chemistry, 546–51. Available at: http://www.cabdirect.org/abstracts/20013171422.html (accessed 31 October 2015).

Srinivasan, M., Sudheer, A. R. and Menon, V. P. (2007). Ferulic acid: Therapeutic potential through its antioxidant property. *J. Clin. Biochem. Nutr.*, 40(2), 92–100.

Sungsopha, J., Moongngarm, A. and Kanesakoo, R. (2009). Application of germination and enzymatic treatment to improve the concentration of bioactive compounds and antioxidant activity of rice bran. *Aust. J. Basic Appl. Sci.*, 3(4), 3653–61.

Tabaraki, R. and Nateghi, A. (2011). Optimization of ultrasonic-assisted extraction of natural antioxidants from rice bran using response surface methodology. *Ultrason. Sonochem.*, 18(6), 1279–86. doi: 10.1016/j.ultsonch.2011.05.004.

Tan, Z. and Shahidi, F. (2011). Chemoenzymatic synthesis of phytosteryl ferulates and evaluation of their antioxidant activity. *J. Agric. Food Chem.*, 59(23), 12375–83. http://doi.org/10.1021/jf2034237.

Tan, Z. and Shahidi, F. (2014). Antioxidant activity of phytosteryl phenolates: phytosteryl phenolates as antioxidants. *Eur. J. Lipid Sci. Technol.*, 116(12), 1701–7. http://doi.org/10.1002/ejlt.201400251.

Tian, S., Nakamura, K. and Kayahara, H. (2004). Analysis of phenolic compounds in white rice, brown rice, and germinated brown rice. *J. Agric. Food Chem.*, 52(15), 4808–13.

Tuncel, N. B. and Yılmaz, N. (2011). Gamma-oryzanol content, phenolic acid profiles and antioxidant activity of rice milling fractions. *Eur. Food Res. Technol.*, 233(4), 577–85. http://doi.org/10.1007/s00217-011-1551-4.

United States Department of Agriculture (2015). Nutrient data laboratory.

US. Food and Drug Administration, Code of Federal Regulations Title 21, 2015.

Valko, M., Leibfritz, D., Moncol, J., Cronin, M. T. D., Mazur, M. and Telser, J. (2007). Free radicals and antioxidants in normal physiological functions and human disease. *Int. J. Biochem. Cell Biol.*, 39(1), 44–84.

Wang, L., Li, X. X. and Chen, Z. X. (2009). Sulfated modification of the polysaccharides obtained from defatted rice bran and their antitumor activities. *Int. J. Biol. Macromol.*, 44(2), 211–14.

Watson, R., Preedy, V. and Zibadi, S. (eds) (2014). *Wheat and Rice in Disease Prevention and Health*. Academic Press, New York, USA.

Xu, Z., Hua, N. and Godber, J. S. (2001). Antioxidant activity of tocopherols, tocotrienols, and γ-oryzanol components from rice bran against cholesterol oxidation accelerated by 2,2′-Azobis (2-methylpropionamidine) dihydrochloride. *J. Agric. Food Chem.*, 49(4), 2077–81. http://doi.org/10.1021/jf0012852.

Yang, F. (2001). Studies on germination conditions and antioxidant contents of wheat grain. *Int. J. Food Sci. Nutr.*, 52(4), 319–30.

Ye, X., Al-Babili, S., Klöti, A., Zhang, J., Lucca, P., Beyer, P. and Potrykus, I. (2000). Engineering the provitamin A (β-carotene) biosynthetic pathway into (carotenoid-free) rice endosperm. *Science*, 287(5451), 303–5.

Zhang, L., Hu, P., Tang, S., Zhao, H. and Wu, D. (2005). Comparative studies on major nutritional components of rice with a giant embryo and a normal embryo. *J. Food Biochem.*, 29(6), 653–61.

Zhang, Q., Xiang, J., Zhang, L., Zhu, X., Evers, J., van der Werf, W. and Duan, L. (2014). Optimizing soaking and germination conditions to improve gamma-aminobutyric acid content in japonica and indica germinated brown rice. *J. Funct. Foods*, 10, 283–91. doi: 10.1016/j.jff.2014.06.009.

Zhang, Z. Y., Lei, Z. F., Lv, Y., Lv, Z. Z. and Chen, Y. (2008). Chemical composition and bioactivity changes in stale rice after fermentation with Cordyceps sinensis. *J. Biosci. Bioeng.*, 106(2), 188–93.

Zhou, Z., Robards, K., Helliwell, S. and Blanchard, C. (2004). The distribution of phenolic acids in rice. *Food Chem.*, 87, 401–6.

Zhu, C., Naqvi, S., Gomez-Galera, S., Pelacho, A. M., Capell, T. and Christou, P. (2007). Transgenic strategies for the nutritional enhancement of plants. *Trends Plant Sci.*, 12(12), 548–55.

Zigoneanu, I. G., Williams, L., Xu, Z. and Sabliov, C. M. (2008). Determination of antioxidant components in rice bran oil extracted by microwave-assisted method. *Bioresour. Technol.*, 99(11), 4910–18.

Chapter 9

Biofortified Golden Rice: an additional intervention for vitamin A deficiency

Adrian Dubock, Golden Rice Humanitarian Board, Switzerland

1 Introduction

For more than a quarter of a century, vitamin A deficiency (VAD) has been recognized by the United Nations as a significant public health problem. Key milestones included the:

- **1990 UN World Summit for Children,** where 50 heads of government and senior government officials committed their governments to the virtual elimination of VAD by the year 2000 (UNICEF, 1990).
- **1992 UN International Conference on Nutrition,** which concluded that
  - VAD control is the most cost-effective child health/survival strategy governments can pursue.
  - All sectors of society should support the virtual elimination of VAD.
  - Strategies should include promoting breast-feeding, dietary diversification, vitamin A supplementation and food fortification.
  - Locally available food-based strategies are the first priority. Vitamin A capsule supplementation is only an interim measure. (FAO, 1992)
- **2004 UNICEF and the Micronutrient Initiative Report 'Vitamin and Mineral Deficiency',** which concluded that 'controlling vitamin and mineral deficiency is an affordable opportunity to improve the lives of two billion people and strengthen the pulse of economic development' and that 'probably no other technology available today offers as large an opportunity to improve lives and accelerate development at such low cost' (UNICEF, 2004).

http://dx.doi.org/10.19103/AS.2016.0003.32

Table 1 Prioritized results from the Copenhagen Consensus 2012, 2008 and 2004

2012		Challenge	Solution
	1	**Hunger and education**	Bundled **micronutrient interventions** to fight hunger and improve education
	2	Infectious disease	Expand subsidy for malaria combination therapy
	3	Infectious disease	Expand childhood immunization coverage
	4	Infectious disease	Deworming of schoolchildren to improve education and health
	5	Infectious disease	Expand tuberculosis treatment
	6	**Hunger**, biodiversity and climate change	R&D for yield enhancement to reduce **hunger**, fight biodiversity destruction, lessen climate change effects
2008			
	1	**Malnutrition**	**Micronutrient supplements** – children (vitamin A and zinc)
	2	*Trade*	*The Doha Development Agenda*
	3	**Malnutrition**	**Micronutrient fortification** (iron and salt iodization)
	4	Diseases	Expanded immunization coverage for children
	5	**Malnutrition**	**Biofortification**
2004			
	1	Diseases	Control HIV/AIDS
	2	**Malnutrition**	**Providing micronutrients**
	3	*Subsidies and trade barriers*	*Trade liberalization*
	4	Diseases	Control of malaria
	5	**Malnutrition**	**Development of new agricultural technologies**

It has been estimated that a universally available source of vitamin A could save 23–34% of all deaths of children under five years globally and reduce measles mortality by up to 50% (West et al., 2010; Mayo-Wilson et al., 2011; WHO, 2013). Combatting micronutrient deficiencies has been judged as 'the best bang for a buck', that is, the most cost-effective way to solve 30 major problems faced by the world. This fact was established at all three separate meetings by different panels of Nobel Laureate Economists as part of the Copenhagen Consensus process (Table 1).

Addressing **micronutrient malnutrition** always scores highly amongst the 30 most pressing global challenges analysed. Of these most cost-effective to address Challenges, only the two *italicised Challenges* would not benefit from populations having improved immune response capability, the most major benefit of vitamin A sufficiency.

'The most recent estimates from FAO indicate that 840 million people do not receive enough energy from their diets to meet their needs. The overwhelming majority of these people – 799 million – live in developing countries. The global toll of people affected by micronutrient deficiency is estimated to be even higher and probably exceeds two billion' (Kennedy et al., 2003; Anon., 2016c).

Table 2 Annual mortality of <5 year children estimated by the UN, various years

<5 y deaths	1990	2009	2010	2012	2013	2014	2015
Per year(m)	12.7	8.1	7.6	6.6	6.3	5.9	5.9
Per day (k)	~35	~22	~21	~18	~17	~16	~16
% reduction (1990–2015)	0%	36%	40%	48%	50%	53%	53%

Table 3 Global mortality figures

Global mortality (millions)	2010	2014
Vitamin A deficiency	1.9–2.8	1.4–2.1*
HIV/AIDS	1.8	1.2**
Tuberculosis	1.4	1.1***
Malaria	0.7	0.6

Notes: *Mostly children <5 years; **1 in 3 HIV deaths is due to TB; ***mortality from TB of HIV negative people.

Data sources: For 2010 data: Dubock, 2013a; for 2014 data: UNICEF, 2016; West et al., 2010; UNAIDS, 2015; WHO, 2014a,b.

By 2014, the latest year for which a full set of comparative data is available, and despite good progress made in reducing child mortality, 5.9 million children under five years still died (Table 2).[1]

Important Millennium Development Goals: 1, to eradicate extreme poverty and hunger, and Goals 4 to 'Reduce by two thirds, between 1990 and 2015, the under-five mortality rate' and 5, to reduce significantly maternal mortality, all between 1990 and 2015, despite significant progress, were not met (Table 2).

The majority of these <5y deaths in both 2010 and 2014 were caused by VAD.[2] Even in 2014, 24 years after the first commitment to 'virtually eliminate VAD by the year 2000', 1–2 million children were still dying annually from this preventable vitamin deficiency (Table 3). The total annual VAD mortality is of the same magnitude as mortality caused by HIV/AIDS and malaria together, or HIV/AIDS and TB together (Dubock, 2013a; Table 3).

2 The problem of vitamin A deficiency (VAD)

Micronutrients are required in much smaller amounts than macronutrients (e.g. carbohydrates, proteins, fats) but are nevertheless essential for a properly functioning human body. Micronutrients exist either as minerals, taken up from the soil and accumulated by plants, and vitamins that are synthesized either by plants or by animals.

1. The UNICEF committee changed its mortality calculation methods from 2011 onwards in 2012: 'Therefore, the estimates presented in this report may differ from and are not necessarily comparable with previous sets of IGME estimates or the most recent underlying country data.'
2. Only a properly conducted Randomized Clinical Trial changes the single parameter(s) of interest (Sommer, pers. comm.).

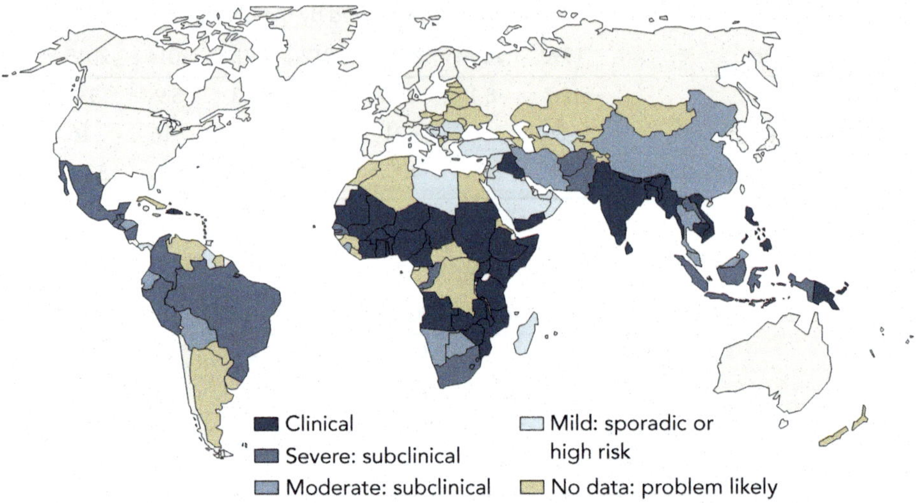

Figure 1 Public health importance of vitamin A deficiency by country.

Source: WHO. Redrawn and reproduced courtesy Banson.

Human populations deficient in the minerals iron and zinc, and of vitamin A, suffer from a variety of serious developmental and public health issues, and these deficiencies are more widespread globally than other important micronutrient deficiencies (Dubock, 2013b). Vitamin A deficiency, which mostly affects children less than 5 years old, and to a lesser extent their mothers, is widespread and well documented by the World Health Organisation (WHO) (Fig. 1).

By the 1990s, VAD was known as the major cause of childhood blindness globally, with about 500 000 cases annually, of which about two-thirds die if not treated (Whitcher et al., 2001). What was not known at the time was that VAD also suppresses the human immune system: VAD is 'a nutritionally acquired immune deficiency syndrome' (Semba, 2012). Increased susceptibility to disease as a result of VAD results in the majority of the millions of preventable <5y child deaths, mainly among children, annually (West et al., 2010; Dubock, 2013a; Table 3).

The Smart Development Goals of the Post-2015 Consensus for 'People' relate to, *inter alia*, providing nutritional supplements, halving malaria infections, reducing tuberculosis deaths, reducing new-born mortality and increasing protection from forms of influenza, pneumonia and diarrhoeal disease (Anon., 2015). Again, all of these objectives would benefit from fully functional immune systems of vulnerable people being enabled by vitamin A sufficiency.

Why is combatting VAD such a challenge? Only animals, including human beings, synthesize vitamin A. The only direct sources of vitamin A itself are animal products: liver, butter, milk and eggs are excellent sources. Animals and humans synthesize vitamin A from carotenoid chemicals naturally occurring in coloured fruits, vegetables and leaves. Beta-carotene (β-carotene) is a particularly important pro-vitamin A carotenoid found in all coloured plant foods. In industrialized countries, VAD is not a significant problem because a varied diet includes animal products and various sources of pro-vitamin A carotenoids.

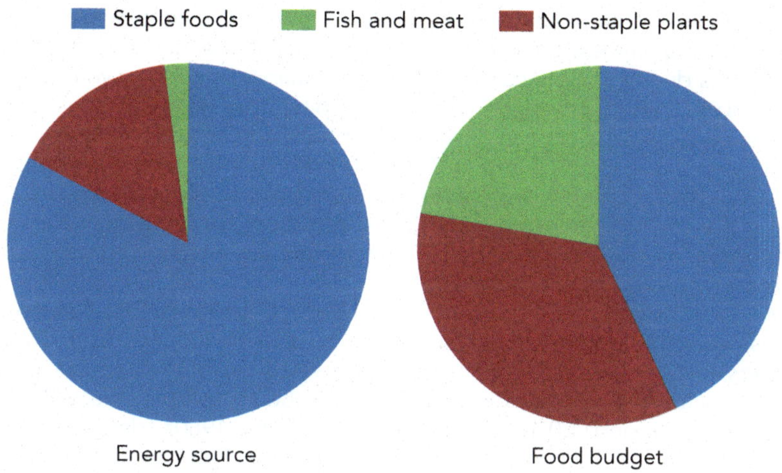

Figure 2 Share of energy source and food budget in rural Bangladesh.

Source: www.harvestplus.org/content/food-crisis.

However, many populations in the developing world are not able to access such balanced diets. In the poorer sections of developing countries, more than 70% of family income may be taken for food. Food sources that are most valuable in terms of micronutrients – animal products including milk, eggs, butter, liver, fish – are usually more expensive and 'beyond the reach of poor families' (Semba, 2012). Food security staple crops such as rice are cheaper and therefore make up most of the diet (Fig. 2). Three and a half billion people daily obtain most of their calories from rice, which can, in the case of countries like Bangladesh, account for up to 80% of their calorie intake (Bouis and Islam, 2012). Countries where the diet lacks sufficient vitamin A suffer from relatively high infant and maternal mortality as a consequence (Dubock, 2014a). The problem of VAD is exacerbated by the limited bioavailability of vitamin A from fruit and vegetables (Tang, 2013). It has been estimated that young children between ages one year and three would need to eat eight servings of dark green leafy vegetables per day in order to meet the recommended dietary allowance ('RDA') for vitamin A. These facts have resulted in the conclusion of 'the virtual impossibility for most poor, young children to meet their vitamin A requirements through vegetable and fruit intake alone' (Semba, 2012).

Supplementation may be used to alleviate micronutrient deficiencies. A good example is vitamin A capsules introduced to vulnerable populations starting in the early 1990s (West et al., 2010; Semba, 2012). At the same time it was recognized by the 1992 UN Conference on Nutrition that 'locally available food based strategies are first priority. Vitamin A capsule supplementation is only an interim measure' (FAO, 1992). Ideally, VAD-vulnerable children between six months and five years should receive a capsule every six months (WHO, 2011, 2013). Although individually costing only a few cents, provision of the capsules is costly – around $1.0 each costing $1.0 billion annually from aid agencies and mostly paid for by Canadian and US tax payers (Tang et al., 2012), and would be more if the infrastructure costs of provision in remote communities were not shared by other public health programmes. Such capsules are not recommended for children younger than six months due to toxicity concerns from the vitamin A, yet very young children are most vulnerable to VAD (Dubock, 2013a). And despite these generous donations over the last 25 years, and millions of young lives undoubtedly saved, 250 million children a year still remain vulnerable to VAD, and in 2014 ~1–2 million <5y preventable child deaths occurred. Even when vitamin A capsules are used, such use does not change the underlying vitamin A status of populations, which therefore remain at risk if the programmes were to be withdrawn for any reason (West, pers. comm.). Supplementation in poor populations is helpful, but is not sufficient on its own.

The use of capsules has also sometimes been controversial, for example, in the Philippines, where a large research trial had to be abandoned due to opposition from some groups. In India – the country with the biggest problem of VAD – political opposition to the use of vitamin A capsules is thought to be the reason that the VAD death rate has remained at around 1.0 million per year since the 1980s, despite neighbouring countries with similar culture and history and challenges, Pakistan and Nepal, reducing the VAD death toll significantly in the same period (Semba, 2012). The relatively high pharmacological concentrations of β-carotene found in some supplements – vitamin pills – may be hazardous to some groups (Mayne, 1996).

Another intervention for micronutrient deficiency is fortification of processed foods. In some countries margarine is fortified by mixing in vitamin A. In Guatemala, sugar has been fortified with vitamin A. Such programmes save lives. But in common with supplementation, fortified food requires sufficient industrial infrastructure to be effective, as well as good distribution networks, both of which can be lacking in poorer countries (Dubock, 2013a).

What promises to be a completely sustainable alternative to supplementation and fortification is biofortification. With biofortification, a crop plant, preferably a staple crop, (so that no cultural change is required in the target population to achieve sufficient consumption), is bred so that the crop contains the necessary micronutrients in sufficient quantities after harvesting, storage, processing and cooking to prevent micronutrient deficiency (Bouis et al., 2009; Dubock, 2013b).

Modern knowledge of both micronutrient deficiencies, and crop genetics allows such biofortified crops to be targeted for production through conventional breeding (which includes random genome alteration through mutagenesis) or through the use of modern biotechnological systems including genetically modified organisms (GMOs), in both cases followed by the selection of beneficial plant breeding stock (Bouis and Islam, 2012; Albabili and Beyer, 2005; Beyer, 2010). When mineral deficiency alleviation is targeted, the crop breeding can only do so much: if a crop plant with the capacity to absorb and possibly accumulate a mineral is grown in a soil that is deficient in that mineral, the plant genetics cannot help much. The situation is different when the targeted deficiency is a vitamin where either the vitamin itself, or a precursor to the vitamin which humans can convert to the vitamin, is synthesized directly within the plant. In this case, soil type is much less relevant: if the crop will grow, it will be usefully biofortified for human consumption.

Staple crops biofortified to contain elevated amounts of pro-vitamin A carotenoids, such as β-carotene, have many potential advantages over vitamin A capsules. At the physiological doses of β-carotene found in food sources (Shekelle et al., 1981; Mayne, 1996; Grodstein et al., 2007; Kang et al., 2009), including biofortified foods (Grune et al., 2010), β-carotene is non-toxic; it is Generally Considered As Safe – 'GRAS' – (U.S. Food and Drug Administration, 1979). Only sufficient β-carotene is converted by the human body to vitamin A as is required for health. After liver stores, and circulating levels of vitamin A become sufficient, excess β-carotene is excreted without conversion. All the cost benefit analyses summarised in Table 1 are based on the delivered cost of vitamin A capsules. People will get better economic returns if a donated pro-vitamin A trait is used, even if the cost, for example of Golden Rice, is exaggerated because of earlier, incorrect assumptions of the bioconversion efficiency of the beta-carotene in Golden Rice to circulating retinol (Stein et al., 2006). Up to six times better efficiency values are provided by Tang et al. (2009, 2012).

In addition, if available as a staple crop for daily consumption, it is expected that mothers will become self-sufficient in vitamin A. Transmission via the placenta for developing foetuses and breast milk for neonates will have a positive effect on the vitamin A status of very young children who are particularly vulnerable to VAD before they can consume solid food or it is safe to administer vitamin A capsules (Dubock, 2013a).

3 The origins of Golden Rice

'Golden' (first known as 'yellow') Rice is the first purposefully created biofortified food. It is a rice that synthesizes and accumulates β-carotene during seed maturation (Ye et al., 2000). Following normal harvesting, grain polishing, storage, cooking and consumption, the human body efficiently converts the β-carotene to vitamin A (Tang et al., 2009, 2012; Anon., 2013; Stokstad, 2015). It was designed specifically as a nutrition and health intervention to combat VAD (Ye et al., 2000; Paine et al., 2005; Giddings et al., 2012; Potrykus, 2013, 2014).

Golden Rice was created after ten years of research by Ingo Potrykus from the Swiss Eidgenössische Technische Hochschule (Federal Institute of Technology, ETH, Zürich, Switzerland) and Peter Beyer from the Albert-Ludwigs-Universität Freiburg, Germany. In 1984 an established rice breeder, Peter Jennings, first suggested developing a yellow rice containing β-carotene as a useful intervention for VAD. Ingo Potrykus initiated research on the genetic engineering required in 1991. The Rockefeller Foundation organized a brainstorming meeting in New York at the end of 1992, inviting world experts on terpenoid biosynthesis. These experts felt that it was possible to develop a yellow rice, though the chances of success were extremely small. Ingo Potrykus and Peter Beyer, who met at that meeting for the first time, decided to work together on the project, supported by the Rockefeller Foundation.[3]

There were four possible alternatives for producing a pro-vitamin A rice:

1 Using natural variation and traditional breeding;
2 Using mutagenesis, through chemicals or radiation, to generate new and usable variation;
3 Identifying and deactivating a regulatory gene acting as a 'switch' to inactivate the biochemical pathway in the endosperm. Rice plants synthesize large amounts of pro-vitamin A in leaves; however, the pathway is selectively turned off in the starch-storing tissue of the endosperm.
4 Harnessing the power of genetic engineering.

3. The Rockefeller Foundation provided funding for two 'Post Docs': one in Ingo Potrykus' laboratory and one in Peter Beyer's. Ingo Potrykus also contributed some of his untied research funding, and later the European Commission a little.

The first option was explored extensively but was not considered further because of lack of sufficient natural variation. No progress was achieved with the second option. In addition, no specific transcriptional activator of the pathway has so far been found to allow the third option to be taken further. This left genetic engineering as the remaining option (Ye et al., 2000; Paine et al., 2005). Ingo Potrykus' team contributed leading expertise in genetic engineering of rice, while Peter Beyer's team provided expertise in the biochemistry and regulation of terpenoid biosynthesis in plants. Both teams complemented each other: Peter's group isolated the genes and Ingo's group introduced them into rice under endosperm-specific regulation.

Their research necessitated developing processes to introduce genes into the genome of rice. It also necessitated understanding the biosynthetic pathway for carotenoids, and understanding why the genome of the rice plant produced carotenoids in other rice plant tissues, including leaves and the alurone layer in the grain, but not in the white endosperm, which is the part of rice that humans consume. Additionally, it involved making new genetic constructs including the necessary genetic components from other organisms, and introducing them to the rice genome with the newly developed genetic engineering techniques for rice. Finally, it required selection of the suitably transformed rice and the growth from the transformed cells of the first (subsequently named Golden Rice) plants (Potrykus, 2014). Success was achieved in 1999 (Fig. 3), and the results published in *Science* (Ye et al., 2000), attracting the world's attention in a new controversial subject – gmo-crops (Fig. 4; Dubock, 2014b).

Since the initial ~800 transformations in 2004, and a selected subset of 6, all Golden Rice work has involved introgressing the nutritional trait through conventional backcross

Figure 3 The first 'proof-of-concept' Golden Rice in April 1999, created by Beyer and Potrykus. Photo courtesy P. Beyer.

Figure 4 Ingo Potrykus and Golden Rice on the cover of the 31 July 2000 US edition of *Time* magazine. (The story also featured in the 12 February 2001 Asian edition of *Time*, but not in the European edition of the magazine.)

breeding and line selection into white indica-type mega varieties of rice common in Asia: IR36, IR64, PSB Rc82 and BR29. The intention is to use these varieties as breeding parents to further introgress the trait into other locally adapted and public sector owned, rice varieties preferred by the populations suffering from VAD.

The current version of Golden Rice, through transformation by *Agrobacterium* in about 2004, received two key genes, one from a commonly ingested soil bacterium and one from maize, to reconstitute in the normally white part of the seed the complex metabolic carotenoid pathway (Paine et al., 2005). This pathway is active in green plant tissues and in many flowers but not otherwise activated in white rice seed (Beyer, 2010). This success has been made possible through modern molecular biology combined with genetic knowledge and techniques (Potrykus, 2014; Ye et al., 2000; Albabili and Beyer, 2005; Beyer, 2010).

4 Developing Golden Rice as a product

Following initial publication in 2000, the author, then working for the company Syngenta (from which he retired in December 2007) approached the inventors, Professors Potrykus and Beyer, requesting the acquisition of commercial rights to the nutritional technology they had created. This approach coincided with the inventors realizing that they needed assistance in navigating what appeared to be a complex set of interacting intellectual property rights associated with their invention, as well as product development expertise.

The inventors wanted to donate their invention and create a free public good so that rice incorporating their nutritional technology would be available free of charge to resource-poor farmers in developing countries. The author saw how to combine the commercial interests of Syngenta with the altruistic interests of the inventors (which he shared and shares) (Potrykus, 2010). In a novel public–private partnership, the author arranged for the definition of several aspects necessary for humanitarian exploitation of the technology in developing countries, the cross-licensing of the inventor's technology and aspects of Syngenta technology to facilitate both applications, and also the in-licensing for the humanitarian project only of third-party intellectual property (Krattiger and Potrykus, 2007). In return for the inventors licensing Syngenta with the nutritional technology for commercial applications, Syngenta undertook to support the inventors in the defined humanitarian applications of the same technology in developing countries (Dubock, 2013a).

I then proposed to the inventors a governance system to turn their technology into a useful product: Golden Rice. A volunteer Golden Rice Humanitarian Board was created by invitation to provide strategic guidance to the project, and a network of 16 public sector rice research institute licensees of the inventors was also created in Bangladesh, China, India, Indonesia, Philippines and Vietnam, all countries whose populations suffer significantly with VAD and where rice is the staple. Subsequently, a website was created to share information concerning the project with the public. The Humanitarian Board has met 17 times since its first meeting in 2000, but has been unable to physically meet since 2013 due to lack of funds.

An important part of the cross-licencing agreement between the inventors and Syngenta was that improvements to the nutritional technology made by either party would be made available to the other. Subsequently, Syngenta scientists improved on the 'proof-of-concept' research of the inventors, and this research was published in *Nature Biotechnology* in 2005 (Paine et al., 2005). Selected versions – transformation events – of the improved materials were then provided to two licensees of Ingo Potrykus for the humanitarian project:

- The Indian Agricultural Research Institute,
- The International Rice Research Institute (IRRI) in the Philippines

Shortly before the 2005 publication, Syngenta announced its decision to cease pursuit of its commercial objectives for Golden Rice but to continue its support for the inventors' humanitarian project: 'Syngenta has supported this public project from its inception and will continue to do so' (Maeder and Michaelis, 2004; Dubock, 2013a).

The potential health benefits of Golden Rice were shown in a 2009 study (Tang et al., 2009). In 2012, a further study on the effectiveness of Golden Rice (GR) involving Chinese children was published (Tang et al., 2012; Anon., 2013; Dubock, 2014a; Stokstad, 2015). The study concluded that 'a bowl of ~100 to 150 g cooked GR (50 g dry weight) can provide ~60% of the Chinese Recommended Nutrient Intake of vitamin A for 6–8-y-old children' and that 'Golden Rice may be as useful as a source of preformed vitamin A from vitamin A capsules, eggs, or milk to overcome VAD in rice-consuming populations' (Tang et al., 2012). Although it was accepted that the data and findings were valid, and that the children had not been caused any harm, there were subsequent complaints, initiated by

Greenpeace, about the way the study had been conducted (notably the issue of consent), which led to the retraction of the paper (Stokstad, 2015).[4]

The detailed 'current status' of the project and its history is described in detail elsewhere (Dubock, 2014a). A key step has been in carrying out open field trials necessary to obtain regulatory approval. At the time of writing, one of the six transformation events is in a third round of field trials, with the first two rounds having demonstrated no agronomic difference between wild-type and Golden Rice versions. If the third field trials are also successfully concluded, the way will be clear for regulatory clearance to be applied for, for selected lines. Except for the open field agronomic data that are awaited, a regulatory data package complying with national regulations has already been created. It has been a long-standing strategy of the Golden Rice Humanitarian project that all regulatory data will be made available free of charge to all relevant national regulatory agencies. Only as it was created using genetic engineering, no cultivation of Golden Rice by farmers, or consumption by populations in any country, can occur until regulatory clearance for food and environmental safety has been achieved in that country.

5 Current challenges to establishing Golden Rice as an additional VAD intervention

Genetic engineering technology to harness the power of genetics for man has been embraced fully in the pharmaceutical sector, including the production of insulin. It is also very common in food production: bread, beer, wine and cheese production makes use of genetically modified enzymes (Renneberg, 2008). Feed meal from genetically modified GM soya and GM maize, imported from the Americas, is widely used by the European livestock industry (James, 2015; Klümper and Qaim, 2014; Eenennaam and Young, 2013). Concerns about the safety of GM crops highlight issues such as the potential to cause allergic reactions, transfer of genes from GM food to cells in the body and 'outcrossing', the migration of genes from GM plants into conventional plants or related wild species with potential adverse health (as well as environmental) effects.

The safety of GMOs and GM crops has been well established by independent scientific authorities globally for more than a decade (Dubock, 2009). As an example, a European Commission report concluded: 'The main conclusion to be drawn from the efforts of more than 130 research projects, covering a period of more than 25 years of research, and involving more than 500 independent research groups, is that biotechnology, and in particular GMOs, are not per se more risky than e.g. conventional plant breeding technologies' (European Commission, 2010). The World Health Organisation (WHO) has concluded that 'GM foods currently available on the international market have passed safety assessments and are not likely to present risks for human health. In addition, no effects on human health have been shown as a result of the consumption of such foods where they have been approved' (WHO, nd , see also Mantell, 2002). A recent analysis of

4. Tang et al. (2012) were criticized by Greenpeace and then Tufts University for procedural reasons. Tufts University also confirmed the research results, their public health significance and lack of harm to subjects (Anon., 2013). At the time of writing, the Office of Inspector General, US Department of Health and Social Security (DHSS), Investigations and Fraud office is reviewing the evidence that Tufts University reviewers violated ethical, due diligence and process standards in reaching their conclusions. The DHSS has a 2016 budget of US$1.093 billion – and 80000 staff – which include the US NIH (National Institutes of Health) who were funders, together with the Chinese Government, of the Tang et al. (2012) research.

almost 1800 studies on GM crop safety also concluded that no significant hazards were connected with the use of such crops (Nicolia et al., 2014). The US National Academics of Sciences, Engineering and Medicine recently published a comprehensive review of GM crops (Anon., 2016a; Pollack, 2016) which concluded: 'While recognising the inherent difficulty of detecting subtle or long-term effects in health or the environment, the study committee found no substantiated evidence of a difference in risks to human health between currently commercialized genetically-engineered (GE) crops and conventionally bred crops, nor did it find conclusive cause-and-effect evidence of environmental problems from the GE crops.' Similarly, the UK's Royal Society has reported that there are no human health or environmental issues specific to GM crops and that 'different GM crops have different characteristics and it is impossible, from a scientific point of view, to make a blanket statement that all GM is good or bad' (Anon., 2016b). Recently, more than 100 Nobel Laureates have called upon the UN, Governments and anti-GMO activists to adopt only science, not ideology, in considerations of the utility of GMO-crops in general and Golden Rice in particular (Anon., 2016d).

However, continued opposition to Golden Rice on the grounds that it involves genetic modification has proved a serious impediment to its final approval and use (Dubock, 2013a, 2014; Wesseler and Zilberman, 2014; Wesseler et al., 2014). There have been a number of objections to GM crops such as Golden Rice. As well as concerns about safety, these have included the claims that GM crops have been developed only to improve the profitability of intensive farming and provide no benefits to consumers or poorer farmers in the developing world.

When available through public sector rice breeding systems used to disseminate new varieties in Asian countries, Golden Rice will cost no more than white rice, and farmers, seed processors and consumers will use it in the same way. There will be no charge for the nutritional trait, and farmers can, without any limitation, plant, harvest, save seed and locally sell Golden Rice, which, in the homozygous form which will be released, breeds true generation to generation. Most rice is grown very close to where it is consumed, and resource-poor farmers, the definition of which includes almost all rice farmers in developing countries, will not require any license. As such, the adoption of Golden Rice will be completely sustainable, not requiring any further institutional support, nor cost, after initial introduction and government-assisted adoption.

The current impasse in making Golden Rice available to farmers can be seen in the trials being conducted by the IRRI in the Philippines. IRRI was the inventors' first licensee institution in 2001. The destruction of a Golden Rice field trial by anti-GMO activists in the Philippines in August 2013 led to an online petition and editorial in *Science* by a number of eminent scientists, which stated: 'If ever there was a clear-cut cause for outrage, it is the concerted campaign by Greenpeace and other nongovernmental organizations, as well as by individuals, against Golden Rice. ... We, and the thousands of other scientists who have signed the statement of protest, stand together in staunch opposition to the violent destruction of required tests on valuable advances such as Golden Rice that have the potential to save millions of impoverished fellow humans from needless suffering and death' (Alberts et al., 2013).

Greenpeace and local activists have also sought to challenge GM trials through local courts with some success. In 2013 the Court of Appeal in Manila upheld a legal challenge to one GM-crop trial (of Bt-aubergine) on the grounds that 'it destroys the balance of our biodiversity. Because it violates the conjunct right of our people to a balanced ecology, the whole constitutional right of our people (as legally and logically construed) is violated'

(Dubock, 2014b). Drawing on the 'Precautionary Principle' of the UN Cartagena Protocol (UN, 2000), the Supreme Court of the Philippines then put a stop to all new GMO research until the Philippines Government redrafted its relevant 'administrative order' governing the conduct of such research (Supreme Court of the Philippines, 2015). This change was achieved within three months by the Philippine authorities (ISAAA, 2016), allowing the trials to continue, and attracted some support in the local press for the potential benefits of genetically modified crops to Philippine farmers and the Philippine economy (Shotkoski, pers. comm.).

Conversely, Rajiv Shah, Administrator of USAID, has written: 'It's taken 25 years of ingenuity and perseverance to bring Golden Rice from vision to reality. Who will stand between it and millions of undernourished children?' (Shah, 2013). Nevertheless, continuing opposition from groups such as Greenpeace, which has had a significant impact on public opinion, and the resulting caution from many governments and institutions, in limiting further trials and practical deployment of gm-crops, including Golden Rice , still present significant obstacles, especially for public sector projects with much more limited resources to impact public opinion. The definition of modern biotechnology in the Cartagena Protocol,[5] which is the basis of almost all GMO crop regulatory requirements (Dubock, 2014b), is actually not limited to transgenesis. Nevertheless, in future, it may be pragmatic to consider that new genome editing techniques produce crop varieties that fall outside the 'Precautionary Principle' in national rules currently governing GMO crop regulatory clearance (Uauy et al., 2015; Specter, 2015; Porterfield, 2015; Lawrenson et al., 2015). A recent decision in this direction taken by the Swedish Board of Agriculture with respect to CRISPR-Cas9 (Jansson and Sundström, 2016) could also be adopted in other jurisdictions. Notwithstanding the above, it is considered that such technologies may be useful more for eliminating undesirable crop traits than for adding desirable traits (Beyer, pers. comm.).

6 Conclusions

All involved with Golden Rice hope that it can be usefully adopted as a sustainable additional intervention for VAD. All involved know, and do not make any contrary claims, that increased dietary variety and poverty elimination are important and enduring goals that Golden Rice cannot completely overcome. All involved are also committed to reducing the misery of VAD for current, as well as future, rice-consuming populations, where preventable deaths and blindness still occur 25 years after widespread UN recognition of the severity of the problem, despite existing interventions, and despite Millennium Development Goals to reduce child and maternal mortality. Wider concerns about GM crops have so far delayed the adoption of Golden Rice as a vitamin A intervention by more than six years. During this time, as many as 6000 children a day have been dying, and many becoming irreversibly blind, from the results of VAD, with many of them in countries where rice is the staple food.

GM techniques have been necessary to achieve what conventional crop breeding cannot achieve in rice: nutritionally beneficial levels of high folate (100x WT, ~3xRDA in 100 g, (Storozhenko et al., 2007; Blancquaert et al., 2015), iron (15 μg/g) and zinc (46 μg/g) (Slamet-Loedin et al., 2016; Ebron, 2016), as well as Golden Rice (Ye et al., 2000; Paine et al., 2005).

5. Where precautions overules experience, and Greenpeace claim partial credit for creating: http://www.greenpeace.org/international/en/campaigns/agriculture/solution-ecological-farming/biosafety-protocol/ (accessed 4 July 2016).

For the important food security crops like cassava, cooking banana (plantain) and sorghum, genetic modification is also a requirement for introducing biofortification traits. Although HarvestPlus has made progress in biofortifying crops by avoiding the use of crops created by genetic modification, Howarth Bouis, HarvestPlus director, has still underlined the important role of genetic techniques in making further progress: 'The research results [from research funded by Harvest Plus], which have exceeded target levels of both Fe and Zn, speak of the immense potential of using transgenic techniques in pursuing biofortification to improve the nutritional value of food crops. This demonstrates how scientific innovations can expand the range of solutions to curbing micronutrient deficiencies' (Ebron, 2016).

At the Second International Conference on Nutrition in 2014, the WHO and the Food and Agriculture Organisation (FAO) (WHO and FAO, 2014) led, *inter alia*, to a technical consultation: 'Staple Crops biofortified with vitamins and minerals: considerations for public health strategies',[6] which occurred in early April 2016 (WHO and FAO, 2016). Following the WHO/FAO consultation, it is to be hoped that biofortified crops – from all available technologies – can be embraced as part of the fight against the impacts of poverty, as poverty itself continues to be reduced by all available means.

7 Acknowledgements

Apart from the cross-licencing deal between the inventors and Syngenta, and the donation of their nutritional trait intellectual property to the resource poor of the developing world by Professors Potrykus and Beyer, many other parties have contributed to the project.

The continuing support of the volunteer Golden Rice Humanitarian Board has been of significant assistance, and especially the network of public sector rice research institute licensees and their skilled plant breeders and other staff in Bangladesh, China, India, Indonesia, Philippines (both the Philippine Rice Research Institute and the IRRI) and Vietnam thus far.

Free licences were secured for technology, possibly necessary for humanitarian applications of the inventors' patented technology from Bayer, Mogen, Novartis, Monsanto, Orynova and Zeneca, together with a Japanese company wishing to remain anonymous. Thus, it was possible to provide free licences to the full technological support package to national research institutions in developing countries for use only in public sector-owned rice germplasm and (if any) only to be stacked with other public sector traits for humanitarian use controlled by the public sector. It is a licence term, agreed with government licensees, that there will be no additional charge for the nutritional trait: Golden Rice will cost the same as the equivalent white rice variety.

Financial support has been gratefully received at different times from the European Commission, Prof Potrykus, The Rockefeller Foundation, USAID, Syngenta, The Syngenta Foundation, The Bill & Melinda Gates Foundation (including also Mr Warren Buffett's contribution to it) and the national budgets of the countries involved.

Dr Jorge Mayer's management of the Golden Rice website, also as an unpaid volunteer, is very welcome.

6. Going to press, 19 July 2016, landraces of rice with up to 21 ppm (=μg/g) Fe and up to 35 ppm Zn have been reported following 3 seasons field trials in India. It is too early to know if these traits can be transferred to high yielding rice varieties: http://www.indialivetoday.com/eat-your-carbs-study-finds-indigenous-varieties-of-rice-with-higher-nutrient-content-says-biofortification-the-way-to-combat-nutritional-deficiencies-in-community/13969.html.

8 Where to look for further information

The Golden Rice website: www.goldenrice.org or contact the author, Adrian Dubock, at contact@goldenrice.org . A summary of more recent developments can be found in A. C. Dubock, 2014a and b.

9 References

Albabili, S. and Beyer, P., 2005. Golden Rice – five years on the road – five years to go? *Trends in Plant Science*, 10(12), pp. 565–73. Available at: http://linkinghub.elsevier.com/retrieve/pii/S1360138505002578.

Alberts, B., Baulcombe, D., Blobel, G., Datta, S., Fedoroff, N., Kennedy, D., Khush, G., Peacock, J., Rees, M. and Sharp, P., 2013. Standing up for GMOs. *Science (New York, NY)*, 341(6152), p. 1320. Available at: http://www.sciencemag.org/content/341/6152/1320.full.

Anon., 2013. Tufts University statement on Golden Rice Research. http://academicsreview.org/2013/09/tufts-university-statement-on-golden-rice-research/.

Anon., 2015. Copenhagen Consensus Center. Available at: http://www.copenhagenconsensus.com/projects (Accessed 28 December 2015).

Anon., 2016a. National Academies of Sciences, Engineering, and Medicine. *Genetically Engineered Crops: Experiences and Prospects*. Washington, DC: The National Academies Press. doi: 10.17226/23395. http://www.nap.edu/23395.

Anon., 2016b. GM plants: Questions and answers. The Royal Society. Issued: May 2016 DES3710. https://royalsociety.org/~/media/policy/projects/gm-plants/gm-plant-q-and-a.pdf.

Anon., 2016c. G20 Agriculture Ministers Communiqué. http://www.g20.org/English/Documents/Current/201606/t20160608_2301.html (accessed 4 July 2016).

Anon., 2016d. Support GMOs and Golden Rice. http://supportprecisionagriculture.org/ (accessed 4 July 2016).

Beyer, P., 2010. Golden Rice and 'Golden' crops for human nutrition. *New Biotechnology*, 27(5), pp. 478–81. Available at: http://linkinghub.elsevier.com/retrieve/pii/S1871678410004450.

Blancquaert, D., VanDaele, J., Strobbe, S., Kiekens, F., Storozhenko, S., DeSteur, H., Gellynck, X., Lambert, W., Stove.C. and VanDerStraeten D., 2015. Improving folate (vitamin B9) stability in biofortified rice through metabolic engineering. *Nature Biotechnology* 33, 1076–8. doi:10.1038/nbt.3358.

Bouis, H. and Islam, Y., 2012. Chapter 10: Biofortification. In Shenggen Fan and Rajul Pandya-Lorch, eds. Leveraging Agriculture to Reduce Hidden Hunger in Reshaping Agriculture for Nutrition and Health (An IFPRI 2020 book). Available at: https://www.ifpri.org/publication/reshaping-agriculture-nutrition-and-health (Accessed 19 May 2016).

Bouis, H., Hotz, C., McClafferty, B., Meenakshi J. and Pfeiffer, W., 2009. Biofortification: A new tool to reduce micronutrient malnutrition. *Food and Nutrition Bulletin*, 32(1), pp. 31–40.

Dubock, A. C., 2009. Crop conundrum. *Nutrition Reviews*, 67(1), pp. 17–20. Available at: http://www.goldenrice.org/PDFs/Crop_conundrum_Dubock_Nutr_Rev_2009.pdf (Accessed 19 May 2014).

Dubock, A. C., 2013a. Golden Rice: A long-running story at the watershed of the GM debate. *www.Goldenrice.Org*, pp. 1–12. Available at: http://b4fa.org/wp-content/uploads/2013/10/Viewpoints-Dubock.pdf\nhttp://www.www.goldenrice.org/PDFs/GR_A_long-running_story.pdf.

Dubock, A. C., 2013b. Nutritional enhancement by biofortification of staple crops. In D. J. Bennett and R. C. Jennings (eds), *Successful Agricultural Innovation in Emerging Economies: New Genetic Technologies for Global Food Production*. Cambridge: Cambridge University Press, pp. 119–220.

Dubock, A. C., 2014a. The present status of Golden Rice. *Journal of Huazhong Agricultural University*, 33(6), pp. 69–84.

Dubock, A. C., 2014b. The politics of Golden Rice. *GM Crops & Food 5:3, 210–222; July/August/ September 2014*. Available at: http://www.goldenrice.org/PDFs/Dubock-The_present_status_of_Golden_Rice-2014.pdf.

Ebron, G., 2016. New GMO rice could fight iron, Zinc deficiencies in developing world Start the discussion... . *Genetic Literacy Project*, pp. 2015–16. Available at: https://www.geneticliteracyproject.org/2016/02/16/newgmoricecouldfightironzincdeficienciesindevelopingworld/ (Accessed 29 April 2016).

European Commission Directorate-General for Research and Innovation; Biotechnologies (2010). A Decade of EU-Funded GMO Research (2001–2010). https://ec.europa.eu/research/biosociety/pdf/a_decade_of_eu-funded_gmo_research.pdf (Accessed 19 May 2016).

FAO, 1992. *The International Conference on Nutrition*. Available at: http://www.fao.org/docrep/v7700t/v7700t02.htm.

Giddings, L. V., Giddings, L., Potrykus, I., Ammann, K. and Fedoroff, N., 2012. Confronting the Gordian knot. *Nature biotechnology*, 30(3), pp. 208–9. Available at: http://www.ncbi.nlm.nih.gov/pubmed/22398608.

Grodstein, F.,Grodstein, F., Kang, J., Glynn, R., Cook, N. and Gaziano, J., 2007. A randomized trial of beta carotene supplementation and cognitive function in men: The Physicians' Health Study II. *Archives of Internal Medicine*, 167, pp. 2184–90.

Grune T., Lietz, G., Palou, A., Ross, C., Stahl, W., Tang, G., Thurnham, D., Yin, S. and Biesalski, H., 2010. β-Carotene is an important vitamin A source for humans. Supplement: *The Journal of Nutrition*. http://jn.nutrition.org/content/140/12/2268S.full (Accessed19 May 2016).

ISAAA, 2016. Philippine Government Approves Revised GM, p. 14218. Available at: http://www.isaaa.org/kc/cropbiotechupdate/article/default.asp?ID=14218.

James, C., 2015. ISAAA Briefs, Executive Summary, Brief 49, Global Status of Commercialized Biotech/GM Crops: 2014. http://www.isaaa.org/resources/publications/briefs/49/executive-summary/pdf/b49-execsum-english.pdf (Accessed 19 May 2016).

Jansson, S. and Sundström, J., 2016. 'Green light in the tunnel'! Swedish Board of Agriculture: a crispr-cas9-mutant but not a gmo. http://www.upsc.se/about-upsc/news/4815-green-light-in-the-tunnel-swedish-board-of-agriculture-a-crispr-cas9-mutant-but-not-a-gmo.html (Accessed 19 May 2016).

Kang, J. H., Cook, N., Manson, J., Buring, J., Albert, C. and Grodstein, F., 2009. Vitamin E, vitamin C, beta-carotene, and cognitive function among women with or at risk of cardiovascular disease. *Circulation*, 119, pp. 2772–80.

Kennedy, G., Nantel, G. and Shetty, P., 2003. The scourge of 'hidden hunger': Global dimensions of micronutrient deficiencies. *Food Nutrition and Agriculture*, 32, pp. 8–16. Available at: ftp://ftp.fao.org/docrep./fao/005/y8346m/y8346m01.pdf.

Klümper, W. and Qaim, M., 2014. A meta-analysis of the impacts of genetically modified crops. *PLoS ONE*, 9(11), p. e111629.

Krattiger, A. and Potrykus, I., 2007. Golden Rice: A Product-Development Partnership in Agricultural Biotechnology and Humanitarian Licensing. In A. Krattiger, R. Mahoney, L. Nelsen, J. Thomson, A. Bennett, K. Satyanarayana, G. Graff, C. Fernandez and S. Kowalski (eds), *Executive Guide to Intellectual Property Management in Health and Agricultural Innovation: A Handbook of Best Practices*. MIHR: Oxford, UK, and PIPRA: Davis, USA. Available www.ipHandbook.org. Oxford, UK and PIPRA: Davis, USA.

Lawrenson, T., Shorinola, O., Stacey, N., Li, C., Østergaard, L., Patron, N., Uauy, C. and Wendy Harwood, W., 2015. Induction of targeted, heritable mutations in barley and Brassica oleracea using RNA-guided Cas9 nuclease. *Genome Biology*,16, 258.

Maeder, C. and Michaelis, D., 2004. *Syngenta to donate Golden Rice to Humanitarian Board*, Form 6-K, Securities and Exchange Commission Washington DC 20549. Available at: http://www.sec.gov/Archives/edgar/data/1123661/000095010304001433/oct1504_6k.htm.

Mantell, K., 2002. WHO urges Africa to accept GM food aid. *Science and Development Network*, http://www.scidev.net/News, 30 August 2002. http://www.scidev.net/global/gm/news/who-urges-africa-to-accept-gm-food.html (Accessed 19 May 2016).

Mayne, S. T., 1996. Beta-carotene, carotenoids, and disease prevention in humans. *The FASEB Journal*, 10, 690–701.

Mayo-Wilson, E., Imdad, A., Herzer, K., Yakoob, M. and Bhutta, Z., 2011. Vitamin A supplements for preventing mortality, illness, and blindness in children aged under 5: Systematic review and meta-analysis. *BMJ*, 343, pp. d5094. Available at: http://www.bmj.com/cgi/doi/10.1136/bmj.d5094 (25 August 2011).

Nicolia, A., Manzo, A., Veronesi, F. and Rosellini, D., 2014. An overview of the last 10 years of genetically engineered crop safety research. *Critical Reviews in Biotechnology*, 34(1), 77–88.

Paine, J., Shipton, C., Chaggar, S., Howells, R., Kennedy, M., Vernon, G., Wright, S., Hinchliffe, E., Adams, J., Silverstone, A. and Drake, R. 2005. Improving the nutritional value of Golden Rice through increased pro-vitamin A content. *Nature Biotechnology*, 23(4), pp. 482–7. Available at: http://www.ncbi.nlm.nih.gov/pubmed/15793573 (Accessed 28 April 2014).

Pollack, A., 2016. Genetically engineered crops are safe, analysis finds. *The New York Times*. http://nyti.ms/24Xdtcr?version=meter+at+1&module=meter-Links&pgtype=article&contentId=&mediaId=&referrer=https%3A%2F%2Fwww.google.co.uk%2F&priority=true&action=click&contentCollection=meter-links-click (17 May 2016).

Porterfield, A., 2015. Next generation crop precision editing can avoid marketing pitfalls of GMOs.

Post-2015 Consensus, 2015. Prioritizing 19 targets instead of the UN's 169 targets is equivalent to doubling or quadrupling foreign aid. Available at: http://www.copenhagenconsensus.com/post-2015-consensus/nobel-laureates-guide-smarter-global-targets-2030.

Potrykus, I., 2010. The private sectors role for public sector genetically engineered crop projects. *Pontifical Academy of Sciences*, 27(43660). Available at: http://www.sciencedirect.com/science/article/B8JG4-50J4MRM-5/2/b25d03d8d135f1ce871ce16f02aa1522.

Potrykus, I., 2013. Unjustified regulation prevents use of GMO technology for public good. *Trends in biotechnology*, 31(3), pp. 131–3. Available at: http://www.ncbi.nlm.nih.gov/pubmed/23273413 (Accessed 26 June 2014).

Potrykus, I., 2014. From the concept of totipotency to biofortified cereals. *Annual Review of Plant Biology*, 66(1), pp. 1–22. Available at: http://www.annualreviews.org/doi/abs/10.1146/annurev-arplant-043014-114734.

Renneberg, R., 2008. *Biotechnology for Beginners*. Elsevier Inc., p. 345.

Semba, R. D., 2012. *The Vitamin A Story: Lifting the Shadow of Death*. In B. Koletzko (ed.), *World Review of Nutrition and Dietetics*, p. 104.

Shah, R., 2013. Genetic Engineering. Letter to the Editor, *The New York Times*, from Rajiv Shah, Administrator, United States Agency for International Development. Washington. http://www.nytimes.com/2013/09/10/opinion/genetic-engineering.html?_r=0 (28 August 2013).

Shekelle, R., Lepper, M., Liu, S., Maliza, C., Raynor, W., Rossof, A., Paul, O., Shryock, A. and Stamler, J., 1981. Dietary vitamin A and risk of cancer in the Western Electric study. *Lancet*, pp. 1185–90.

Slamet-Loedin, I., Trijatmiko, K., Dueñas, C., Tsakirpaloglou, T., Torrizo, L., Arines, F., Adeva, C., Balindong, J., Oliva, N., Sapasap, M., Borrero, J., Rey, J., Francisco, P., Nelson, A., Nakanishi, H. and Lo, E., 2016. Biofortified indica rice attains iron and zinc nutrition dietary targets in the field. *Nature Scientific Reports* (19792).

Specter, B. Y. M., 2015. The Gene Hackers. *The New Yorker*, pp. 1–21.

Stein, A., Sachdev, H. and Qaim, M., 2006. Potential impact and cost-effectiveness of Golden Rice. *Nature Biotechnology*, 24(10), 200–1. Available at : https://www.uni-hohenheim.de/uploads/media/GR-suppl.pdf.

Stokstad, E., 2015. Golden rice paper retracted after legal bid fails. *Science*, pp. 7–9. Available at: http://news.sciencemag.org/asiapacific/2015/07/golden-rice-paper-retracted-after-legal-bid-fails.

Storozhenko, S., De Brouwer, V., Volckaert, M., Navarrete, O., Blancquaert, D., Zhang, G., Lambert, W. and Van Der Straeten, D., 2007. Folate fortification of rice by metabolic engineering. *Nature Biotechnology*, 11, pp. 1277–9.

Supreme Court of the Philippines, 2015. *International Service for the Acquisition of Agri-Biotech Applications, Inc. Vs. Greenpeace Southeast Asia (Philippines)*. http://www.lawphil.net/judjuris/juri2015/dec2015/pdf/gr_209271_velasco.pdf (Accessed 19 May 2016).

Tang, G., Qin, J., Dolnikowski, G., Russell, R. and Grusak, M., 2009. Golden Rice is an effective source of vitamin A. *American Journal of Clinical Nutrition*, 89, pp. 1–8. Available at: http://ajcn.nutrition.org/content/89/6/1776.short (Accessed 24 July 2014).

Tang, G., et al., 2012. β-Carotene in Golden Rice is as good as β-carotene in oil at providing vitamin A to children. *American Journal Of Clinical Nutrition*, 96, pp. 658–64. Available at: http://ajcn.nutrition.org/content/early/2012/07/31/ajcn.111.030775.abstract.

Tang, G., 2013. Using plant foods rich in β-carotene to combat vitamin A deficiency. *Sight And Life* 27 (1). http://www.sightandlife.org/fileadmin/data/Magazine/2013/27_1_2013/using_plant_foods_rich_in_beta_carotene_to_combat_VAD.pdf.

U.S. Food and Drug Administration, 1979. Select Committee on GRAS Substances (SCOGS). http://www.fda.gov/Food/IngredientsPackagingLabeling/GRAS/SCOGS/ucm261245.htm (Accessed 19 May 2016).

Uauy, C., Harwood, W. and Østergaard, L., 2015. On Biology (Using CRISPR/Cas9 to genetically modify crops). http://blogs.biomedcentral.com/on-biology/2015/11/30/using-crisprcas9-genetically-modify-crops/ (Accessed 19 May 2015).

UN, 2000. Secretariat of the Convention on Biological Diversity. *Cartagena Protocol on Biosafety to the Convention on Biological Diversity*, Montreal. Available at: https://www.cbd.int/doc/legal/cartagena-protocol-en.pdf.

UNAIDS, 2015. Fact sheet 2015 | UNAIDS Fact sheet 2015 Fact sheet 2015 | UNAIDS (June 2015), pp. 2015–16. Available at: http://www.unaids.org/en/resources/campaigns/HowAIDSchanged-everything/factsheet.

UNICEF, 2004. Vitamin & Mineral deficiency. A global damage assessment report. http://www.micronutrient.org/CMFiles/PubLib/Report-67-VMD-A-Global-Damage-Assessment-Report1KSB-3242008-9634.pdf

UNICEF, 1990. World Declaration on the Survival, Protection and ..., pp. 134–6. Available at: http://www.unicef.org/wsc/declare.htm.

UNICEF, 2016. UNICEF Data: Monitoring the Situation of Children and Women TOPIC + COUNTRY Committing to Child Survival: A Promise Renewed – Progress Report 2014 National, regional, and global sex ratios of infant, child, and under-5 mortality and identification, pp. 2014–16. Available at: http://data.unicef.org/child-mortality/under-five.html (Accessed 29 April 2016).

Van Eenennaam, A. L. and Young, A. E., 2013. Prevalence and impacts of genetically engineered feedstuffs on livestock populations 1.

Wesseler, J. and Zilberman, D., 2014. The economic power of the Golden Rice opposition. *Environment and Development Economics*, 19(6), pp. 724–42.

Wesseler, J., Kaplan, S. and Zilberman, D., 2014. The cost of delaying approval of golden rice. *Agricultural and Resource Economics, Giannini Foundation of Agricultural Economics*, 17, pp. 1–3. Available at: http://giannini.ucop.edu/media/are-update/files/articles/V17N3_1.pdf.

West, K. P., Klemm, R. D. W. and Sommer, A., 2010. Sound science, sound policy. *Journal of the World Public Health Nutrition Association*, 1(5), pp. 211–29.

Whitcher, J. P., Srinivasan, M. and Upadhyay, M. P., 2001. Corneal blindness: A global perspective. *Bull World Health Organ*, 79(3), pp. 214–21.

WHO, 2011. Guideline: Vitamin A supplementation in infants and children 6–59 months of age. http://apps.who.int/iris/bitstream/10665/44664/1/9789241501767_eng.pdf (Accessed 19 May 2016).

WHO, 2013. Nutrition, micronutrient deficiencies, vitamin A deficiency. Available at: http://www.who.int/nutrition/topics/vad/en/.

WHO, 2014a. Global Health Observatory (GHO) data How many TB cases and deaths are there ?, pp. 1–2. Available at: http://www.who.int/gho/tb/epidemic/cases_deaths/en/ (Accessed 29 April 2016).

WHO, 2014b. WHO | Fact sheet on the World Malaria Report 2014 (December 2014), pp. 2014–16. Available at: http://www.who.int/malaria/media/world_malaria_report_2014/en/.

WHO Frequently asked questions on genetically modified food, n.d. http://www.who.int/foodsafety/areas_work/food-technology/faq-genetically-modified-food/en/ (Accessed 19 May 2016).

WHO and FAO, 2014. Second International Conference on Nutrition (October), pp. 19–21. Available at: http://www.fao.org/3/a-ml542e.pdf [Accessed April 29, 2016].

WHO and FAO, 2016. Nutrition Staple crops biofortified with vitamins and minerals: Considerations for a public health strategy, pp. 4–5. Available at: http://www.who.int/nutrition/events/2016_consultation_staplecrops_biofortified_vitminerals_5to8april/en/.

Ye, X., et al., 2000. Engineering the provitamin A (beta-carotene) biosynthetic pathway into (carotenoid-free) rice endosperm. *Science (New York, NY)*, 287(5451), pp. 303–5.

Chapter 10

Development of rice varieties with improved iron content in grain

Navreet K. Bhullar, ETH Zurich, Switzerland

1 Introduction: iron deficiency anaemia

The deficiency of different minerals and vitamins, also known as Hidden Hunger, is considered a serious global challenge for human health (White and Broadley, 2009). Iron deficiency anaemia (IDA), together with vitamin A, iodine and zinc deficiencies, is the most common form of micronutrient malnutrition. IDA is widespread in both developing and developed countries, with reports on 43% of children, 38% of pregnant women and 29% of nonpregnant women affected worldwide (Stevens et al., 2013). Although IDA can occur at all stages of life, it is often more prevalent in children, adolescents and women of reproductive age. The deficiency has social and economic implications including negative effects on cognitive/physical development of children and reduced immunity and work productivity (WHO, 2015). Low intake of bioavailable iron, increased iron requirement in circumstances such as pregnancy, excess blood loss as a result of accidents and/or infections or impaired iron absorption caused by chronic diseases are among several reasons leading to IDA. For human adults, daily iron intakes of 8 mg/day (males), 18 mg/day (females) and 27 mg/day (pregnant women) are recommended (National Academy of Sciences, 2001).

Several possible strategies have been suggested for combating the deficiency in affected populations. These include education about appropriate dietary habits, iron supplementation of specific groups at high risk and iron fortification of foods (Abbaspour et al., 2014). However, these approaches are often hindered by socio-economic and/or biological factors (Frossard et al., 2000; Mayer et al., 2008). Alteration

http://dx.doi.org/10.19103/AS.2016.0003.33

in diet is often difficult to implement in affected populations, mainly due to limited access to iron-rich foods. Iron supplementation, particularly when administered in high doses, may lead to side effects such as nausea and epigastric pain (Galloway and McGuire, 1994). Furthermore, there are concerns regarding increased severity of infectious diseases when using iron supplements, particularly in the presence of malaria (Oppenheimer, 2001; Sazawal et al., 2006). Food fortification has been suggested as the most effective and the safest intervention. However, iron fortification of food is more difficult as compared to fortification with other nutrients such as zinc, iodine and vitamin A. The most bioavailable and soluble iron compounds often cause unacceptable colour and flavour changes (Hurrell and Egli, 2010). Considering such bottlenecks, biofortification (i.e. production of nutrient-enriched crops) has been suggested as an inexpensive and sustainable approach to address micronutrient deficiencies worldwide. Several programmes aimed at biofortification are active in different regions globally, such as the HarvestPlus program (http://www.harvestplus.org/), the Biofort Brazil program and the HarvestPlus-China program (http://harvestplus-china.org/). Staple food crops including rice, wheat, maize, beans, sweet potato, cassava and pearl millet are the target crops in these programmes.

2 Biofortification of rice for increased iron content in grain

Cereals are vital to world food security and serve as the most important source of nutrients for human beings. Rice (*Oryza sativa* L.) is a source of more than 20% of the calories consumed worldwide. In many Asian countries, the dietary contribution of rice is substantially higher with people depending on it for 35–59% of their caloric intake (White and Broadley, 2009). In rice grains, starch represents the main component (almost 90%), followed by storage proteins, a small fraction of oils and a very low amount of nutrients (Kennedy et al., 2003). Furthermore, rice grains are polished before long-term storage because the oil-rich aleurone layer turns the seed rancid upon storage. The limited amount of micronutrients that are almost exclusively stored in the bran, aleurone layer and embryo of rice are removed during polishing and are therefore not available for human consumption. Considering these facts, improving the iron content and thus the nutritional quality of the rice endosperm is vital from a human nutrition and health point of view. Germplasm screening efforts have shown that rice germplasm has a very low or nearly no diversity for the key micronutrients, posing difficulties for the development of improved rice lines (Beyer, 2010). The iron content in rice grains varies from 6.3 to 24.4 mg/kg (Gregorio et al., 2000); however, most of this iron is lost during the milling process, leaving only about 2 μg/g in the polished grains (Bouis et al., 2011). Due to this insufficient natural genetic variation in the iron levels of rice endosperm, increasing iron levels through conventional breeding has not been possible (Slamet-Loedin et al., 2015). In such cases, biotechnological approaches offer options that, together with plant breeding, can help to achieve the target iron levels in popular rice varieties. For successful iron biofortification of crops, detailed knowledge about the complex homeostatic mechanisms that tightly control micronutrient concentrations in plants is essential. Among different plant species, Arabidopsis and rice are the ones for which substantial research on iron uptake, translocation and regulation has been conducted.

3 Molecular mechanisms controlling iron uptake and translocation in rice

Broadly, plants have developed two distinct strategies to acquire iron from the rhizosphere: the reduction strategy (Strategy I) in nongraminaceous plants and the chelation strategy (Strategy II) in graminaceous plants that include most of the important staple grain crops. Strategy I involves the induction of three mechanisms at the plasma membrane of root cells. In the first step, protons are released into the soil by H^+-ATPases, which results in acidification that increases the solubility of iron (Fig. 1). The FERRIC CHELATE REDUCTASES reduce the Fe(III) to the Fe(II) form; in Arabidopsis, the membrane-bound FERRIC REDUCTASE OXIDASE 2 (FRO2) serves this function (Robinson et al., 1999). Finally, transportation of solubilized Fe(II) into root cells is carried out by an IRON-REGULATED TRANSPORTER (IRT1), a member of the ZINC-REGULATED TRANSPORTER (ZRT)/

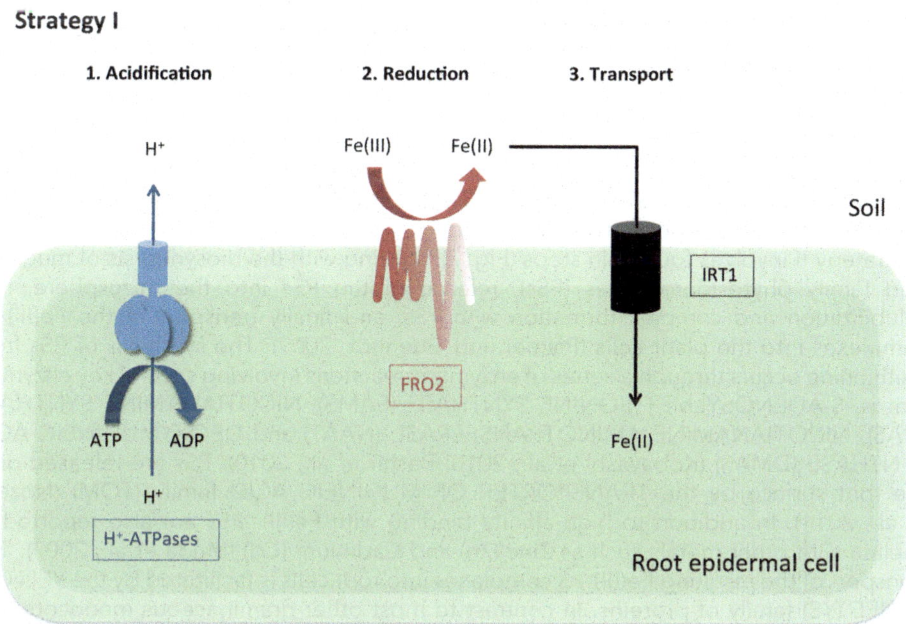

Figure 1 Iron acquisition mechanism used by nongraminaceous plants through reduction strategy (Strategy I).

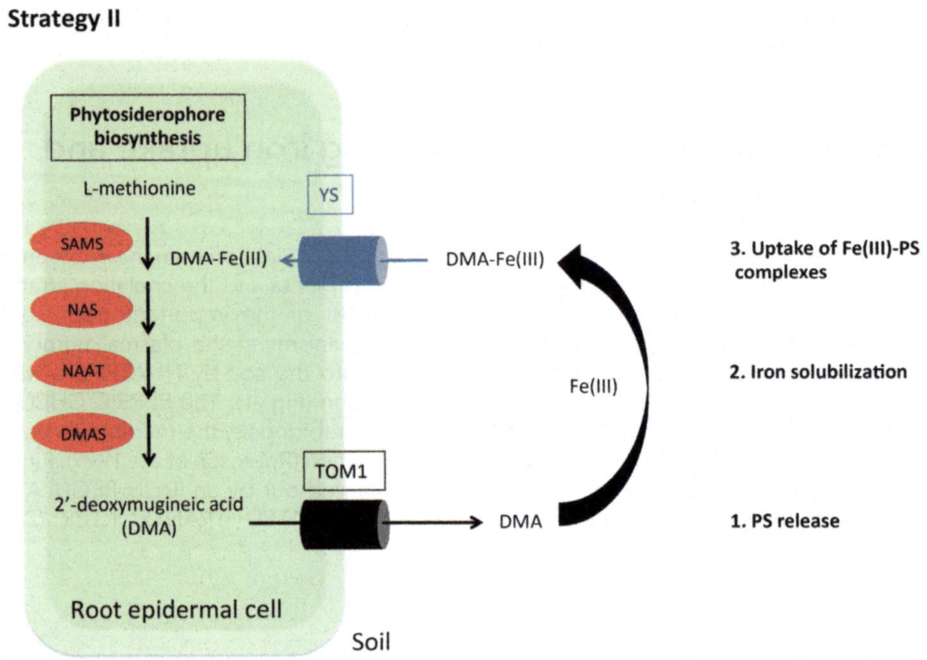

Figure 2 Iron acquisition mechanism used by graminaceous plants through chelation strategy (Strategy II).

IRON-REGULATED TRANSPORTER (ZRT/IRT1)-RELATED PROTEIN (ZIP) divalent metal transporter family (Connolly et al., 2002).

Strategy II involves four main steps (Fig. 2), starting with the biosynthesis of mugineic acid family phytosiderophores (PSs), release of the PSs into the rhizosphere, iron solubilization and complex formation with PSs, and finally transport of the Fe(III)-PS complexes into the plant cells (Palmer and Guerinot, 2009). The synthesis of PSs from methionine occurs through a series of enzymatic reactions involving several key enzymes, that is, S-ADENOSYLMETHIONINE SYNTHASE (SAMS), NICOTIANAMINE SYNTHASE (NAS), NICOTIANAMINE AMINOTRANSFERASE (NAAT) and DEOXYMUGINEIC ACID SYNTHASE (DMAS) (Kobayashi et al., 2010; Bashir et al., 2010). PSs are released onto the root surface by the TRANSPORTER OF MUGINEIC ACID family (TOM) (Nozoye et al., 2011). In addition to high-affinity binding with Fe(III), PSs are also reported to chelate with other metals such as Zinc (Zn) and Cadmium (Cd) (Meda et al., 2007). The transport of the resulting Fe(III)-PS complexes into root cells is facilitated by the YELLOW STRIPE (YS) family of proteins. In contrast to most other graminaceous monocots, rice possesses an efficient Fe(II) uptake system, although it apparently cannot accomplish ferric reduction as Strategy I plants do (Cheng et al., 2007). Rice *NAAT* mutants having an impaired ability to synthesize PSs could be rescued with external Fe(II) supply, while these mutants showed strong growth defects upon Fe(III) supply (Cheng et al., 2007).

The iron-deficiency-inducible genes that were upregulated when these mutants were grown on Fe(III) supply were not upregulated when mutants were grown on Fe(II), indicating that the *NAAT* mutants were not experiencing iron deficiency in the latter situation. This implied that rice can use Fe(II) as the sole source of iron unlike other grasses that require the Fe(III)-PS uptake system, which is in accordance with abundance of Fe(II) in the water-logged paddy fields where rice is typically grown. The homologs of the *Arabidopsis thaliana IRT1* (*AtIRT1*) in rice, that is, *OsIRT1* and *OsIRT2*, were shown to be induced upon iron deficiency (Ishimaru et al., 2006).

Iron translocation and maintenance of homeostasis in general have been researched extensively at the molecular level. Several genes that coordinate iron uptake, translocation and storage are reported to be transcriptionally regulated in response to iron availability. The principal transcription factors involved in iron homeostasis in plants include the IRON DEFICIENCY RESPONSE (FER) and the FER-LIKE IRON-DEFICIENCY-INDUCED TRANSCRIPTION FACTOR (FIT). In rice, *O. sativa*, IRON-DEFICIENCY-INDUCIBLE BASIC HELIX-LOOP-HELIX (bHLH) TRANSCRIPTION FACTOR 2 (IRO2) has a major role in the regulation of iron-responsive genes, such as *OsNAS1*, *OsNAAT1*, *OsDMAS1*, *OsTOM1* and *OsYSL15* (Ogo et al., 2007), and other transcription factors – IRON-DEFICIENCY-RESPONSIVE CIS-ACTING ELEMENT (IDE)-BINDING FACTOR (IDEF 1 and 2) – interact with IDE 1 and 2 (Kobayashi et al., 2007; Ogo et al., 2008). The IDEF1 has been suggested to play an important role in sensing cellular metal status by binding directly to the divalent metal ions, including iron and zinc (Kobayashi et al., 2012).

Different chelators including citrate, nicotianamine (NA) and deoxymugineic acid (DMA) facilitate long-distance iron transport within the plant. Citrate is predicted to chelate 99.5% of the iron present in the xylem (Durrett et al., 2007). The transporter FERRIC REDUCTASE DEFECTIVE 3 (FRD3), a member of the MULTIDRUG AND TOXIC COMPOUND EXTRUSION (MATE) family of transporters, facilitates efflux of citrate into the root vasculature in Arabidopsis in order to translocate iron to the other plant parts, including leaves (Durrett et al., 2007). The FRD3-LIKE PROTEIN (FRDL1) in rice is involved in iron-citrate translocation from rice roots to shoots (Yokosho et al., 2009). The presence of NA and DMA in xylem sap under iron deficiency suggested the function of these compounds in long-distance iron translocation (Kakei et al., 2009; Rellán-Álvarez et al., 2008). In addition, YELLOW STRIPE1-LIKE (YSL) family transporters are also suggested to play a role in iron transport, with the YSL2 thought to facilitate lateral movement of metals (Curie et al., 2009; Koike et al., 2004). Although relatively little is yet known about intracellular metal ion transport and storage, it has been demonstrated that the vacuole is an essential cell compartment for metal ion storage. In Arabidopsis, the VACUOLAR IRON TRANSPORTER 1 (VIT1), located in the vacuolar membrane, was suggested to mediate iron sequestration into vacuoles (Kim et al., 2006). Conversely, the NATURAL-RESISTANCE-ASSOCIATED MACROPHAGE PROTEIN (NRAMP) family members are required for export of iron from the vacuoles to the cytosol. In Arabidopsis, NRAMP3 and NRAMP4 are induced under iron deficiency conditions and mobilize vacuolar iron stores during early development (Lanquar et al., 2005). FERRITIN (FER), a complex protein capable of storing up to 4500 iron atoms, facilitates storage of iron in different plant organelles (Carrondo, 2003). The majority of FER is localized in chloroplasts, while mitochondria and other plastids also host FER for iron sequestration. Plant FERs comprise different isoforms, for instance FER1 and FER2, encoded by *OsFER1* and *OsFER2*, respectively, in rice (Stein et al., 2009). During seed germination, iron needs to be released from FER and/or phytate (Fu et al., 2010). In terms of human nutrition however, iron stored by FER is more bioavailable as compared to that stored by phytate (Murray-Kolb et al., 2003; Hoppler et al., 2008).

Together, these studies have increased our knowledge on transporters and chelators necessary for effective iron translocation within the plant. However, for each translocation step, the speciation of iron forms and chelators, the contribution of each type of transporter and the precise iron flux still need to be clarified.

4 Genetic engineering for increasing iron content in rice grain

Gene technology approaches aimed at increasing the iron content in rice grains focused on transformation with genes involved in iron uptake, transport and storage in the endosperm (Table 1). Different strategies are summarized here.

4.1 Increasing the iron stores in the rice endosperm and improved bioavailability

Due to high iron storage capacity, FER has been widely used in rice iron biofortification programmes. The FER-bound iron is bioavailable at gastric pH and is efficiently released during cooking (Hoppler et al., 2008). Endosperm-specific promoters have been used to drive the expression in order to target *FER* expression to the endosperm of transformed lines. Expression of *Glycine max* (soybean) *FERRITIN H1 (GmFERRITIN H1)* under the control of the rice-seed-specific promoter GLUTELIN B1 (*pGLUB1*) led to an increased iron and zinc content in the transformed seeds (Goto et al., 1999; Vasconcelos et al., 2003). Likewise, overexpression of *Phaseolus vulgaris* (common bean) *FERRITIN* (*PvFERRITIN*) in rice under the control of the *GLUTELIN* promoter also led to a twofold increase in iron content of the transformed rice seeds (Lucca et al., 2002). More recently, *PHYTOFERRTIN* genes expressed under the control of the rice *GLUTELIN* promoter have been transformed into rice mega variety IR64, leading to iron levels of up to 7.6 μg/g in the transgenic polished grains (Oliva et al., 2014). There have been efforts to increase the endosperm expression of *FER* by producing double *FER* lines by transforming rice with *FER* expressed under two different seed storage gene promoters: *pGLUB1* and *pGLOBULIN1* (*pGLB1*) promoters. Although *FER* expression could be elevated in these lines, there were no significant differences in the iron content of rice grains of single and double *FER* lines (Qu et al., 2005).

In the past, efforts have also been made to improve the bioavailability of iron through overexpression of proteins promoting iron absorption in humans or by elimination of antinutrients, like phytic acid. Rice plants that either overexpressed *CYSTEINE-RICH METALLOTHIONEIN-LIKE PROTEIN* or *PHYTASE* genes were produced (Lucca et al., 2001a). Significant increases in cysteine content of the soluble seed protein were reported, but no evidences on bioefficacy in reducing phytic acid were reported. Lucca and colleagues (Lucca et al., 2001b) reported a low activity of the *in planta* produced PHYTASE protein in the transgenic rice seeds upon cooking. This field of research has not

Table 1 Biotechnological approaches used to improve the iron content in rice grains. Different reports are grouped by the strategies used

Promoter::gene combination	Transformed rice cultivar	Increase of iron content in polished grains as compared to control[a]	Increase of iron content in unpolished grains as compared to control[a]	Increase of zinc content as compared to control	Reference
Endosperm-specific expression of *FERRITIN* gene encoding for iron storage protein					
pOsGLUB1::GmFERRITIN	*Japonica* cv. Kitaake	2-fold (T_2 seeds)	Up to 3-fold (T_1 seeds)	Not reported	Goto et al., 1999
pOsGLUB1::PvFERRITIN	*Japonica* cv. Taipei309	Not reported	Up to 2-fold (T_1 seeds)	Not reported	Lucca et al., 2002
pOsGLUB1::GmFERRITIN	*Indica* cv. IR68144	Up to 3.7-fold (T_2 seeds)	Up to 2.2-fold (T_1 seeds)	Up to 1.65-fold (T_1 unpolished seeds)	Vasconcelos et al., 2003
pOsGLUB1::GmFERRITIN H1, and *pOsGLB1::GmFERRITIN H1*	*Japonica* cv. Kitaake and FK22 line (Goto et al., 1999)	Not reported	Up to 2.4-fold (seeds of T_5 plants)[b]	Up to 1.1-fold (unpolished seeds of T_6 plants)[b]	Qu et al., 2005
pOsGLUA2::OsFERRITIN2	*Indica* cv. Pusa-Sugandhi II	Up to 2.09-fold (T_3 seeds)	Not reported	Up to 1.37-fold (T_3 polished seeds)	Paul et al., 2012
pOsGLUB4::GmFERRITIN H1	*Indica* cv. IR64	Up to 3.4-fold (T_3 seeds)	Not reported	Not reported	Oliva et al., 2014
Overexpression of genes encoding iron chelators and/or transporters, i.e. *NAS, YSL, DIOXYGENASE, IRT*					
pCaMV35S::HvNAS1	*Japonica* cv. Tsukinohikari	2.3-fold (T_2 seeds)	1.4-fold (T_2 seeds)	1.5-fold (T_2 polished seeds)	Masuda et al., 2009
pOsACTIN1::HvNAS1	*Japonica* cv. Tsukinohikari	Up to 4.5-fold (T_1 seeds) Up to 3.4-fold (T_2 seeds)	Up to 1.2-fold (T_2 seeds)	Up to 2.5 fold (T_1 polished seeds)	Masuda et al., 2009
OsNAS3 with downstream *35S* enhancer element	*Japonica* cv. Dongjin	2.6-fold	Up to 2.9-fold	2.2-fold (polished & unpolished seeds)	Lee et al., 2009a
pCaMV35S::OsNAS1	*Japonica* cv. Nipponbare	2.2-fold (T_1 seeds)	Up to 2.4-fold (T_1 seeds)	1.9-fold (T_1 unpolished seeds)	Johnson et al., 2011
pCaMV35S::OsNAS2	*Japonica* cv. Nipponbare	Up to 4.2-fold (T_1 seeds)	Up to 3.5-fold (T_1 seeds)	2.5-fold (T_1 unpolished seeds)	Johnson et al., 2011
pCaMV35S::OsNAS3	*Japonica* cv. Nipponbare	2.2-fold (T_1 seeds)	Up to 2.7-fold (T_1 seeds)	2.1-fold (T_1 unpolished seeds)	Johnson et al., 2011

(Continued)

Table 1 (Continued)

Promoter::gene combination	Transformed rice cultivar	Increase of iron content in polished grains as compared to control[a]	Increase of iron content in unpolished grains as compared to control[a]	Increase of zinc content as compared to control	Reference
pOsACTIN1::OsYSL15	*Japonica* cv. Dongjin	Not reported	Up to 1.2-fold	No increase	Lee et al., 2009b
pOsSUT1::OsYSL2	*Japonica* cv. Tsukinohikari	Up to 4.4-fold	Not reported	Not reported	Ishimaru et al., 2010
HvIDS3 genome fragment	*Japonica* cv. Tsukinohikari	1.4-fold	1.25-fold	1.35-fold (polished seeds)	Masuda et al., 2008
HvIDS3 genome fragment	*Japonica* cv. Tsukinohikari	Not reported	1.26-fold	Up to 1.36-fold (unpolished seeds)	Suzuki et al., 2008
pZmUBIQUITIN::OsIRT1		Not reported	1.1-fold	1.1-fold (unpolished seeds)	Lee and An, 2009
pCaMV35S::MxIRT1	*Japonica* cv.	Not reported	Up to 2.4-fold (T_3 seeds)	Up to 3-fold (T_3 unpolished seeds)	Tan et al., 2015
Overexpression of *OsIRO2* encoding transcription factor					
pCaMV35S::OsIRO2	*Japonica* cv. Tsukinohikari	Not reported	Up to 2.7-fold	Up to 1.4-fold (unpolished seeds)	Ogo et al., 2011
Knockdown of genes encoding vacuolar iron transporters, i.e. *OsVIT* 1 or 2					
OsVIT1 or	*Japonica* cv. Zhonghua11	Not reported	1.2-fold	1.2-fold (unpolished seeds)	Zhang et al., 2012
OsVIT2 T-DNA insertion	*Japonica* cv. Dongjin	Not reported	1.5-fold	1.4-fold (unpolished seeds)	
OsVIT2 T-DNA insertion		1.7-fold	1.3-fold	1.6-fold (unpolished seeds)	Bashir et al., 2013
Strategies focusing on concerted expression of different genes					
pOsGLB1::PvFERRITIN, *pCaMV35S::AtNAS1*, and *pOsGLB1::AfPHYTASE*	*Japonica* cv. Taipei 309	Up to 6.3-fold (Seeds of T_3 plants)	Up to 2.0-fold (Seeds of T_3 plants)	Up to 1.6-fold (unpolished seeds of T_3 plants)	Wirth et al., 2009

pOsGLUB1::GmFERRITIN H2, *pOsGLB1::GmFERRITIN H2*, *pOsACTIN1::HvNAS1*, *pOsSUT1::OsYSL2*, and *pOsGLB1::OsYSL2*	*Japonica* cv. Tsukinohikari	Up to 6-fold (T_2 seeds) and up to 4.4-fold (T_3 seeds)	1.6-fold (T_3 seeds)	Up to 1.6-fold (T_3 polished seeds)	Masuda et al., 2012
pOsGLUB1::GmFERRITIN H2, *pOsGLB1::GmFERRITIN H2*, *pOsACTIN1::HvNAS1*, *pOsSUT1::OsYSL2*, and *pOsGLB1::OsYSL2*	*Japonica* cv. Paw San Yin	Up to 3.4-fold (T_2 seeds)	Up to 1.37-fold (T_2 seeds)	Up to 1.3-fold (T_2 polished seeds)	Aung et al., 2013
pOsGLUB1::GmFERRITIN H2, *pOsGLB1::GmFERRITIN H2*, *HvNAS1*, *HvNAAT-A,-B*, and *IDS3* genome fragment	*Japonica* cv. Tsukinohikari	Up to 3.6-fold (T_3 seeds)	Up to 1.5-fold (T_3 seeds)	Up to 1.3-fold (T_3 polished seeds)	Masuda et al., 2013
pMsENOD12B::AtIRT1, *pOsGLB1::PvFERRITIN*, *pCaMV35S::AtNAS1*, and *pOsGLB1::AfPHYTASE*	*Japonica* cv. Taipei 309 and NFP rice (Wirth et al., 2009)	Up to 4.24-fold (T_3 seeds)[b]	Up to 1.47-fold (T_3 seeds)[b]	Up to 1.5-fold (T_3 polished seeds)[b]	Boonyaves et al., 2015

[a] Plant generation of tested grains written in parentheses, if reported.

[b] Fold change values are provided in comparison to the non-transformed rice variety and not to the transgenic background used for supertransformation.

Abbreviations used in the table are explained as follows: *pCaMV35S, CAULIFLOWER MOSAIC VIRUS 35S* promoter; *pMsENOD12B, Medicago sativa EARLY NODULIN 12B promoter; pOsACTIN1, Oryza sativa ACTIN1* promoter; *pOsGLB1, Oryza sativa GLOBULIN* promoter; *pOsGLUA2, Oryza sativa GLUTELINA2* promoter; *pOsGLUB1, Oryza sativa GLUTELIN* promoter; *pOsGLUB4, Oryza sativa GLUTELIN-4* promoter; *pOsSUT1, Oryza sativa SUCROSE TRANSPORTER 1* promoter; *pZmUBIQUITIN, Zea mays UBIQUITIN* promoter; *AfPHYTASE, Aspergillus fumigatus PHYTASE; AtIRT1, Arabidopsis thaliana IRON-REGULATED TRANSPORTER 1; AtNAS1, Arabidopsis thaliana NICOTIANAMINE SYNTHASE 1; GmFERRITIN H1, Glycine max FERRITIN H1; GmFERRITIN H2, Glycine max FERRITIN H2; HvIDS3, Hordeum vulgare IDS3; HvNAAT-A, -B, Hordeum vulgare NICOTIANAMINE AMINOTRANSFERASE-A, -B; HvNAS1, Hordeum vulgare NICOTIANAMINE SYNTHASE 1; MxIRT1, Malus xiaojinensis IRON-REGULATED TRANSPORTER 1; OsFERRITIN2, Oryza sativa FERRITIN 2; OsIRO2, Oryza sativa IRON-DEFICIENCY-INDUCIBLE BASIC HELIX-LOOP-HELIX (bHLH) TRANSCRIPTION FACTOR 2; OsIRT1, Oryza sativa IRON-REGULATED TRANSPORTER 1; OsNAS1, Oryza sativa NICOTIANAMINE SYNTHASE 1; OsNAS2, Oryza sativa NICOTIANAMINE SYNTHASE 2; OsNAS3, Oryza sativa NICOTIANAMINE SYNTHASE 3; OsVIT1, Oryza sativa VACUOLAR IRON TRANSPORTER 1; OsVIT2, Oryza sativa VACUOLAR IRON TRANSPORTER 2; OsYSL2, Oryza sativa YELLOW STRIPE1-LIKE 2; OsYSL15, Oryza sativa YELLOW STRIPE1-LIKE 15; PvFERRITIN, Phaseolus vulgaris FERRITIN.*

been extended much further, particularly due to shift of focus towards FER that stores iron in a bioavailable form.

4.2 Transformation of rice with iron uptake- and translocation-related genes

Among several other genes involved in iron uptake and translocation, *NAS*, one of the key genes involved in PS biosynthesis, has received maximum focus for rice transformation with respect to iron biofortification. In addition to promoting an increased production of PSs, the overexpression of *NAS* could enhance the production of NA, which is one of the main chelators required for effective iron translocation within the plants. Expression of the *Hordeum vulgare* (barley) *NAS* gene (*HvNAS*) under the control of *CAULIFLOWER MOSAIC VIRUS 35S* (*CaMV35S*) and rice *ACTIN 1* (*OsACTIN1*) promoters led to an increased iron and zinc content in the polished rice grains (Masuda et al., 2009). The activation of *OsNAS2* led to a threefold iron increase in mature rice seeds (Lee et al., 2012). Similarly, increased iron, zinc and copper concentrations in the grains of activation-tagged mutant rice lines overexpressing *OsNAS3* were reported (Lee et al., 2009a). Rice lines overexpressing *OsNAS2* contained a fourfold and more than twofold higher iron and zinc content in the polished rice grains, respectively (Johnson et al., 2011).

Besides the focus on PS synthesis related genes, iron transporters have also been tested. Expression of *OsYSL2* under the control of *SUCROSE TRANSPORTER 1* (*OsSUT1*) promoter led to 4.4-fold increased iron content in the polished rice grains (Ishimaru et al., 2010). *OsACTIN 1* promoter driven expression of *OsYSL15* also exhibited increased iron content in unpolished grains (Lee et al., 2009b). Rice engineered with *HvYS1* showed increased iron accumulation in leaves, but no such increases were reported in the seeds (Gómez-Galera et al., 2012). Knockdown of the *OsVIT1* and *OsVIT2* elevated iron levels up to 1.7-fold in the polished grains, most likely due to enhanced iron translocation from flag leaves to rice grains (Bashir et al., 2013; Zhang et al., 2012). Overexpression of the *OsIRT1* in rice plants led to an increased iron concentration in shoots, roots and mature seeds, as well as increased iron deficiency tolerance at the seedling stage in the transformed plants (Lee and An, 2009). Introduction of *Arachis hypogaea* (peanut) *IRT1* (*AhIRT1*) in rice resulted in high tolerance to low iron availability when grown on calcareous soils (Xiong et al., 2014). The overexpression of *Malus xiaojinensis* (apple) *IRT1* (*MxIRT1*) in rice led to a 2.4-fold increased iron content in the mature grains (Tan et al., 2015). Besides the transporters, constitutive overexpression of *OsIRO2*, a transcription factor responsible for regulating iron-deficiency-inducible genes, led to a 2.7-fold higher iron content in the rice grains (Ogo et al., 2011).

4.3 Strategies focusing on concerted expression of multiple genes

A strategy expressing genes for effective iron uptake and storage from a single construct was pioneered by Wirth and colleagues (Wirth et al., 2009), where *AtNAS1* and *PvFERRITIN* exhibited a synergistic effect. The transformed plants had more than sixfold increase in iron content in the polished grains, with accompanied increases in zinc content. Later, a number of studies followed and transformed different combinations of genes responsible for iron uptake, translocation and/or storage in the endosperm. Rice engineered with *HvNAS1*,

OsYSL2 and *GmFERRITIN* showed significantly increased iron content in the polished grains of two different rice cultivars (Aung et al., 2013; Masuda et al., 2012). Similarly, overexpression of different genes involved in PS production, including *HvNAS1*, *HvNAAT* and *IDS3* genome fragments, together with the endosperm-specific expression of *FER*, elevated the iron content in rice endosperm by nearly fourfold (Masuda et al., 2013). More recently, the concerted expression of Arabidopsis *IRT1* (*AtIRT1*), together with Arabidopsis *NAS1* and *PvFERRITIN*, was shown to significantly increase iron content in both polished and unpolished rice grains, yielding rice lines containing around 9.6 µg/g DW iron in the polished grains (Boonyaves et al., 2015). In this study, the *AtIRT1* was expressed under control of the *Medicago sativa EARLY NODULIN 12B* promoter and was transformed into previously developed high-iron NFP rice lines expressing *NICOTIANAMINE SYNTHASE* (*AtNAS1*) and *FER* (Wirth et al., 2009). The strategies involving combination of genes appear relatively more promising as compared to those focusing on single genes at a time. However, further iron increases need to be achieved in order to generate rice lines with the target levels of around 15 µg/g dry weight iron in the polished grains that could meet 30% of the recommended daily dietary allowance (Bouis et al., 2011).

5 Future trends

Iron biofortification of rice endosperm has been very challenging. The majority of approaches tested so far have not generated rice lines that meet the target iron levels set by the HarvestPlus program. So far, there is only a single published report on rice lines transformed with *OsNAS2* where iron concentrations of 14 µg/g and 19 µg/g were detected in the endosperm of T_1 grains (Johnson et al., 2011). However, this report used cultivar Nipponbare for transformation with around 4 µg/g baseline iron content in endosperm, which is higher than most other rice cultivars ranging between 1 and 2 µg/g for iron content in their endosperm. In terms of fold changes, approximately fourfold increase in iron content was obtained, which is in a similar range to that achieved by using several other approaches. Reaching an iron content of up to 15 µg/g from a baseline of 2 µg/g in the rice endosperm has proven difficult. This is mainly due to complex regulation of micronutrient homeostasis in plants. Although a lot of research has focused on understanding iron uptake, translocation and storage in plants, including rice, our understanding on the precise role of each of these genes/transporters is still far from sufficient. The choice of *NAS* genes facilitating effective iron uptake and of FER proteins for storing the iron in the endosperm has worked excellently, but in order to further raise the iron levels, additional candidate genes are required. With all these studies, it has also become evident that increased amounts of iron transported into the plant do not necessarily reach the grain. It is worth noting that iron levels in rice leaves are much higher than those within the grain, and in the seed much more iron accumulates in the aleurone layer and seed coat (which are removed during polishing) than in the endosperm. The rice lines with constructs expressing the *FER* gene twice in the endosperm did not increase the iron content as compared to lines expressing *FER* only once (Qu et al., 2005). These independent studies suggest the presence of strong physiological/molecular barrier(s) that limits iron transport into the grain and/or endosperm part. One such barrier may be the formation of iron-phytate complexes that become deposited in the protein storage vacuoles (PSVs) of the aleurone layer and embryo scutellum. Therefore, the biofortification strategies need to extend to candidate genes that can increase the availability of free iron for transport and storage into the grains, in addition to promoting effective translocation within the plant.

The broad-range metal transport ability of different candidate transporters is another important aspect that should be addressed by the biofortification programmes. It is known that changes in content of one element within a plant ionome can lead to changes in the concentrations of other elements. The overexpression of a *NAS* gene in order to increase the iron levels often also resulted in accompanied increase in zinc content (Johnson et al., 2011; Masuda et al., 2012; Tan et al., 2015; Wirth et al., 2009). Likewise, the IRT1 is capable of transporting iron, zinc, manganese and cadmium (Vert et al., 2002). While increases in zinc and manganese levels could be beneficial, increases in cadmium concentrations of rice grains are not desirable. Therefore, it is also important that the efforts aimed at iron biofortification comprehensively screen the profiles of different elements in the transformed grains. This is particularly important for avoiding increases in levels of toxic metals such as cadmium. When using broad-range transporters as candidates, consideration should be given to metal selectivity and transport abilities of these transporters. Furthermore, the majority of the iron biofortification efforts involved transformation of tissue culture competent genotypes. Future efforts need to target transformation to elite cultivars that are widely grown for broader and faster accessibility to breeding programmes. Field studies are also currently very limited and should be performed to examine if the iron levels achieved in the greenhouse grown plants are maintained under field conditions. To facilitate product development, precise molecular data on transgene copy number, insertion site and expression levels are required.

6 Where to look for further information

Breeding of rice varieties with increased iron content has met with significant scientific interest and has been discussed vividly (for review, see Bouis et al., 2011; Mayer et al., 2008). Recent research has provided insight into the complexity of iron homeostasis in plants (for review, see Bashir et al., 2010; Curie et al., 2009; Kobayashi et al., 2010; Palmer et al., 2009). Several breeding approaches and biotechnological modifications allowed to increase the iron content in the endosperm of rice grains, as reviewed in Slamet-Loedin et al. (2015). Global projects like the HarvestPlus Program (http://www.harvestplus.org/) lead efforts to breed crops with improved nutritional traits.

7 References

Abbaspour, N., Hurrell, R. and Kelishadi, R. (2014), Review on iron and its importance for human health, *J. Res. Med. Sci.*, 19(2), 164–74.

Aung, M. S., Masuda, H., Kobayashi, T., Nakanishi, H., Yamakawa, T. and Nishizawa, N. K. (2013), Iron biofortification of Myanmar rice, *Front. Plant Sci.*, 4, 158. doi: 10.3389/fpls.2013.00158.

Bashir, K., Ishimaru, Y. and Nishizawa, N. K. (2010). Iron uptake and loading into rice grain, *Rice*, 3, 122–30.

Bashir, K., Takahashi, R., Akhtar, S., Ishimaru, Y., Nakanishi, H. and Nishizawa, N. K. (2013), The knockdown of *OsVIT2* and *MIT* affects iron localization in rice seed, *Rice*, 6(31), 1–7.

Beyer, P. (2010), Golden Rice and 'Golden' crops for human nutrition, *N. Biotechnol.*, 27(5), 478–81.

Boonyaves, K., Gruissem, W. and Bhullar, N. K. (2015), NOD promoter-controlled *AtIRT1* expression functions synergistically with *NAS* and *FERRITIN* genes to increase iron in rice grains, *Plant Mol. Biol.*, in press. doi:10.1007/s11103-015-0404-0.

Bouis, H. E., Hotz, C., McClafferty, B., Meenakshi, J. V. and Pfeiffer, W. H. (2011), Biofortification: A new tool to reduce micronutrient malnutrition, *Food Nutr. Bull.*, 32(Supplement 1), 31S–40S.

Carrondo, M. A. (2003), Ferritins, iron uptake and storage from the bacterioferritin viewpoint, *EMBO J.*, 22(9), 1959–68.

Cheng, L., Wang, F., Shou, H., Huang, F., Zheng, L., He, F., Li, J., Zhao, F. J., Ueno, D., Ma, J. F. and Wu, P. (2007), Mutation in nicotianamine aminotransferase stimulated the Fe(II) acquisition system and led to iron accumulation in rice, *Plant Physiol.*, 145(4), 1647–57.

Connolly, E. L., Fett, J. P. and Guerinot, M. L. (2002), Expression of the IRT1 metal transporter is controlled by metals at the levels of transcript and protein accumulation, *Plant Cell*, 14(6), 1347–57.

Curie, C., Cassin, G., Couch, D., Divol, F., Higuchi, K., Le Jean, M., Misson, J., Schikora, A., Czernic, P. and Mari, S. (2009), Metal movement within the plant: Contribution of nicotianamine and yellow stripe 1-like transporters, *Ann. Bot.*, 103(1), 1–11.

Durrett, T. P., Gassmann, W. and Rogers, E. E. (2007), The FRD3-mediated efflux of citrate into the root vasculature is necessary for efficient iron translocation, *Plant Physiol.*, 144(1), 197–205.

Frossard, E., Bucher, M., Mächler, F., Mozafar, A. and Hurrell, R. (2000), Potential for increasing the content and bioavailability of Fe, Zn and Ca in plants for human nutrition, *J. Sci. Food Agric.*, 80(7), 861–79.

Fu, X., Deng, J., Yang, H., Masuda, T., Goto, F., Yoshihara, T. and Zhao, G. (2010), A novel EP-involved pathway for iron release from soya bean seed ferritin, *Biochem. J.*, 427, 313–21.

Galloway, R. and McGuire, J. (1994), Determinants of compliance with iron supplementation: Supplies, side effects, or psychology?, *Soc. Sci. Med.*, 39(3), 381–90.

Gómez-Galera, S., Sudhakar, D., Pelacho, A. M., Capell, T. and Christou, P. (2012), Constitutive expression of a barley Fe phytosiderophore transporter increases alkaline soil tolerance and results in iron partitioning between vegetative and storage tissues under stress, *Plant Physiol. Biochem.*, 53, 46–53.

Goto, F., Yoshihara, T., Shigemoto, N., Toki, S. and Takaiwa, F. (1999), Iron fortification of rice seed by the soybean ferritin gene, *Nat. Biotechnol.*, 17(3), 282–6.

Gregorio, G. B., Senadhira, D., Htut, H. and Graham, R. D. (2000), Breeding for trace mineral density in rice, *Food Nutr. Bull.*, 21(4), 382–6.

Hoppler, M., Schönbächler, A., Meile, L., Hurrell, R. F. and Walczyk, T. (2008), Ferritin-iron is released during boiling and in vitro gastric digestion, *J. Nutr.*, 138(5), 878–84.

Hurrell, R. and Egli, I. (2010), Iron bioavailability and dietary reference values, *Am. J. Clin. Nutr.*, 91(5), 1461S–7S.

Ishimaru, Y., Suzuki, M., Tsukamoto, T., Suzuki, K., Nakazono, M., Kobayashi, T., Wada, Y., Watanabe, S., Matsuhashi, S., Takahashi, M., Nakanishi, H., Mori, S. and Nishizawa, N. K. (2006), Rice plants take up iron as an Fe^{3+} phytosiderophore and as Fe^{2+}, *Plant J.*, 45(3), 335–46.

Ishimaru, Y., Masuda, H., Bashir, K., Inoue, H., Tsukamoto, T., Takahashi, M., Nakanishi, H., Aoki, N., Hirose, T., Ohsugi, R. and Nishizawa, N. K. (2010), Rice metal nicotianamine transporter, *OsYSL2*, is required for the long distance transport of iron and manganese, *Plant J.*, 62(3), 379–90.

Johnson, A. A., Kyriacou, B., Callahan, D. L., Carruthers, L., Stangoulis, J., Lombi, E. and Tester, M. (2011), Constitutive overexpression of the *OsNAS* gene family reveals single-gene strategies for effective iron-and zinc-biofortification of rice endosperm, *PLoS One*, 6(9), e24476.

Kakei, Y., Yamaguchi, I., Kobayashi, T., Takahashi, M., Nakanishi, H., Yamakawa, T. and Nishizawa, N. K. (2009), A highly sensitive, quick and simple quantification method for nicotianamine and 2′-deoxymugineic acid from minimum samples using LC/ESI-TOF-MS achieves functional analysis of these components in plants, *Plant Cell Physiol.*, 50(11), 1988–93.

Kennedy, G., Burlingame, B. and Nguyen, V. (2003), Nutritional contribution of rice and impact of biotechnology and biodiversity in rice-consuming countries, *Food Chem.*, 80(4), 589–96.

Kim, S. A., Punshon, T., Lanzirotti, A., Li, L., Alonso, J. M., Ecker, J. R., Kaplan, J. and Guerinot, M. L. (2006), Localization of iron in *Arabidopsis* seed requires the vacuolar membrane transporter VIT1, *Science*, 314(5803), 1295–8.

Kobayashi, T., Ogo, Y., Nakanishi Itai, R., Nakanishi, H., Takahashi, M., Mori, S. and Nishizawa, N. K. (2007), The transcription factor IDEF1 regulates the response to and tolerance of iron deficiency in plants, *Proc. Natl. Acad. Sci. USA*, 104(48), 19150–5.

Kobayashi, T., Nakanishi, H. and Nishizawa, N. K. (2010), Recent insights into iron homeostasis and their application in graminaceous crops, *Proc. Jpn. Acad., Series B*, 86(9), 900–13.

Kobayashi, T., Nakanishi Itai, R., Aung, M. S., Senoura, T., Nakanishi, H. and Nishizawa, N. K. (2012), The rice transcription factor IDEF1 directly binds to iron and other divalent metals for sensing cellular iron status, *Plant J.*, 69(1), 81–91.

Koike, S., Inoue, H., Mizuno, D., Takahashi, M., Nakanishi, H., Mori, S. and Nishizawa, N. K. (2004), *OsYSL2* is a rice metal-nicotianamine transporter that is regulated by iron and expressed in the phloem, *Plant J.*, 39(3), 415–24.

Lanquar, V., Lelièvre, F., Bolte, S., Hamès, C., Alcon, C., Neumann, D., Vansuyt, G., Curie, C., Schröder, A., Krämer, U., Barbier-Brygoo, H. and Thomine, S. (2005), Mobilization of vacuolar iron by AtNRAMP3 and AtNRAMP4 is essential for seed germination on low iron, *EMBO J.*, 24(23), 4041–51.

Lee, S. and An, G. (2009), Over expression of *OsIRT1* leads to increased iron and zinc accumulations in rice, *Plant Cell Environ.*, 32(4), 408–16.

Lee, S., Jeon, U. S., Lee, S. J., Kim, Y.-K., Persson, D. P., Husted, S., Schjørring, J. K., Kakei, Y., Masuda, H., Nishizawa, N. K. and An, G. (2009a), Iron fortification of rice seeds through activation of the *nicotianamine synthase* gene, *Proc. Natl. Acad. Sci. USA*, 106(51), 22014–19.

Lee, S., Chiecko, J. C., Kim, S. A., Walker, E. L., Lee, Y., Guerinot, M. L. and An, G. (2009b), Disruption of *OsYSL15* leads to iron inefficiency in rice plants, *Plant Physiol.*, 150(2), 786–800.

Lee, S., Kim, Y.-S., Jeon, U. S., Kim, Y.-K., Schjoerring, J. K. and An, G. (2012), Activation of rice *nicotianamine synthase 2* (*OsNAS2*) enhances iron availability for biofortification, *Mol. Cells*, 33(3), 269–75.

Lucca, P., Hurrell, R. and Potrykus, I. (2001a), Genetic engineering approaches to improve the bioavailability and the level of iron in rice grains, *Theor. Appl. Genet.*, 102(2–3), 392–7.

Lucca, P., Hurrell, R. and Potrykus, I. (2001b), Approaches to improving the bioavailability and level of iron in rice seeds, *J. Sci. Food Agric.*, 81(9), 828–34.

Lucca, P., Hurrell, R. and Potrykus, I. (2002), Fighting iron deficiency anemia with iron-rich rice, *J. Am. Coll. Nutr.*, 21(3), 184S–90S.

Masuda, H., Suzuki, M., Morikawa, K. C., Kobayashi, T., Nakanishi, H., Takahashi, M., Saigusa, M., Mori, S. and Nishizawa, N. K. (2008), Increase in iron and zinc concentrations in rice grains via the introduction of barley genes involved in phytosiderophore synthesis, *Rice*, 1(1), 100–8.

Masuda, H., Usuda, K., Kobayashi, T., Ishimaru, Y., Kakei, Y., Takahashi, M., Higuchi, K., Nakanishi, H., Mori, S. and Nishizawa, N. K. (2009), Overexpression of the barley nicotianamine synthase gene *HvNAS1* increases iron and zinc concentrations in rice grains, *Rice*, 2(4), 155–66.

Masuda, H., Ishimaru, Y., Aung, M. S., Kobayashi, T., Kakei, Y., Takahashi, M., Higuchi, K., Nakanishi, H. and Nishizawa, N. K. (2012), Iron biofortification in rice by the introduction of multiple genes involved in iron nutrition, *Sci. Rep.*, 2, 543.

Masuda, H., Kobayashi, T., Ishimaru, Y., Takahashi, M., Aung, M. S., Nakanishi, H., Mori, S. and Nishizawa, N. K. (2013), Iron-biofortification in rice by the introduction of three barley genes participated in mugineic acid biosynthesis with soybean *ferritin* gene, *Front. Plant Sci.*, 4, 132. doi: 10.3389/fpls.2013.00132.

Mayer, J. E., Pfeiffer, W. H. and Beyer, P. (2008), Biofortified crops to alleviate micronutrient malnutrition, *Curr. Opin., Plant Biol.*, 11(2), 166–70.

Meda, A. R., Scheuermann, E. B., Prechsl, U. E., Erenoglu, B., Schaaf, G., Hayen, H., Weber, G. and von Wirén, N. (2007), Iron acquisition by phytosiderophores contributes to cadmium tolerance, *Plant Physiol.*, 143(4), 1761–73.

Murray-Kolb, L. E., Welch, R., Theil, E. C. and Beard, J. L. (2003), Women with low iron stores absorb iron from soybeans, *Am. J. Clin. Nutr.*, 77(1), 180–4.

National Academy of Sciences, Institute of Medicine (2001), Dietary Reference Intakes for Vitamin A, Vitamin K, Arsenic, Boron, Chromium, Copper, Iodine, Iron, Manganese, Molybdenum, Nickel, Silicon, Vanadium, and Zinc, 290–393 [Online]. Available: http://www.nap.edu/read/10026/chapter/1 [Accessed 29 October 2015].

Nozoye, T., Nagasaka, S., Kobayashi, T., Takahashi, M., Sato, Y., Sato, Y., Uozumi, N., Nakanishi, H. and Nishizawa, N. K. (2011), Phytosiderophore efflux transporters are crucial for iron acquisition in graminaceous plants, *J. Biol. Chem.*, 286(7), 5446–54.

Ogo, Y., Nakanishi Itai, R., Nakanishi, H., Kobayashi, T., Takahashi, M., Mori, S. and Nishizawa, N. K. (2007), The rice bHLH protein *OsIRO2* is an essential regulator of the genes involved in Fe uptake under Fe deficient conditions, *Plant J.*, 51(3), 366–77.

Ogo, Y., Kobayashi, T., Nakanishi Itai, R., Nakanishi, H., Kakei, Y., Takahashi, M., Toki, S., Mori, S. and Nishizawa, N. K. (2008), A novel NAC transcription factor, IDEF2, that recognizes the iron deficiency-responsive element 2 regulates the genes involved in iron homeostasis in plants, *J. Biol. Chem.*, 283(19), 13407–17.

Ogo, Y., Nakanishi Itai, R., Kobayashi, T., Aung, M. S., Nakanishi, H. and Nishizawa, N. K. (2011), *OsIRO2* is responsible for iron utilization in rice and improves growth and yield in calcareous soil, *Plant Mol. Biol.*, 75(6), 593–605.

Oliva, N., Chadha-Mohanty, P., Poletti, S., Abrigo, E., Atienza, G., Torrizo, L., Garcia, R., Dueñas Jr, C., Poncio, M. A., Balindong, J., Manzanilla, M., Montecillo, F., Zaidem, M., Barry, G., Hervé, P., Shou, H. and Slamet-Loedin, I. H. (2014), Large-scale production and evaluation of marker-free *indica* rice IR64 expressing phytoferritin genes, *Mol. Breed.*, 33(1), 23–37.

Oppenheimer, S. J. (2001), Iron and its relation to immunity and infectious disease, *J. Nutr.*, 131(2), 616S–35S.

Palmer, C. M. and Guerinot, M. L. (2009), Facing the challenges of Cu, Fe and Zn homeostasis in plants, *Nat. Chem. Biol.*, 5(5), 333–40.

Paul, S., Ali, N., Gayen, D., Datta, S. K. and Datta, K. (2012), Molecular breeding of *Osfer2* gene to increase iron nutrition in rice grain, *GM Crops Food*, 3(4), 310–16.

Qu, L. Q., Yoshihara, T., Ooyama, A., Goto, F. and Takaiwa, F. (2005), Iron accumulation does not parallel the high expression level of ferritin in transgenic rice seeds, *Planta*, 222(2), 225–33.

Rellán-Álvarez, R., Abadía, J. and Álvarez-Fernández, A. (2008), Formation of metal-nicotianamine complexes as affected by pH, ligand exchange with citrate and metal exchange. A study by electrospray ionization time-of-flight mass spectrometry, *Rapid Commun. Mass Spectrom.*, 22(10), 1553–62.

Robinson, N. J., Procter, C. M., Connolly, E. L. and Guerinot, M. L. (1999), A ferric-chelate reductase for iron uptake from soils, *Nature*, 397(6721), 694–7.

Sazawal, S., Black, R. E., Ramsan, M., Chwaya, H. M., Stoltzfus, R. J., Dutta, A., Dhingra, U., Kabole, I., Deb, S., Othman, M. K. and Kabole, F. M. (2006), Effects of routine prophylactic supplementation with iron and folic acid on admission to hospital and mortality in preschool children in a high malaria transmission setting: Community-based, randomised, placebo-controlled trial, *Lancet*, 367(9505), 133–43.

Slamet-Loedin, I. H., Johnson-Beebout, S. E., Impa, S. and Tsakirpaloglou, N. (2015), Enriching rice with Zn and Fe while minimizing Cd risk, *Front. Plant Sci.*, 6, 121.

Stein, R. J., Ricachenevsky, F. K. and Fett, J. P. (2009), Differential regulation of the two rice ferritin genes (*OsFER1* and *OsFER2*), *Plant Sci.*, 177(6), 563–9.

Stevens, G. A., Finucane, M. M., De-Regil, L. M., Paciorek, C. J., Flaxman, S. R., Branca, F., Peña-Rosas, J. P., Bhutta, Z. A., Ezzati, M. and Nutrition Impact Model Study Group (2013), Global, regional, and national trends in haemoglobin concentration and prevalence of total and severe anaemia in children and pregnant and non-pregnant women for 1995-2011: A systematic analysis of population-representative data, *Lancet Glob. Health*, 1(1), e16–25.

Suzuki, M., Morikawa, K. C., Nakanishi, H., Takahashi, M., Saigusa, M., Mori, S. and Nishizawa, N. K. (2008), Transgenic rice lines that include barley genes have increased tolerance to low iron availability in a calcareous paddy soil, *J. Soil Sci. Plant Nutr.*, 54(1), 77–85.

Tan, S., Han, R., Li, P., Yang, G., Li, S., Zhang, P., Wang, W. B., Zhao, W. Z. and Yin, L.P. (2015), Over-expression of the *MxIRT1* gene increases iron and zinc content in rice seeds, *Transgenic Res.*, 24(1), 109–22.

Vasconcelos, M., Datta, K., Oliva, N., Khalekuzzaman, M., Torrizo, L., Krishnan, S., Oliveira, M., Goto, F. and Datta, S. K. (2003), Enhanced iron and zinc accumulation in transgenic rice with the *ferritin* gene, *Plant Sci.*, 164(3), 371–8.

Vert, G., Grotz, N., Dédaldéchamp, F., Gaymard, F., Guerinot, M. L., Briat, J. F. and Curie, C. (2002), IRT1, an Arabidopsis transporter essential for iron uptake from the soil and for plant growth, *Plant Cell*, 14(6), 1223–33.

White, P. J. and Broadley, M. R. (2009), Biofortification of crops with seven mineral elements often lacking in human diets-iron, zinc, copper, calcium, magnesium, selenium and iodine, *New Phytol.*, 182(1), 49–84.

Wirth, J., Poletti, S., Aeschlimann, B., Yakandawala, N., Drosse, B., Osorio, S., Tohge, T., Fernie, A. R., Günther, D., Gruissem, W. and Sautter, C. (2009), Rice endosperm iron biofortification by targeted and synergistic action of nicotianamine synthase and ferritin, *Plant Biotechnol. J.*, 7(7), 631–44.

World Health Organisation (2015), Micronutrient deficiencies: Iron deficiency anaemia [Online]. Available: http://www.who.int/nutrition/topics/ida/en/ [Accessed 31 October 2015].

Xiong, H., Guo, X., Kobayashi, T., Kakei, Y., Nakanishi, H., Nozoye, T., Zhang, L., Shen, H., Qiu, W., Nishizawa, N. K. and Yuanmei, Z. (2014), Expression of peanut iron regulated transporter 1 in tobacco and rice plants confers improved iron nutrition, *Plant Physiol. Biochem.*, 80, 83–9.

Yokosho, K., Yamaji, N., Ueno, D., Mitani, N. and Ma, J. F. (2009), OsFRDL1 is a citrate transporter required for efficient translocation of iron in rice, *Plant Physiol.*, 149(1), 297–305.

Zhang, Y., Xu, Y. H., Yi, H. Y. and Gong, J. M. (2012), Vacuolar membrane transporters *OsVIT1* and *OsVIT2* modulate iron translocation between flag leaves and seeds in rice, *Plant J.*, 72(3), 400–10.

Chapter 11

Quality parameters and testing methods in rice cultivation

Rachelle Ward, NSW Department of Primary Industries, Australia

1 Introduction

Rice quality parameters are designed to assess the suitability of particular rice for a desirable suite of processing and/or consumer traits. Rice quality characteristics typically assessed on the uncooked grain can relate to grain dimension, chalkiness, colour, uniformity, composition and millout, and on the cooked grain the traits may include elongation, texture and digestibility. Rice quality traits are assessed along the rice production pathway from farm to fork. They are used to make selection decisions on which breeding lines progress to the next generation. They are also used to inform research on genotype–structure–property and genotype by environment investigations. Regardless of pathway or stage within the pathway, there are several common challenges to ensure rice quality parameters and testing are relevant and robust enough to meet the broader rice industry needs. These include, but are not limited to, technology that satisfies needs across the industry, an understanding and response to consumer needs and the ability for the industry to deliver a consistent product to the consumer.

Selection pressures for quality attributes in a new variety are similar to a consumer making a purchase. At a basic level the choices are grain shape, texture, aroma and health claim that are reflected in the cost of research or product. As a cultivar progresses through parallel quality and breeding programs, sample size increases and so the ability to measure a broader range of phenotypes becomes possible. By understanding the genotype and structure that explains the phenotype of interest, breeders can then select for these quality traits with greater confidence in earlier generations. The viability of this research approach

http://dx.doi.org/10.19103/AS.2016.0003.34

is dependent on good rice quality parameters and testing and the ability to explain the traits with a suite of molecular markers. The research challenge here is that many tools for measuring phenotype rice quality parameters are at the crossroads of technology and knowledge.

Rice consumers have a sophisticated awareness of the various rice quality types in the market. The quality type originates in the varietal development through rice breeding, the growing conditions and rice processing. Premium rice varieties with unique quality traits, such as low glycemic index (GI) or coloured rice, tend to be segregated on varietal basis and, at times, geographical basis. Common quality types such as medium grain tend to be comprised of various varietal types that are grown in regions that optimize their grain quality and yield. Once a variety is established and grown on a commercial scale, it is the food labelling laws and the product specification agreements between producer and customer that dictate the quality measurements and requirements during production. In Australia, product labelling must adhere to the Australia New Zealand Food Standards Code governed by Food Standards Australia New Zealand. From 2016, any health claims made on packaging need to be substantiated and included in the Nutritional Information Panel. There are also voluntary labelling options such as the Health Star Ratings that rate the nutritional profile of a food. These laws drive the evaluation of the end product.

The next stage to satisfying consumer needs is to ensure that natural variation between seasons is within tolerance of the various policies and consumer acceptances. Global rice production areas will also gradually shift towards a more reliable water supply, and away from adverse climate change, pests and disease. Some varieties can maintain grain quality across environments, while others respond adversely. Trait by trait we are beginning to understand these changes, although how the traits combine into a functional property is unknown. For example, we may know that amylose content in some varieties might decrease in high growth temperature, but we do not know how this impacts GI. Developing a base level of quality data across environments is vital for future planning from both research and commercial perspectives to provide a consistent product to the consumer. With a deeper understanding of variety by environment interactions, the development of new varieties should be prioritized to withstand an increasingly volatile growing environment.

This chapter focuses on the three key research areas identified as having the greatest impact on the rice production pathway including the development of new rice varieties. There is a need for technology that can satisfy both scientific and practical requirements, options to meet consumer needs and the importance of predicting grain quality in the face of changing growing environments.

2 Technology for assessing rice quality: core parameters measured and the requirements of research and commercial settings

Core quality parameters measured and selected for the quality and breeding program are foremost the yield of whole milled grain after milling paddy (millout), followed by physical characteristics including chalk content, grain dimension and colour (Pallas 2016). Composition of the grain is limited to amylose content and nitrogen content as a proxy for protein content. Sulphur content is also measured to monitor the N:S ratio. The fine structure of these components is gaining attention because composition alone is unable to explain enough variation in quality. Functional parameters typically measured are viscosity, fragrance, texture and gelatinization temperature and, more recently, GI and antioxidants. Greater use of genetic marker technology in early generations has increased the probability of these traits progressing through the breeding program.

Fundamental to all grain quality evaluation is the need to continually improve all aspects of evaluation with aims to fast-track the breeding process from 10 years to 6–7 years. The ability to genotype vast numbers of samples across a wide slice of genome has become increasingly cheaper and faster in the past decade, to the extent that genotyping has shifted from single-marker screens in-house to outsource for microarray analysis of hundreds of markers. Advances in phenotyping have shifted more towards the use of sophisticated equipment and techniques where the premium is on information rather than throughput. For example, branching frequency was once measured by debranching starch and measuring the reducing ends (Takeda et al. 1987) whereas NMR is now the preferred choice (Dona et al. 2007). The cost differentiation between these two techniques can stem breeding advances. Nonetheless, renewal of existing equipment, of international standard methods and of marker validation, needs to catch up.

Within Australia's rice breeding and grain quality program, there is a 4-month period between rice harvest and rice sowing available to conduct all rice quality evaluations and to make selection decisions on the next generation of cultivars. Not turning data around within this tight time frame can mean that the number of cultivars sown in the next crop may be inflated, as selection decisions for that season occur after sowing. With finite land and water allocation, this can also stem the potential to have a multiplicity of concurrent breeding objectives. This inevitable lag time can be managed to some extent; however, robust and informative technology to improve throughput is required.

In a commercial setting, a crude grain analysis begins with delivery of a truck of paddy rice to the receival weighbridge from which a subsample of paddy is taken for a thorough analysis at a later date, up to several months. This time delay has implications for double-handling the subsample, limits the ability to segregate incoming grain on a quality basis and interrupts the payment process to growers if the variety attracts a grain premium or penalty on its quality. Concurrently, the tools to develop a new variety are research scale and do not necessarily translate linearly to commercial scale. This can create surprises when a new variety is released and the performance at receivals is viewed to be different. Once again, these shortcomings could be overcome by equipment and methods that can equally withstand the receival shed environment and the accuracy of a research environment.

The physical appearance of rice is the basis of the initial choice a consumer makes, is a primary selection tool for breeding a new variety and is a key classifier of grain quality. Physical attributes of uncooked rice quality can include millout, grain dimensions (length and width), chalk and colour; and when cooked, elongation. Rice dimensions have traditionally been classified as either long or medium grain. This is even reflected in most image analysis systems with sample delivery trays/wheels/shakers being either medium or long grain. In the past ten years especially, a greater understanding of existing and emerging markets both domestically and globally means that grain dimension qualifiers are now extended to include bold, short and intermediate grains. The problem is these newer quality types are not physically compatible with sample delivery platforms; as examples, bold, medium grains such as the Australian Reiziq can be too broad for a long grain application, yet too long for the medium grain application, and two short grains can fit into a medium grain slot. A more complicated trait to investigate is fissures in the grain that may affect millout, indicate how a grain elongates during cooking, and even alter the ability of the variety to withstand processing (Tuyen et al. 2012; Li et al. 2014; Grigg and Siebenmorgen 2015).

Amylose content has long been held responsible for many grain quality attributes from texture to digestibility (Bhattacharya and Sowbhagya 1979; Fitzgerald et al. 2011). Methods to measure amylose content are based on measuring the colour produced when iodine binds to the internal single helix of amylose. An AACC method exists for this measure, although in reality the method has been modified to suit the equipment in each laboratory (including grinders), source of biological standard (starch, amylose only or mixture of amylose and amylopectin) and time frame (quick dissolution of flour in sodium hydroxide by heat reflux vs. overnight incubation at room temperature) (Fitzgerald et al. 2009). Alternative methods are gaining attention through a cross-disciplinary approach to develop new methods. For example, the use of capillary electrophoresis to determine the amylose: amylopectin ratio (Herrero-Martinez et al. 2004). This method avoids the need for a biological standard, the process can be largely automated and the time per sample is comparable to that of iodine binding.

Together the protein and lipid component of a rice grain is around 7–14% of the milled grain, and is often overlooked in favour of starch content and structure to explain grain quality. Changes in growing conditions, increased popularity of different quality types (e.g., fragrant and sushi) and recognition that starch can only explain up to 70% of grain variability (Kharabian-Masouleh et al. 2012) has meant that research into protein and lipid content and structure is back in vogue. The hiatus from rice protein research has meant that new methods to quantify rice are required. A refined and simplified sample preparation and HPLC separation technique is able to distinguish varieties of different quality classes, as well as explore differences within a quality class (Balindong et al. Submitted). With this tool available, there is now renewed opportunity to tease out subtle differences in functional qualities within a narrow quality class such as sushi rice or long grain rice. As Australia's rice-growing region expands out of the traditional rice-growing region of Southeast Australia, the inevitable fine changes in grain protein composition have the potential to impact on key grain quality parameters such as millout and texture. Capturing this knowledge and translating it into farm management techniques can empower growers to have a hand in the final grain quality of the rice produced.

With affordable options to screen extensively across the genome, the bottleneck in marker adoption into a grain quality and breeding program now resides with the ability to effectively utilize that information. A comprehensive understanding of how well a marker or, more likely, a marker combination can explain the variability of a trait again depends on the integrity of the phenotype (as well as the diversity of the population). Strategically selecting for a suite of non-conflicting quality traits without inadvertently creating weakness in other attributes can also be an issue. As an example, to achieve low chalk levels in rice grown in warm climates requires grain size to be maintained but grain number to be sacrificed (Fitzgerald and Resurreccion 2009). A loss in yield is unacceptable to both growers and primary breeding goals, and so perhaps by selecting for high yield, chalk has inadvertently been selected for. Chalk is also believed to be associated with higher activity of alpha-amylase during grain filling (Hakata et al. 2012; Shiraya et al. 2015), and so overlaying complementary studies will likely yield a closer resolution to combating chalk.

In considering the rice industry from all stakeholder perspectives, it becomes clear that the need to renew instrumentation around the assessment of physical grain quality would benefit key stakeholders from farm to fork. Although with a limited choice of equipment in the marketplace that can fulfil the broader rice industry needs, researchers are either turning to alternate industries to fill this gap or developing in-house alternatives. While this is a short- to medium-term solution, without any benchmarking or correlation between methods, there may be negative implications for trade, especially grain destined for processing. Method development is following suit with cereal chemists working with polymer chemists to develop innovative methods that surpass existing methods. A great example of this is utilizing capillary electrophoresis to determine the amylose to amylopectin ratio.

Technology adoption, upkeep and transfer of grain quality testing methods across the broader rice industry can be improved by promoting a cross-disciplinary and needs based approach towards the development of new and improved instrumentation, standardised methods and capacity to pyramid the information.

3 Consumer health needs: fortification and digestibility

There is increased pressure on food producers and manufacturers to satisfy a diverse range of consumer needs. Grain quality, at its most fundamental level, is to be free of pest and disease. To some, grain quality attributes begin with the appearance of the cooked grain and how the rice suits a particular cuisine. To others, clean and good quality rice are expected and the need is for health attributes such as low GI, high fibre and conservation of nutrients and protein. Still others prefer rice in its processed form in products like rice cakes, microwave packs and as an ingredient in prepared meals. In terms of rice processing and testing, the challenge to meet consumer needs is centred on understanding the goal and developing strategies to reach that goal.

In 2008, the World Health Organisation predicted that 347 million people had diabetes, and in 2020 the disease is likely to be the seventh leading cause of death worldwide and double in incidence in the next 20 years. In 2014, the World Health Authority identified six Global Nutrition Targets for 2025 to improve the maternal, infant and young child nutrition. The 2015 estimates of child malnutrition showed that among children under five:

- stunting (caused by chronic malnutrition) has steadily declined to 159 million children or 23.8% of that population;
- overweight (caused by consuming more calories than are expended) has increased to 41 million children or 6.1% of that population; and
- wasting (caused by acute malnutrition) affects 50 million children with 2.4% of that population classified as severely wasted.

Over 70% of malnourished children live in Asia and Africa, the same regions that also have a high level of rice consumption, and so rice has the real potential to be a vehicle to combat a range of these health issues. Domestically in social media and populist cookbooks, complex carbohydrates like rice can be misunderstood as being unhealthy. To resolve this misbelief, a deeper understanding of the digestibility of cooked rice, more low GI options available to consumers and a general education program is essential.

3.1 Rice fortification

The practice of fortifying rice with micronutrients is supported by mandatory legislation in six countries and voluntarily by another six countries. According to the Food Fortification Initiative, the limiting adoption rate is related to the higher cost of production and limited technology to fortify rice. Recognition that the cooked rice remains fortified and is palatable are essential criteria for consumer adoption of fortified rice.

There are several avenues to increase the micronutrient levels in rice, and each option targets a different stage of the rice production pathway from farm to fork. At the farm end, biofortification via transgenics has been shown to significantly increase the level of zinc and iron (Trijatmiko et al. 2016). Foliar spray is an alternate method (Fang et al. 2008; Joy et al. 2015). Coloured rice is also an option to improve the nutritional load in rice (Deepa et al. 2008; Bett-Garber et al. 2013). As mentioned, the question remains if the micronutrient is bioavailable once the food is prepared and on the plate (Fang et al. 2008; Saha et al. 2015).

At the fork end, many countries parboil rice at an industry or household scale as a means to combat pests and disease as well as increase the micronutrient load of the rice. Parboiling involves paddy rice to be soaked, steamed and dried before milling and subsequent cooking. Recent studies have shown that adding water-soluble micronutrients such as folic acid to the soaking water can increase the nutrient content of the rice without disturbing the appearance or palatability of the rice (Kam et al. 2012a,b). Migration of minerals such as iron and zinc into the grain are also possible under certain parboiling conditions (Prom-u-thai et al. 2011). Using coloured rice as the base for parboiling is another option to preserve the free phenolics (Paiva et al. 2016).

3.2 Digestibility

Topical in Australia is the consumer-driven and evidence-based need for low GI rice varieties across different quality types and across different growing environments (Augustin et al. 2016). GI (ISO 26642:2010) is a standardized in vivo measure of the digestion and absorption rate of a food relative to a readily digested standard such as white bread. A diet of low GI foods is recommended for the general population but also for specific populations for example as a preventative measure for a range of chronic diseases such as type 2 diabetes and coronary heart disease (Augustin et al. 2016). The popularity of a low GI diet is reflected in recently updated food labelling laws (ANZFSC Food Standard 1.2.7 and GIF policy for declaration of low GI Foods). Selection for new Australian rice varieties already encompasses a vast array of rice quality selection tools that include molecular screening, amylose content, gelatinization temperature and texture, and these have been used to successfully select for candidates with low GI. However, there are opportunities to improve the precision, and thus fast-track, the selection process by adopting or developing more direct quality testing.

Identifying the target GI and accounting for seasonal variability around that target is an essential part of the breeding objective to satisfy current GI labelling legislation and endorsement. Relying on in vivo methodology as a development tool is cost-prohibitive, so a cost-effective, high-throughput, reproducible and simple method to predict GI is essential. The question then becomes which method and/or modification of a method to choose (Englyst et al. 1992; Shrestha et al. 2010; Fitzgerald et al. 2011). For this, a comprehensive correlation between in vivo evaluation and in vitro measurements needs to be established and standardized (Ek et al. 2014).

With a good broad spectrum of in vitro digestion profiles of cultivars, it then becomes possible to pinpoint the drivers of GI so that the selection process can occur even earlier in the breeding program. Understanding the drivers of low GI has largely centred on the concept that long amylose chains are difficult to digest (Miller et al. 1992; Fitzgerald et al. 2011). While certainly promising, this is not always the case (Ek et al. 2014; Dhital et al. 2015), and so higher orders of starch organization have been suggested to explain slower digestibility (Tester et al. 2004; Dhital et al. 2015). This introduces a need to explore another swathe of instrumentation for routine analysis.

Research is shifting towards mapping the digestion process at a higher level of starch organization (Blazek and Gilbert 2010; Dona et al. 2010; Shrestha et al. 2015) with the realization that digestion is determined largely by the accessibility and binding site of the relatively large alpha-amylase (radial width is 6 nm) (Casset et al. 1995). For example, double helix amylose cannot physically fit into the binding site of alpha-amylase (Casset et al. 1995; Zhang et al. 2015). So characterizing the lamellar repeat length and regularity using single angle X-ray scattering (Donald et al. 2001; Cardoso and Westfahl 2010) and determining the ratio of amorphous to crystalline regions by using X-ray diffraction (Lopez-Rubio et al. 2008), as well as the composition and arrangement of the crystalline regions including degree of branching by using nuclear magnetic resonance (Dona et al. 2007), will enable researchers to establish how conserved these features are across different rice quality types and how they change when varieties are grown under different environmental conditions. Exactly how the more traditional measures of amylose content and gelatinization temperature (that characterizes the melting of A-chains of amylopectin (Waters et al. 2006)) contribute to the higher order structure and digestibility is yet another research tangent that needs clarification (Witt et al. 2012).

Knowledge gleaned from studying water migration into brown rice and parboiled rice also has an advantage in that it can be applied to understand the process of cooking rice (Oli et al. 2014; Wu et al. 2016). The ways to cook rice are vast: rinse or no rinse, excess or limited water (absorption method), microwave, stove or rice cooker, same or different method for all rice types and, of course, the time needed to cook. Understanding cooking, or more specifically the degree of gelatinization and hydration, is also a key driver of digestibility, because the pathway for hydration creates the pathway for digestion enzymes.

Stepping closer again towards the initial crosses within a breeding program introduces the need for molecular markers to explain these variations in starch structure at all levels of organization. Much work has focused on the combined role of Granule Bound Starch Synthase I (GBSSI) and Starch Synthase IIa genes to explain starch structure (Fitzgerald et al. 2011; Kharabian-Masouleh et al. 2012; Wang et al. 2015). If indeed the digestion of the amorphous growth rings and amorphous lamellae are the first regions to be digested (Blazek and Gilbert 2010) and as such create the distinction between rapidly and slowly digested starch, then more work relating starch synthesis enzymes to higher order architecture is required.

Meeting consumer needs through rice processing parameters and testing means we can contribute to the Global Nutrition Targets for 2025 (http://www.who.int/nutrition/global-target-2025/en/). Shown here, the quality parameters to measure digestion and micronutrients against are still being clarified. Evident, however, is the need to understand opportunities across the broader rice industry all along the pathway from farm to fork.

4 Quality parameters and testing methods to address the problem of natural variation in rice crops

With new and revised technologies available and a clearer recognition of consumer needs and labelling requirements, the challenge to deliver a consistent product will depend on the integration and management of the natural variation in rice quality traits. This natural variation should then be used to inform reasonable processing tolerances and end-product specifications.

One of the greatest challenges for the future of Australian rice production will be to understand and overcome the potential variation between crop year and the slow impact of climate change. The rice production area of Australia is in the temperate southeast regions of the Murray and Murrumbidgee Irrigation Areas. Water availability for rice in these regions is increasingly challenged by both irregular rainfall and restricted water allocation in preference for competing high-security permanent crops like grapes and oranges. Anticipated changes in solar radiation, diurnal periodicity, and changes in carbon dioxide levels all influence ambient temperature. We know that water and temperature combinations at critical stages of rice impact on both rice yield and quality (Peng et al. 2004; Deng et al. 2015).

There are two broad opportunities to build rice grain quality consistency into a changing environment: development of new varieties that can withstand abiotic stress and identification of a collection of varieties that can be grown in different environments but deliver the same quality. Knowing the key criteria to select for in new varieties largely comes from identifying those varieties that display resilience to variable climates. So, either way there is a need to capture the performance of varieties across locations and years to develop a level of understanding that can be communicated from farm to fork. In Australia, extensive research trials are conducted annually on current and upcoming varieties to capture grain variation in grain quality as well as fine-tune recommendations for sowing and harvest dates to maximize yield and quality.

Often the impact of environment on grain quality is measured on the milled, whole grain. It has been long established that high growth temperatures have been shown to alter almost every grain quality attribute including increased chalk (Lisle et al. 2000), protein content and gelatinization temperature (Chun et al. 2015), and decreased amylose content (Sano 1984). Some of these compositional and functional traits have been partially explained by single nucleotide polymorphisms (SNP) and microsatellites that are used as a selection tool early in the breeding program.

Amylose content, to use a well-studied example, is characterized by a combination of CT repeats and SNP at the intron 1 splice site (Ayres et al. 1997), deletion in exon 2 (Teng et al. 2012), an SNP in exon 6 and exon 10 (Chen et al. 2008), and an SNP within intron 1 (Ward 2007) of the waxy (*Wx*) gene. Different combinations of these SNPs, within defined populations, are able to withstand varying environmental conditions.

Amylose structure can be defined more extensively as chain length, the ratio of single to double helices, location in the grain, contribution to crystallinity, lamellar periodicity and more (Juliano 1979; Hizukuri et al. 1981; Ball et al. 1998; Cardoso and Westfahl 2010). It is often these structures that have a greater influence on functional properties (Gidley 1989; Syahariza et al. 2013; Ek et al. 2014; Dhital et al. 2015; Shrestha et al. 2015; Teng et al. 2016). In 2015 alone, 110 journal articles were published on a search of amylose, structure and rice! When 'enzyme' is added to the search, the number of references is reduced to 26 articles with many highlighting that amylose structure is controlled by more than the *Wx* gene. Starch branching enzyme activity is linked to the growth temperature sensitive T at the intron 1 spice site of the *Wx* gene (Hirano et al. 1998), has a role in the synthesis of short amylose chains (Li et al. 2015) and may explain the suggestion that GBSSI does not control the relative ratio of short to long chains of amylose (Wang et al. 2015). Together with starch synthase IIa (Kharabian-Masouleh et al. 2012; Wang et al. 2015), the number of possible genotype combinations to explain and predict amylose content and structure is vast.

We have long surpassed the single-marker-trait validation that has worked well (Waters et al. 2006) and the focus has now shifted into the realm of determining how these enzymes work in concert, and under different abiotic stress, and their ability to explain the observed variation in structure at a molecular, lamellar and granular level of order. In turn, these structures define the properties that consumers identify with and expect to be consistent. Coupling the complexity of the genotype by environment interactions along the rice production pathway becomes more an exercise in data management, to inform reasonable end-product specifications (and labelling requirements).

In reaction to the overwhelming amount of information available in the developing grain, from a quality parameter and processing perspective, it may even prove worthwhile to simplify much of it with the introduction of a new measure for carbohydrate content (Albalasmeh et al. 2013). This measure could potentially gauge the varietal stability across environments, observe the interplay between protein and starch demands on carbon translocated into the grain, and screen for any observed change in grain quality (Chun et al. 2015). Practically, the carbohydrate content is also required to be reported on the Nutritional Information Panel of every packaged good, and it is necessary for GI calculations.

Even though much of this work is conducted on milled whole grain, there is also a need to investigate broken grains and grains that disintegrate during milling. Largely, broken grains are studied from a physical perspective, with a focus on fissures: their origin of pre- or post-harvest, drivers at a cellular or molecular level, wet or dry milling, grain moisture, grain shape and more (Buggenhout et al. 2013). Poorly formed grains are considered as an interruption to source–sink dynamics during grain filling. Typically, the longer the grain-filling duration, the slower the grain-filling rate (Yang et al. 2008); if the two are not synchronized then adverse yield and quality ensue. High growth temperature has consistently shown to decrease the grain-filling duration (Kim et al. 2011) which lowers the ability to add to the cumulative mean temperature and solar radiation required for effective grain-filling (Yang et al. 2008) and leads to a faster grain growth rate (Morita et al. 2005). Explanations for this have seen studies on leaf senescence (Kim et al. 2011), regulation of the sucrose transporter gene (Phan et al. 2013), activity of ADP glucose pyrophosphorylase (Tuncel and Okita 2013; Zhang et al. 2013) and cell number (Morita et al. 2005) under different conditions (Zhang et al. 2012), with many leading to an associated explanation for chalkiness (Fitzgerald and Resurreccion 2009; Siebenmorgen et al. 2013).

Grain quality is also dictated on farm (Champagne et al. 2004; Champagne et al. 2005), with head rice yield influenced by nitrogen fertilization (Siebenmorgen et al. 2013), harvest moisture (Kester et al. 1963) and wet/dry cycles (Dunn et al. 2014). To optimize recommendations to Australian growers, regular trials that capture these three

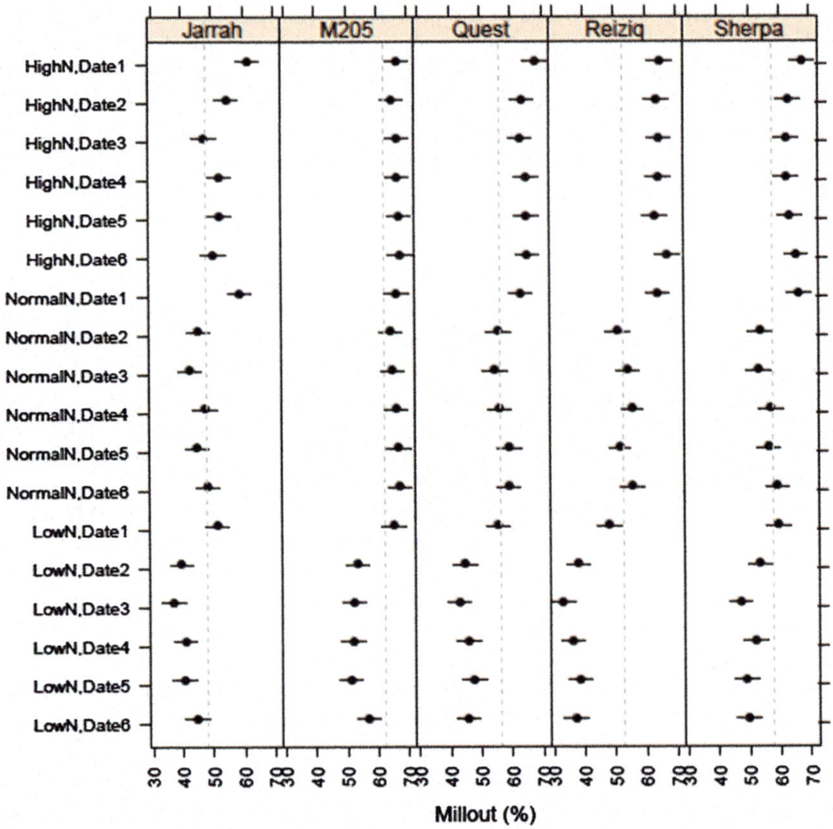

Figure 1 Contribution of nitrogen application and harvest date (Date 1-early to Date 6-late) on millout (%) of five Australian varieties. Data point represents the average and distribution.

Source: Martin et al. 2013.

components are conducted and communicated to growers via field days and fact sheets (Beecher et al. 2006; Martin et al. 2013). To illustrate the point, Fig. 1 shows different responses in millout to differing nitrogen and harvest dates (sowing date was the same in this trial). A similar spectrum of information is collected commercially via farm location and management data coupled with grain analysis at receivals. Utilization of this data to estimate the beneficial value of regional trials and to capture a more commercial picture of genotype by environment would be sensible.

Grain quality measured from genotype by environment interactions provides an initial evaluation of the spectrum of quality expected within a variety. This knowledge then provides a platform to identify key traits that can then add resilience into new varieties or match qualities with varieties grown in different locations. Better understanding the drivers of grain filling and the influence on the starch synthesis enzymes is one opportunity. Another is recognition of the need to use this information to support growers in their on-farm decision making.

5 Summary and future trends

The need for quality parameter and testing resides within the genotype–structure–function research paradigm, and at various stages along the pathway from farm to fork. Developing good phenotype screening across a broadening range of parameters suitable for implementation in both research and commercial settings is the first step. Understanding how we can utilize this information to fast-track the development of new and resilient varieties that exceed in agronomic and market-ready quality attributes is an even greater challenge. Strategies and opportunities to achieve this are vast as evidenced by the mounting (and at times overwhelming) volume of publications and in-house developments.

Research and developments within rice quality parameters and testing are reliant on evolving methodology and the introduction of entirely new techniques adapted through a cross-disciplinary approach. Cereal chemists need to continue to work with polymer chemists, physiologists, rheologists, biochemists, dieticians, physicists and geneticists to characterize the components of the rice grain. At the same time, we need to maintain an industry relevance to the adaption of new technologies and methodologies that can withstand different analytical environments, are suitable for fast and high-throughput analysis, can be substituted for a simpler evaluation method, can help define the link between genotype and phenotype, and much more.

Consumers are increasingly interested in the functionality of the grain, both culinary and health wise. These needs are typically defined through global agencies, local industry marketing intelligence, trade history, legislation, and even populist dietary trends. Responding to these needs to maintain market share ultimately requires a greater investment in understanding options to meet these needs. Fortification of rice with various micronutrients has been tackled at a transgenic and post-harvest level. In contrast, slower digestibility has, so far, focused on the level of variety development. So opportunities to meet consumer needs can and should be explored along the entire rice supply chain from farm to fork.

Parallel to the development of a new variety with improved agronomic and quality performance is the need to build resilience into the grain so that desired traits are maintained across environments. Typically, the focus has been placed on the milled whole grain to provide the answer; however, there is also merit in isolating the causes behind poorly milled grain. Marker-assisted screening for grain-filling traits may become as important as markers that predict amylose content in different growth temperatures.

Overall, the future of rice quality evaluation and testing methods needs to be addressed and integrated throughout the supply chain. Existing and new equipment and methodologies need to satisfactorily characterize structural and functional grain qualities. These tools can then be used to deliver on customer requirements via improved varieties and farming practices.

6 Where to look for further information

World Health

http://www.who.int/nutrition/global-target-2025/en/
http://www.fao.org/home/en/

Food labelling

http://www.foodstandards.gov.au/consumer/labelling/Pages/default.aspx
http://www.foodstandards.gov.au/code/Pages/default.aspx
http://healthstarrating.gov.au/internet/healthstarrating/publishing.nsf/content/home
http://www.gisymbol.com/

Methods

http://methods.aaccnet.org/toc.aspx
http://www.iso.org/iso/home/store/catalogue_ics.htm

Projects

http://www.rirdc.gov.au/research-project-details/

7 References

Albalasmeh, A. A., A. A. Berhe and T. A. Ghezzehei (2013). 'A new method for rapid determination of carbohydrate and total carbon concentrations using UV spectrophotometry'. *Carbohydrate Polymers* **97**(2): 253–61.

Augustin, L. S. A., L. Chiavaroli, J. Campbell, A. Ezatagha, A. L. Jenkins, A. Esfahani and C. W. C. Kendall (2016). 'Post-prandial glucose and insulin responses of hummus alone or combined with a carbohydrate food: a dose-response study'. *Nutrition Journal* **15**: 13–13.

Ayres, N. M., A. M. McClung, P. D. Larkin, H. F. J. Bligh, C. A. Jones and W. D. Park (1997). 'Microsatellites and a single-nucleotide polymorphism differentiate apparent amylose classes in an extended pedigree of US rice germ plasm'. *Theoretical and Applied Genetics* **94**: 773–81.

Balindong, J., L. Liu, R. Ward and D. L. E. Waters (Submitted). 'Optimisation and standardisation of extraction and HPLC analysis of rice grain protein'.

Ball, S. G., M. van de Wal and R. Visser (1998). 'Progress in undertanding the bioynthesis of amylose'. *Trends in Plant Science* **3**(12): 462–7.

Beecher, H. G., B. W. Dunn, J. A. Thompson, E. Humphreys, S. K. Mathews and J. Timsina (2006). 'Effect of raised beds, irrigation and nitrogen management on growth, water use and yield of rice in south-eastern Australia'. *Australian Journal of Experimental Agriculture* **46**(10): 1363–72.

Bett-Garber, K. L., J. M. Lea, A. M. McClung and M.-H. Chen (2013). 'Correlation of sensory, cooking, physical, and chemical properties of whole grain rice with diverse bran color'. *Cereal Chemistry* **90**(6): 521–8.

Bhattacharya, K. R. and C. M. Sowbhagya (1979). 'Pasting behaviour of rice: A new method of viscography'. *Journal of Food Science* **44**: 797–804.

Blazek, J. and E. P. Gilbert (2010). 'Effect of enzymatic hydrolysis on native starch granule structure'. *Biomacromolecules* **11**(12): 3275–89.

Buggenhout, J., K. Brijs, I. Celus and J. A. Delcour (2013). 'The breakage susceptibility of raw and parboiled rice: A review'. *Journal of Food Engineering* **117**(3): 304–15.

Cardoso, M. B. and H. Westfahl, Jr. (2010). 'On the lamellar width distributions of starch'. *Carbohydrate Polymers* **81**(1): 21–8.

Casset, F., A. Imberty, R. Haser, F. Payan and S. Perez (1995). 'Molecular modeling of the interaction between the catalytic site of pig pancreatic alpha-amylase and amylose fragments'. *European Journal of Biochemistry* **232**(1): 284–93.

Champagne, E. T., K. L. Bett-Garber, A. M. McClung and C. Bergman (2004). 'Sensory characteristics of diverse rice cultivars as influenced by genetic and environmental factors'. *Cereal Chemistry* **81**(2): 237–43.

Champagne, E. T., K. L. Bett-Garber, J. Thompson, R. Mutters, C. C. Grimm and A. M. McClung (2005). 'Effects of drain and harvest dates on rice sensory and physicochemical properties'. *Cereal Chemistry* **82**(4): 369–74.

Chen, M.-H., C. Bergman, S. Pinson and R. Fjellstrom (2008). 'Waxy gene haplotypes: Associations with apparent amylose content and the effect by the environment in an international rice germplasm collection'. *Journal of Cereal Science* **47**(3): 536–45.

Chun, A., H.-J. Lee, B. R. Hamaker and S. Janaswamy (2015). 'Effects of ripening temperature on starch structure and gelatinization, pasting, and cooking properties in rice (oryza sativa)'. *Journal of Agricultural and Food Chemistry* **63**(12): 3085–93.

Deepa, G., V. Singh and K. A. Naidu (2008). 'Nutrient composition and physicochemical properties of Indian medicinal rice - Njavara'. *Food Chemistry* **106**(1): 165–71.

Deng, N., X. Ling, Y. Sun, C. Zhang, S. Fahad, S. Peng, K. Cui, L. Nie and J. Huang (2015). 'Influence of temperature and solar radiation on grain yield and quality in irrigated rice system'. *European Journal of Agronomy* **64**: 37–46.

Dhital, S., L. Dabit, B. Zhang, B. Flanagan and A. K. Shrestha (2015). 'In vitro digestibility and physicochemical properties of milled rice'. *Food Chemistry* **172**: 757–65.

Dona, A., C.-W. W. Yuen, J. Peate, R. G. Gilbert, P. Castignolles and M. Gaborieau (2007). 'A new NMR method for directly monitoring and quantifying the dissolution kinetics of starch in DMSO'. *Carbohydrate Research* **342**(17): 2604–10.

Dona, A. C., G. Pages, R. G. Gilbert and P. W. Kuchel (2010). 'Digestion of starch: In vivo and in vitro kinetic models used to characterise oligosaccharide or glucose release'. *Carbohydrate Polymers* **80**(3): 599–617.

Donald, A. M., K. L. Kato, P. A. Perry and T. A. Waigh (2001). 'Scattering studies of the internal structure of starch granules'. *Starch* **53**: 504–12.

Dunn, B. W., T. S. Dunn and H. G. Beecher (2014). 'Nitrogen timing and rate effects on growth and grain yield of delayed permanent-water rice in south-eastern Australia'. *Crop and Pasture Science* **65**(9): 878–87.

Ek, K. L., S. Wang, L. Copeland and J. C. Brand-Miller (2014). 'Discovery of a low-glycaemic index potato and relationship with starch digestion in vitro'. *British Journal of Nutrition* **111**(4): 699–705.

Englyst, H. N., S. M. Kingman and J. H. Cummings (1992). 'Classification and measurement of nutritionally important starch fractions'. *European Journal of Clinical Nutrition* **46**: S33–S50.

Fang, Y., L. Wang, Z. Xin, L. Zhao, X. An and Q. Hu (2008). 'Effect of foliar application of zinc, selenium, and iron fertilizers on nutrients concentration and yield of rice grain in China'. *Journal of Agricultural and Food Chemistry* **56**(6): 2079–84.

Fitzgerald, M. A., C. J. Bergman, A. P. Resurreccion, J. Moeller, R. Jimenez, R. F. Reinke, M. Martin, P. Blanco, F. Molina, M.-H. Chen, V. Kuri, M. V. Romero, F. Habibi, T. Umemoto, S. Jongde, E. Graterol, K. R. Reddy, P. Z. Bassinello, R. Sivakami, N. S. Rani, S. Das, Y. J. Wang, S. D. Indrasari, A. Ramli, R. Ahmad, S. S. Dipti, L. Xie, N. T. Lang, P. Singh, D. Castillo Toro, F. Tavasoli and C. Mestres (2009). 'Addressing the Dilemmas of Measuring Amylose in Rice'. *Cereal Chemistry* **86**(5): 492–8.

Fitzgerald, M. A., S. Rahman, A. P. Resurreccion, J. Concepcion, V. D. Daygon, S. S. Dipti, K. A. Kabir, B. Klingner, M. K. Morell and A. R. Bird (2011). 'Identification of a major genetic determinant of glycaemic index in rice'. *Rice* **4**(2): 66–74.

Fitzgerald, M. A. and A. P. Resurreccion (2009). 'Maintaining the yield of edible rice in a warming world'. *Functional Plant Biology* **36**(12): 1037–45.

Gidley, M. J. (1989). 'Molecular mechanisms underlying amylose aggregation and gelation'. *Macromolecules* **22**: 351–8.

Grigg, B. C. and T. J. Siebenmorgen (2015). 'Impacts of kernel thickness and associated physical properties on milling yields of long-grain rice'. *Applied Engineering in Agriculture* **31**(3): 505–11.

Hakata, M., M. Kuroda, T. Mitashita, T. Yamaguchi, M. Kojima, H. Skakibra, T. Mitsui and H. Yamakawa (2012). 'Suppression of α-amylase genes improves quality of rice grain ripened under high temperature'. *Plant Biotechnology Journal* **10**(9): 1110–17.

Herrero-Martinez, J. M., P. J. Schoenmakers and W. T. Kok (2004). 'Determination of the amylose-amylopectin ratio of starches by iodine-affinity capillary electrophoresis'. *Journal of Chromatography A* **1053**(1–2): 227–34.

Hirano, H.-Y., M. Eiguchi and Y. Sano (1998). 'A single base change altered the regulation of the *Waxy* gene at the posttranscriptional level during the domestication of rice'. *Molecular Biology and Evolution* **15**(8): 978–87.

Hizukuri, S., Y. Takeda, M. Yasuda and A. Suzuki (1981). 'Multi-branched nature of amylose and the action of de-branching enzymes'. *Carbohydrate Research* **94**(1981): 205–13.

Joy, E. J. M., A. J. Stein, S. D. Young, E. L. Ander, M. J. Watts and M. R. Broadley (2015). 'Zinc-enriched fertilisers as a potential public health intervention in Africa'. *Plant and Soil* **389**(1–2): 1–24.

Juliano, B. O. (1979). 'Amylose analysis in rice - a review'. *Proceedings of the workshop on Chemical Aspects of Rice Grain Quality. Review of Methodolgy*: 251–60.

Kam, K., J. Arcot and R. Ward (2012a). 'Fortification of rice with folic acid using parboiling technique: Effect of parboiling conditions on nutrient uptake and physical characteristics of milled rice'. *Journal of Cereal Science* **56**(3): 587–94.

Kam, K., J. M. Murray, J. Arcot and R. Ward (2012b). 'Fortification of parboiled rice with folic acid: Consumer acceptance and sensory evaluation'. *Food Research International* **49**(1): 354–63.

Kester, E. B., H. C. Lukens, R. E. Ferrel, A. Miohammad and D. C. Finrock (1963). 'Influence of maturity on properties of western rices'. *Cereal Chemistry* **40**(4): 323–36.

Kharabian-Masouleh, A., D. L. E. Waters, R. F. Reinke, R. Ward and R. J. Henry (2012). 'SNP in starch biosynthesis genes associated with nutritional and functional properties of rice'. *Scientific Reports* **2**.

Kim, J., J. Shon, C.-K. Lee, W. Yang, Y. Yoon, W.-H. Yang, Y.-G. Kim and B.-W. Lee (2011). 'Relationship between grain filling duration and leaf senescence of temperate rice under high temperature'. *Field Crops Research* **122**(3): 207–13.

Li, E., A. C. Wu, J. Li, Q. Liu and R. G. Gilbert (2015). 'Improved understanding of rice amylose biosynthesis from advanced starch structural characterization'. *Rice* **8**: 20–8.

Li, Y.-N., K. Li, W.-M. Ding, K.-J. Chen and Q.-S. Ding (2014). 'Correlation between head rice yield and specific mechanical property differences between dorsal side and ventral side of rice kernels'. *Journal of Food Engineering* **123**: 60–6.

Lisle, A., M. Martin and M. A. Fitzgerald (2000). 'Chalky and translucent rice grains differ in starch composition and structure and cooking properties'. *Cereal Chemistry* **77**(5): 627–32.

Lopez-Rubio, A., B. M. Flanagan, E. P. Gilbert and M. J. Gidley (2008). 'A novel approach for calculating starch crystallinity and its correlation with double helix content: A combined XRD and NMR study'. *Biopolymers* **89**(9): 761–8.

Martin, M., N. Coombes, P. Snell and R. Ward (2013). 'Spectrum of rice grain quality at the farm gate'. *Australian Cereal Chemistry Conference*. Perth.

Miller, J. B., E. Pang and L. Bramall (1992). 'Rice – a high or low glycemic index food'. *American Journal of Clinical Nutrition* **56**(6): 1034–6.

Morita, S., J. Yonemaru and J. Takanashi (2005). 'Grain growth and endosperm cell size under high night temperatures in rice (Oryza sativa L.)'. *Annals of Botany* **95**(4): 695–701.

Oli, P., R. Ward, B. Adhikari and P. Torley (2014). 'The Diffusion of Moisture in Paddy During Hydration and Dehydration Processes'. *Drying Technology* **32**(12): 1423–34.

Paiva, F. F., N. L. Vanier, J. D. J. Berrios, V. Z. Pinto, D. Wood, T. Williams, J. Pan and M. C. Elias (2016). 'Polishing and parboiling effect on the nutritional and technological properties of pigmented rice'. *Food Chemistry* **191**: 105–12.

Pallas, L. (2016). 'Rice processing: beyond the farm gate'. In C. Wrigley, H. Corke, K. Seetharaman and J. Faubion (eds), *Encyclopedia of Food Grains*, 446–52. Oxford: Academic Press.

Peng, S., J. Huang, J. E. Sheehy, R. C. Laza, R. M. Visperas, X. Zhong, G. S. Centeno, G. S. Khush and K. G. Cassman (2004). 'Rice yields decline with higher night temperature from global warming'. *PNAS* **101**(27): 9971–5.

Phan, T. T. T., Y. Ishibashi, M. Miyazaki, H. T. Tran, K. Okamura, S. Tanaka, J. Nakamura, T. Yuasa and M. Iwaya-Inoue (2013). 'High temperature-induced repression of the rice sucrose transporter (OsSUT1) and starch synthesis-related genes in sink and source organs at milky ripening stage causes chalky grains'. *Journal of Agronomy and Crop Science* **199**(3): 178–88.

Prom-u-thai, C., L. Huang, I. Cakmak and B. Rerkasem (2011). 'Simultaneous fortification of iron and zinc in parboiled rice kernel'. *Scienceasia* **37**(4): 296–302.

Saha, S., B. Mandal, G. C. Hazra, A. Dey, M. Chakraborty, B. Adhikari, S. K. Mukhopadhyay and R. Sadhukhan (2015). 'Can agronomic biofortification of zinc be benign for iron in cereals?' *Journal of Cereal Science* **65**: 186–91.

Sano, Y. (1984). 'Differential regulation of waxy gene expression in rice endosperm'. *Theor. Appl. Genet.* **68**: 467–73.

Shiraya, T., T. Mori, T. Maruyama, M. Sasaki, T. Takamatsu, K. Oikawa, K. Itoh, K. Kaneko, H. Ichikawa and T. Mitsui (2015). 'Golgi/plastid-type manganese superoxide dismutase involved in heat-stress tolerance during grain filling of rice'. *Plant Biotechnology Journal* **13**(9): 1251–63.

Shrestha, A. K., J. Blazek, B. M. Flanagan, S. Dhital, O. Larroque, M. K. Morell, E. P. Gilbert and M. J. Gidley (2015). 'Molecular, mesoscopic and microscopic structure evolution during amylase digestion of extruded maize and high amylose maize starches'. *Carbohydrate Polymers* **118**: 224–34.

Shrestha, A. K., C. S. Ng, A. Lopez-Rubio, J. Blazek, E. P. Gilbert and M. J. Gidley (2010). 'Enzyme resistance and structural organization in extruded high amylose maize starch'. *Carbohydrate Polymers* **80**(3): 699–710.

Siebenmorgen, T. J., B. C. Grigg and S. B. Lanning (2013). 'Impacts of Preharvest Factors During Kernel Development on Rice Quality and Functionality'. *Annual Review of Food Science and Technology, Vol 4* **4**: 101–15.

Siebenmorgen, T. J., B. C. Grigg and S. B. Lanning (2013). 'Impacts of preharvest factors during kernel development on rice quality and functionality'. In M. P. Doyle and T. R. Klaenhammer (eds), *Annual Review of Food Science and Technology, Vol 4* **4**: 101–15.

Syahariza, Z. A., S. Sar, J. Hasjim, M. J. Tizzotti and R. G. Gilbert (2013). 'The importance of amylose and amylopectin fine structures for starch digestibility in cooked rice grains'. *Food Chemistry* **136**(2): 742–9.

Takeda, Y., S. Hizukuri, C. Takeda and A. Suzuki (1987). 'Structures of branched molecules of amyloses of various origins, and molar fractions of branched and unbranched molecules'. *Carbohydrate Research* **165**: 139–45.

Teng, A., T. Witt, K. Wang, M. Li and J. Hasjim (2016). 'Molecular rearrangement of waxy and normal maize starch granules during in vitro digestion'. *Carbohydrate Polymers* **139**: 10–19.

Teng, B., R. Zeng, Y. Wang, Z. Liu, Z. Zhang, H. Zhu, X. Ding, W. Li and G. Zhang (2012). 'Detection of allelic variation at the Wx locus with single-segment substitution lines in rice (Oryza sativa L.)'. *Molecular Breeding* **30**(1): 583–95.

Tester, R. F., J. Karkalas and X. Qi (2004). 'Starch structure and digestibility enzyme-substrate relationship'. *Worlds Poultry Science Journal* **60**(2): 186–95.

Trijatmiko, K. R., C. Duenas, N. Tsakirpaloglou, L. Torrizo, F. M. Arines, C. Adeva, J. Balindong, N. Oliva, M. V. Sapasap, J. Borrero, J. Rey, P. Francisco, A. Nelson, H. Nakanishi, E. Lombi, E. Tako, R. P. Glahn, J. Stangoulis, P. Chadha-Mohanty, A. A. T. Johnson, J. Tohme, G. Barry and I. H. Slamet-Loedin (2016). 'Biofortified indica rice attains iron and zinc nutrition dietary targets in the field'. *Scientific Reports* **6**: 19792–805.

Tuncel, A. and T. W. Okita (2013). 'Improving starch yield in cereals by over-expression of ADPglucose pyrophosphorylase: Expectations and unanticipated outcomes'. *Plant Science* **211**: 52–60.

Tuyen, T., T. Vinh, S. Fukai and B. Bhandari (2012). 'Changes in cracking behavior and milling quality of selected Australian rice varieties due to postdrying annealing and subsequent storage'. *Drying Technology* **30**(16): 1831–43.

Wang, K., J. Hasjim, A. C. Wu, E. Li, R. J. Henry and R. G. Gilbert (2015). 'Roles of GBSSI and SSIIa in determining amylose fine structure'. *Carbohydrate Polymers* **127**: 264–74.

Ward, R. (2007). *Potential Impact of Temperature and Carbon Dioxide Levels on Rice Quality*. PhD, USyd.

Waters, D. L. E., R. J. Henry, R. F. Reinke and M. A. Fitzgerald (2006). 'Gelatinization temperature or rice explained by polymorphisms in starch synthase'. *Plant Biotechnology Journal* **4**(1): 115–22.

Witt, T., J. Doutch, E. P. Gilbert and R. G. Gilbert (2012). 'Relations between Molecular, Crystalline, and Lamellar Structures of Amylopectin'. *Biomacromolecules* **13**(12): 4273–82.

Wu, J., J. Chen, W. Liu, C. Liu, Y. Zhong, D. Luo, Z. Li and X. Guo (2016). 'Effects of aleurone layer on rice cooking: A histological investigation'. *Food Chemistry* **191**: 28–35.

Yang, W., S. Peng, M. L. Dionisio-Sese, R. C. Laza and R. A. Visperas (2008). 'Grain filling duration, a crucial determinant of genotypic variation of grain yield in field-grown tropical irrigated rice'. *Field Crops Research* **105**(3): 221–7.

Zhang, B., S. Dhital, B. M. Flanagan, P. Luckman, P. J. Halley and M. J. Gidley (2015). 'Extrusion induced low-order starch matrices: Enzymic hydrolysis and structure'. *Carbohydrate Polymers* **134**: 485–96.

Zhang, H., H. Li, L. Yuan, Z. Wang, J. Yang and J. Zhang (2012). 'Post-anthesis alternate wetting and moderate soil drying enhances activities of key enzymes in sucrose-to-starch conversion in inferior spikelets of rice'. *Journal of Experimental Botany* **63**(1): 215–27.

Zhang, Z., G. Chu, L. Liu, Z. Wang, X. Wang, H. Zhang, J. Yang and J. Zhang (2013). 'Mid-season nitrogen application strategies for rice varieties differing in panicle size'. *Field Crops Research* **150**: 9–18.

Chapter 12

Agronomic and environmental factors affecting rice grain quality

Chuan Tong and Jinsong Bao, Zhejiang University, China

1 Introduction

Rice (*Oryza sativa* L.) is one of the world's most important crops and is a fundamental part of the diet of over half the world's population. Rice grain quality is usually evaluated using parameters such as milling and processing performance, appearance, cooking and eating quality and nutritional composition. Recent research has highlighted the importance of both genetic and environmental factors in determining rice quality attributes (Siebenmorgen et al., 2013; Xu et al., 2004).

Among the main chemical components of rice, starch, protein and lipid play a critical role in rice quality. The starch in the rice seed endosperm consists of amylose and amylopectin, the biosynthesis of which requires enzymes such as granule-bound starch synthase (GBSS), starch synthases (SS), starch branching enzymes (SBE) and debranching enzymes (DBE). It has been shown that the bioactivity of these enzymes is significantly influenced by abiotic factors (Denyer et al., 1994; Jiang et al., 2003; Umemoto et al., 1995). Key proteins in the rice endosperm are albumin and globulin (4–10%), prolamin (5–18%) and glutelin (77–90%) (Huebner et al., 1990). Prolamin and glutelin have been shown to affect rice pasting and textural characteristics (Baxter et al., 2004; Martin and Fitzgerald, 2002). They have been shown to be significantly influenced by factors such as temperature during grain ripening (Ashida et al., 2013). Lipids have been shown to affect rice pasting and gelatinization behaviour through interaction with amylose (Gelders et al., 2006; Morrison, 1995). Starch lipids have also been shown to be influenced by combinations of environmental and genetic factors (Tong et al., 2014).

http://dx.doi.org/10.19103/AS.2016.0003.36

It is now widely accepted that rises in global mean surface temperature related to increases in anthropogenic carbon dioxide (CO_2) emissions are likely to result in more unpredictable and extreme weather (IPCC, 2014). It has been suggested that increases in rice yields can be mainly resulted from genetic factors such as improved varieties but that climactic factors are also significant (Song et al., 2015). This correlation highlights the importance of assessing the effects of abiotic factors on rice quality.

2 The effects of sowing, transplantation and harvesting dates

Rice yields and quality are determined by three basic factors: genetics, environmental conditions and cultivation practices (Fig. 1). The latter include sowing, transplantation and harvesting dates, as well as irrigation, fertilizer use and tillage systems. Sowing date is a key step during rice farming, determining subsequent heading and harvesting dates, and significantly influencing efficient heat use and final water content of the rice grain. Average maximum and minimum temperatures during rice development, especially at heading and filling stages, have significant effects on agronomic yield and grain quality.

2.1 Yield, milling performance and appearance

It has been shown that favourable weather conditions such as temperature and sunshine contribute to higher tillers and grain yield more than early transplanting dates (Brar et al., 2012). However, delayed sowing and transplantation times can significantly influence leaf area index, light extinction co-efficient and filled spikelets growth (Ghosh et al., 2004), and thus reduce grain yield (Ghosh et al., 2004; Slaton et al., 2003). Decreases in yields caused by delays in sowing and transplantation dates have been shown to vary from 10.8 to 35.6% and from 11.8 to 33.2%, respectively (Dhaliwal et al., 1986). In India, impacts of four different sowing dates on rice seed quality were evaluated, and it was found that seeding setting, germination, discolouration, leachate conductivity and longevity were

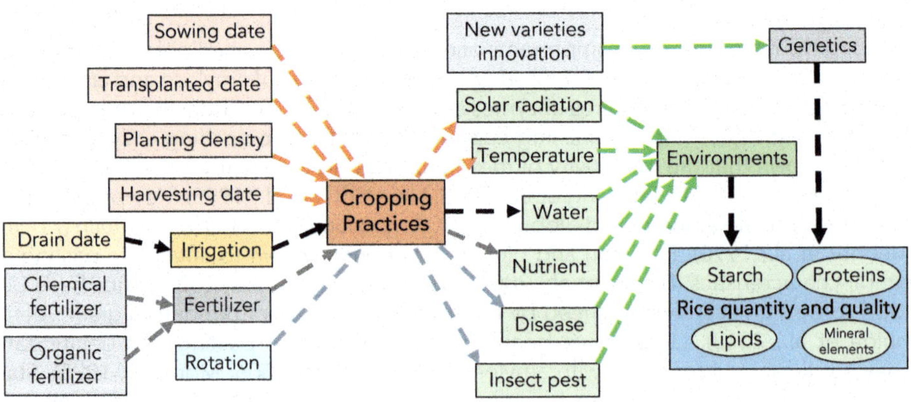

Figure 1 Basic rice cropping system.

more affected by environmental factors such as sowing date than by genotypes (Krishnan and Rao, 2005).

Singh et al. (2014) have reported that a 20-day delay in transplanting rice caused lower grain lightness and whiteness. Chalkiness is one of the key determinants of rice milling quality and appearance. Wide variations in maximum and minimum grain chalkiness across six sowing dates among 40 varieties have also been observed (Chen et al., 2013). These might be explained by the positive relationship between chalkiness and the temperature during the second week after heading (Chen et al., 2013). A delay in rice sowing means delaying heading and flowering resulted from lower temperature after heading. It has been suggested that optimal rice milling quality requires warm weather conditions with high solar radiation but without excessive rain during grain filling, with the best harvest date from 127 to 130 days after planting for Lemont variety (Krishnan and Rao, 2005; McCauley and Way, 2002).

2.2 Nutritional quality

Lower lipids content was also associated with a delay in rice transplanted dates (Singh et al., 2014). Drain dates had a significant effect on protein content (Champagne et al., 2005). Protein profiling has indicated a decreased prolamines content for PR114 and PAU201 cultivars related to late transplanting, though no variations in polypeptide of GBSS, glutelin acidic and globulins (Singh et al., 2014). However, an increase in protein and free fatty acid content with late transplanting in rice was shown by Dhaliwal et al. (1986).

2.3 Eating and cooking quality

Amylose content (or apparent amylose content) is one of most important factors determining the eating and cooking quality of rice. However, the rapid viscosity analyser (RVA) profile can distinguish eating quality rice varieties with similar amylose content. It has been reported that delayed planting and transplanting results in lower amylose content (Dhaliwal et al., 1986; Ghosh et al., 2004). Drain dates also had a significant effect on amylose content, although different harvest dates did not affect amylose content (Champagne et al., 2005).

Decreased peak viscosity (PV), cool paste viscosity (CPV) and setback (SB) of RVA pasting parameters, and lower grain hardness for textural quality, were observed in rice cultivars due to delayed transplantation (Dhaliwal et al., 1986; Singh et al., 2014). Trends towards lower PV, hot paste viscosity (HPV) and CPV of RVA parameters were also observed for long-grain rice cultivars grown in lower temperature conditions (Missouri) compared to higher temperature locations (Arkansas and Texas) in the United States (Aboubacar et al., 2006). These studies showed consistently higher amylose content, the proportion of short amylopectin linear chains and a lower proportion of long amylopectin linear chains of rice cultivars grown at lower temperature locations. Similarly, higher amylose content in cool growing season and lower PV and SB in pasting behaviour were observed in Australia (Dang and Copeland, 2004).

Harvest moisture content in rice related to harvest date is closely related with grain physiochemical properties such as milling performance and pasting properties. An early harvest date can also bring about higher SB and lower breakdown (BD) of RVA parameters, and correspondingly higher hardness and cohesiveness for better texture (Champagne et al., 2005). As harvest moisture content decreased, head rice yield and PV viscosity increased, depending on growing location and rice variety (Dilday, 1989; Wang et al., 2004). Levels of lipid oxidation products and branched chain length of amylopectin also varied across harvest dates (Champagne et al., 2005).

3 The effects of irrigation, fertilizer application and other aspects of cultivation

3.1 Irrigation

The influence of irrigation on rice qualities results from the sensitivity of rice plants to the water content in soils, particularly at the filling stage (Gomez, 1979). Lack of water during the rice filling stage leads to leaf senescence, lower amylose and increased chalkiness contents (Gomez, 1979). Protein content in most types of rice decreased with decreased water in soil (Gomez, 1979). It was inferred that reduced nitrogen uptake induced by the absence of water resulted in lower protein content in rice grain.

Appropriate draining times of floodwater in rice paddy have a critical influence on optimum rice grain yield and quality. It has been shown that, while delayed flood irrigation had no effect on head rice yields, full season flush irrigation reduced grain yields (Grigg et al., 2000). Drain date had no significant effect on cooking texture or flavour, but early drain could result in lower SB and BD (Champagne et al., 2005).

Water-saving practices can be beneficial for rice grain properties. As an example, milling quality, appearance and cooking properties were improved and decreased by the respective use of straw mulching and plastic film management (Zhang et al., 2008). Qualities such as chalkiness also significantly declined with the use of plastic film (Gomez, 1979). In addition, Pan et al. (2009) found that alternately wetting and drying irrigation could improve brown rice grain yield and milling quality.

3.2 Fertilizer

Nitrogen (N) fertilizer has been shown to have a positive effect on tiller number and negative effect on filling rate and grain weight (Jongkaewwattana and Geng, 1991), but no effects on chalk percentage and grain width (Bryant et al., 2015). The type, the amount and period of N fertilizer application are closely correlated with rice yield and quality, including rice grain filling and milling performance (Leesawatwong et al., 2005; Pan et al., 2009; Seetanun and Dedatta, 1973). Split application N fertilizer has been shown to increase grain yield, whereas topdressing of N fertilizer can raise head rice rate (Seetanun and Dedatta, 1973). Late N fertilizer application can improve milled rice translucency but reduce grain whiteness (Alcantara and Cassman, 1996).

There have been inconsistent findings on the role of N fertilizer on amylose in rice grain. Some studies show a high N fertilizer level had a negative influence on amylose content in the early report (Champagne et al., 2009), whereas others showed no great change (Bryant et al., 2015). A positive correlation between protein content (such as glutelin in the endosperm) and N fertilizer has been clearly demonstrated in the literature (Bryant et al., 2015; Champagne et al., 2009; Leesawatwong et al., 2005; Seetanun and Dedatta, 1973). As an example, head rice protein increased significantly with late N fertilizer application at flowering (Alcantara and Cassman, 1996). N fertilizer application has also been positively related to a higher percentage of unbroken rice (Leesawatwong et al., 2005). Lower PV, HPV, CPV and higher SB and consistency (CS) have been associated with increasing use of N fertilizer (Bryant et al., 2015). Decreased initial slickness and stickiness, and improved roughness, hardness and moisture absorption have been linked to increasing N fertilizer supply (Champagne et al., 2009; Leesawatwong et al., 2005).

3.3 Other aspects of cultivation

Other aspects of cultivation can also affect rice quality, such as planting density. Ottis and Talbert (2005) found that differences in rice seeding rate, from 57 to 500 seeds m^{-2} corresponding to 73–373 plants m^{-2} of rice density, did not affect rice yield. Crop rotation (e.g. rice–soya bean and rice–corn systems) can improve protein content and PV, although it does not affect other properties such as amylose and lipid content (Bryant et al., 2009).

4 The effects of elevated temperature and other aspects of climate change

Climate change associated with global warming has major implications for crop physicochemical properties and production (Fig. 2). In the case of rice, extreme temperatures and water scarcity are considered to be the main abiotic stresses to endanger

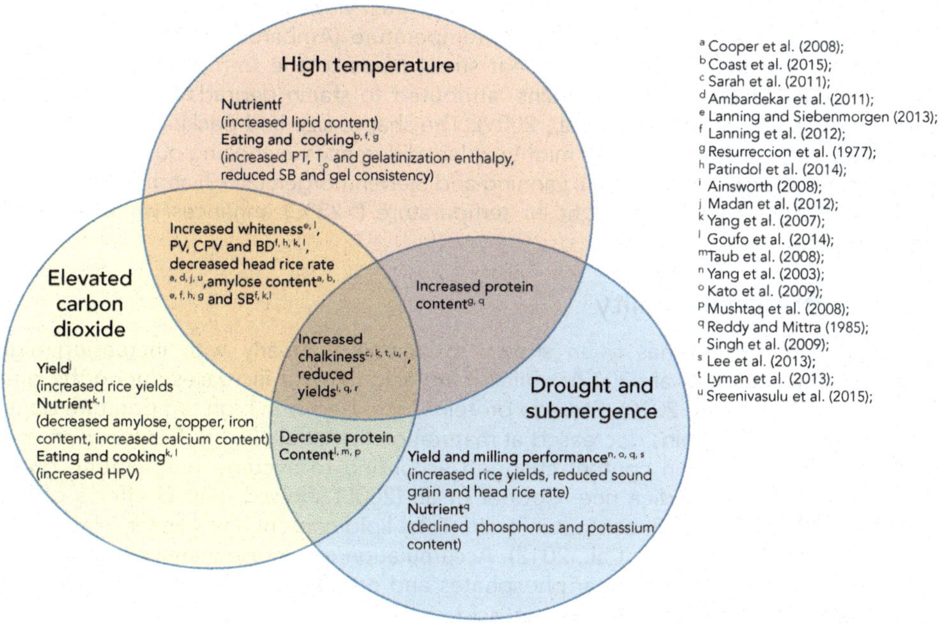

Figure 2 Influences of global environment changing on rice physicochemical properties.

rice quality and production (Lin et al., 2010). Nonetheless, not all quality parameters are equally affected. As an example, chalkiness is more sensitive to climate change than other properties such as head rice rate, gel CS, gelatinization temperature, amylose and protein content (Sreenivasulu et al., 2015).

4.1 High temperature

High temperature has a significant impact on rice grain physicochemical properties (Coast et al., 2015). Generally, head rice yield, grain dimensions, spikelet fertility, amylose content and gel CS are reduced with higher temperatures (Coast et al., 2015; Cooper et al., 2008). Chalkiness seems to increase in relation to increased night temperature during kernel development (Lanning et al., 2011). Decreases in head rice yield resulting from higher temperature have been shown to vary with rice cultivar (Lanning et al., 2011). One study has shown a linear, inverse but no significant correlations between high night temperature for long-grain pure line, long-grain hybrid and medium-grain pure line cultivars (Ambardekar et al., 2011). Interestingly, high temperature has not been shown to influence rice kernel mass (Cooper et al., 2008).

4.1.1 Milling performance and appearance

High temperature during grain filling reduces head rice yield (Ambardekar et al., 2011; Sreenivasulu et al., 2015), but causes increased chalkiness, poorer transparency and more broken kernels which affect milling efficiency and appearance (Lyman et al., 2013; Sreenivasulu et al., 2015). Cooper et al. (2006) found average daily night-time temperature had a strong influence on head rice yield. In cool climates, high temperatures during early vegetative growth can lead to an increase of rice grain yield (Horai et al., 2014).

Chalkiness is positively correlated with temperature (Ambardekar et al., 2011; Chen et al., 2013; Ishimaru et al., 2009). Heat stress leads to the formation of milky-white and white-cored types of chalky grains, attributed to starch degradation by α-amylase (Gregory et al., 2005; Ishimaru et al., 2009). The shape, size and packing of amyloplasts and cells in chalky rice endosperm might explain differences in cooking quality under high temperature (Gregory et al., 2005). Lanning and Siebenmorgen (2013) and Siebenmorgen et al. (2013) found that high night air temperature (>23°C) enhances whiteness while reducing yellowness.

4.1.2 Nutritional quality

Crude protein content has been shown to decrease linearly with increasing night temperature (Lanning et al., 2012), while amino acid content increases during the grain filling stage (Yamakawa, 2010). Storage proteins increased with high temperature during early filling while prolamins decreased at maturation (Lin et al., 2010). Resurreccion et al. (1977) found that protein content changed according to average temperature during filling in japonica and indica rice. Cooper et al. (2008) showed limited effects of high temperature on rice protein content in grain. Total lipid content has a linear relation with night temperature (Lanning et al., 2012). Accumulation of sucrose increased with higher temperature while levels of sugar phosphates and organic acids involved in glycolysis/gluconeogenesis and the tricarboxylic acid cycle decreased (Yamakawa and Hakata, 2010).

4.1.3 Eating and cooking quality

High temperatures have been shown to impair starch deposition (Yamakawa and Hakata, 2010). High temperatures during rice grain ripening can result in lower amylose content (Lanning et al., 2012; Lin et al., 2010; Patindol et al., 2014; Resurreccion et al., 1977). Amylose content in low amylose genotypes correlated negatively, while those in middle and high amylose genotypes correlated positively, with higher temperatures during filling (Resurreccion et al., 1977).

Higher night-time temperatures are associated with decreased amounts of short side chains of amylopectin including unit chains with degree polymerization (DP) 6 and 11–13 and increased amounts of long chains including DP 8, 22–24 and 29 (Inouchi et al., 2000). Works of Patindol et al. (2014) also revealed the decreased percentage of amylopectin short chains (DP $\leq$ 18) and increased long chains (DP $\geq$ 19) were significantly related to high temperature. The increased amount of long chains of amylopectin might be because of the higher activity and expression levels of SSI compared to those in BEI, BEIII and GBSS (Jiang et al., 2003; Wang et al., 2006). Due to changes in amylose, amylopectin and chain-length distribution of amylopectin, greater PV, CPV, BD, pasting temperature (PT), onset temperature (T_o), gelatinization enthalpy and relative crystallinity as well as smaller SB viscosity and gel CS have been observed as a result of higher night-time temperatures during cultivation (Coast et al., 2015; Lanning et al., 2012; Lin et al., 2010; Lisle et al., 2000; Patindol et al., 2014; Resurreccion et al., 1977).

Liao et al. (2015) investigated rice transcriptome changes responsible for high night temperature and found a number of genes involved in regulating processes such as defence response and photosynthetic processes were affected by high night temperatures. One quantitative trait locus (QTL) for tolerant to high temperature located within a 1.9-Mb region on chromosome six was detected (Shirasawa et al., 2013).

4.2 Carbon dioxide (CO_2) and Ozone (O_3)

It is widely accepted that CO_2 and O_3 are two of the most significant anthropogenic greenhouse gases responsible for global warming. Also, it is clear that greenhouse air-induced changes affect not only rice yield but also rice quality. There have been many studies investigating the effects of elevated atmospheric CO_2 and O_3, both separately and in combination, on the physical and chemical composition of rice, and their biochemical mechanisms responsible for the alteration of appearance quality and productivity in rice.

4.2.1 Appearance and milling performance

Seneweera et al. (1996) identified greater average grain weight under high CO_2. A meta-analysis by Ainsworth found that, under high CO_2 conditions, an increase in grain mass, panicle and grain number contributed to a 23% increase in rice yields, while higher O_3 resulted in reduced biomass, grain number and grain mass and a 14% decrease in rice yields (Ainsworth, 2008). Increased grain width, individual kernel weight and seed yield, and a reduced proportion of broken grains with high CO_2 were observed in long-grain genotypes (Madan et al., 2012). Yang et al. (2007) found that, under CO_2 enrichment, milled rice yield and chalkiness degree significantly increased (by 10.7% and 28.3%) while head rice yield markedly reduced (by 13.3%). Other studies have found increased grain whiteness and redness associated with enriched CO_2, together with lower grain filling which makes grains more prone to breakage when milling (Goufo et al., 2014).

Similarly, a significant increase of the proportion of rice immature kernels was observed under O_3 exposure (Sawada et al., 2016). The elevated O_3 result in increased loose pack of endosperm starch granules with large spaces and decreased long-side amylopectin chains, and induce alterations of short side-chains of amylopectin chain length (Sawada et al., 2016). The alteration patterns of amylose and amylopectin induced by O_3 were found to be similar to those in *Starch Synthase IIa* (*SSIIa*)-deficient rice mutants (Sawada et al., 2016). Although the significant deterioration in appearance quality caused by abnormal rounded granules was found, the eating quality of cooked rice exposed to O_3 was not impaired (Sawada et al., 2016).

4.2.2 Nutritional quality

Studies of the influence of elevated CO_2 on rice nutritional quality vary. Goufo et al. (2014) found that increasing levels of CO_2 had no impact on starch and gross energy content, but that they reduced protein and amylose content while improving dietary fibre, sugar and fat content in white rice (Goufo et al., 2014; Yang et al., 2007). A meta-analysis by Taub (2008) also reported a reduction of protein content associated with elevated CO_2. However, Lieffering et al. (2004) did not find any change in micronutrient composition. Yang et al. (2007) reported decreases in iron and copper content with an increase in calcium bioavailability but no effect on phytic acid or zinc content. Goufo et al. (2014) observed reductions in amino acid content but increases in linoleic acid in brown rice. Under high O_3 environments, starch concentration, protein and iron yield in brown rice decreased in response to O_3, while the concentration of protein and lipids increased (Frei et al., 2012). An exciting strategy that increased starch content in grains while suppressed methanogenesis due to addition of a single transcription flux gene, barley *SUSIBA2*, into rice (Su et al., 2015), may be possibly applied to breeding programmes of CO_2-resistant or O_3-resistant rice.

4.2.3 Eating and cooking properties

Studies of the effects of CO_2 levels on amylose content vary. No significant relation between high CO_2 and amylose content was found by Madan et al. (2012). Other studies have suggested both lower and higher amylose content related to high CO_2 levels (Seneweera et al., 1996; Yang et al., 2007). Reduced SB and higher PV and BD have resulted from elevated CO_2, indicating a reduction of hardness and improved cohesiveness and resilience for cooking (Yang et al., 2007). Goufo et al. (2014) also found increased PV, HPV, CPV and BD viscosity under elevated CO_2. However, increased levels of O_3 resulted in lower amylose content, PV, HPV, CPV and BD, as well as higher SB viscosity, though with no change in gelatinization temperature and gel CS of rice starch (Wang et al., 2014).

4.3 Drought and submergence

Water stress at the rice heading and filling stage has been shown to significantly influence rice quality and yield (Bouman et al., 2005; Yang et al., 2003). Mushtaq et al. (2008) found that post-fertilization drought stress resulted in a decrease in abundance of nucleoside diphosphate kinase and storage protein globulins, resulting in a reduction in spikelet fertility. It has been found that *Wx* gene expression, necessary for endosperm amylose synthesis, was lower in developing endosperms during the early stage of grain filling under drought stress conditions (Liu et al., 2010; Wang et al., 2006). In contrast, drought-tolerant rice cultivars have shown higher growth rates (Degenkolbe et al., 2009; Kato, et al., 2009).

Submergence has been shown to delay rice flowering and maturity, reducing the number of grains per panicle, grain weight, head rice yield, grain yield, grain sugar, amylase activity and starch content (Reddy and Mittra, 1985; Singh et al., 2009). Extreme wet during grain filling significantly increases immature chalky grains and protein content, thus degrading rice quality (Lee et al., 2013). Submergence has also been associated with lower levels of phosphorus and potassium (Reddy and Mittra, 1985), though oil concentration, amylose content, gelatinization temperature and gel CS are less affected by short periods of submergence (Singh et al., 2009). As a result there has been extensive research on developing submergence-tolerant rice varieties (Singh et al., 2009; Xu et al., 2006).

5 The impact of organic cultivation on rice quality

Organic agriculture is a rapidly growing sector and one of the effective approaches for sustainable agriculture. It provides potential strategies addressing current challenges in traditional agriculture. The desire to reduce fertilizer and pesticide to make cultivation more sustainable, together with consumers' perceptions that organic rice is of better nutritional quality, have been significant factors in the development of organic rice cultivation, despite some disadvantages such as higher labour costs (Crowder and Reganold, 2015; Finamore et al., 2004; Magkos et al., 2003; Mazzoncini et al., 2015). A recent development is the development of rice–aquaculture systems using prawns or fish. The use of aquatic livestock reduces the need for fertilizers and pesticides, for example by reducing pests (Lansing and Kremer, 2011; Nair et al., 2014; Xie et al., 2011). A key factor in organic cultivation is effective nutrition (e.g. use of composts) which has a major impact on yield (Delmotte et al., 2011; Saha et al., 2007, 2010).

5.1 Appearance and nutritional quality

Great differences of appearance properties of milled rice between organic mode and conventional mode have been reported. Compared with milled rice from conventional farming, higher thousand kernel weight and length/breadth ratio while lower bulk density and lightness have been showed in milled organic rice (Kaur et al., 2015). Studies of variations in nutritional values between organic and conventional cultivation vary in their results (Table 1). Some suggest a lower concentration of macronutrients and higher content of secondary metabolites in organic rice (Mazzoncini et al., 2015). Others have shown a lower concentration of copper and a higher iron and phosphorus content in organic rice grains, with no significant difference in potassium, zinc and phytic acid content compared to conventionally grown rice (Saha et al., 2007). And still others have shown no great differences in mineral content, including phosphorus, potassium, magnesium, calcium, manganese, iron and zinc (Champagne et al., 2007). Chen and McClung (2015)

Table 1 Physicochemical characteristics of rice flour obtained from different cultivars grown under conventional (CF) or organic (OF) farming systems[a]

Cultivars	Systems	Nutrient sources	Rice yield (Mg ha^{-1})[b]	Mineral content P (%)	K (%)	Zn (μg/g)	Cu (μg/g)	Fe (μg/g)	Mn (μg/g)	Amylose content (%)	Protein content (%)	Pasting parameters[d] PV (RVU)	HPV (RVU)	CPV (RVU)	BD (RVU)	SB (RVU)	CS (RVU)	PT (°C)	References
SRSN-18 (aromatic)	CF	Mineral fertilizer	4.61	0.24[c]	0.27	18.7	1.7	33.8	7.8	15.9	6.79	–	–	–	–	–	–	–	Saha, 2010
	OF	Cattle manures	4.82	0.25	0.29	16.4	1.8	43.4	9.3	16.5	6.87	–	–	–	–	–	–	–	
RP3392-79-45-18-8 (aromatic)	CF	Mineral fertilizer	4.00	0.22	0.24	15.5	4.1	42.1	–	–	8.98	–	–	–	–	–	–	–	Saha, 2007
	OF	*Lantana* compost		0.24	0.26	15.5	3.6	49.6	–	–	7.92	–	–	–	–	–	–	–	
	OF	Kudzu compost		0.24	0.26	15.7	3.3	45.4	–	–	8.11	–	–	–	–	–	–	–	
	OF	*Urtica* compost		0.25	0.25	15.2	3.6	46.2	–	–	8.21	–	–	–	–	–	–	–	
	OF	Winter weeds	3.01–3.87	0.23	0.25	15.3	3.8	47.1	–	–	8.12	–	–	–	–	–	–	–	
	OF	Farmyard manure		0.24	0.25	16.4	3.6	47.6	–	–	7.71	–	–	–	–	–	–	–	
	OF	Poultry manure		0.26	0.25	16.7	3.4	45.6	–	–	7.88	–	–	–	–	–	–	–	
Bengal (medium grain)	CF	50% N mineral fertilizer	–	0.28	0.25	24.0	–	19.0	22.0	11.1	7.7	324.0	185.0	297.0	139	−27.0	112.0	79.0	Champagne, 2007
	CF	100% N mineral fertilizer	–	0.29	0.25	26.0	–	22.0	25.0	11.2	9.3	293.0	175.0	285.0	118	−8.0	110.0	81.0	
	OF	Chicken litter	–	0.29	0.25	30.0	–	16.0	20.0	10.9	8.3	331.0	199.0	305.0	133	−27.0	106.0	79.0	
Cypress (long grain)	CF	50% N mineral fertilizer	–	0.28	0.20	28.0	–	16.0	19.0	20.6	8.2	244.0	156.0	312.0	88.0	68.0	156.0	82.0	

	CF	100% N mineral fertilizer	–	0.25	0.18	24.0	–	17.0	20.0	20.7	9.6	207.0	126.0	292.0	82.0	84.0	166.0	85.0
	OF	Chicken litter	–	0.30	0.22	25.0	–	18.0	21.0	20.4	8.6	249.0	159.0	323.0	90.0	74.0	164.0	83.0
Jacinto (long grain)	CF	50% N mineral fertilizer	–	0.31	0.25	24.0	–	15.0	22.0	10.5	8.2	297.0	184.0	306.0	113.0	9.0	122.0	85.0
	CF	100% N mineral fertilizer	–	0.33	0.25	27.0	–	15.0	28.0	10.7	9.7	261.0	170.0	302.0	91.0	41.0	132.0	85.0
	OF	Chicken litter	–	0.30	0.25	22.0	–	23.0	19.0	10.0	8.2	299.0	182.0	306.0	117.0	8.0	124.0	85.0
Jasmine 85 (aromatic, long grain)	CF	50% N mineral fertilizer	–	0.27	0.21	21.0	–	18.0	19.0	14.6	7.9	271.0	163.0	321.0	109.0	49.0	158.0	87.0
	CF	100% N mineral fertilizer	–	0.25	0.19	25.0	–	20.0	18.0	14.6	9.3	277.0	166.0	322.0	110.0	44.0	156.0	87.0
	OF	Chicken litter	–	0.27	0.20	25.0	–	19.0	18.0	14.4	8.3	284.0	167.0	311.0	117.0	27.0	104.0	86.0
Neches (long grain, waxy)	CF	50% N mineral fertilizer	–	0.33	0.29	32.0	–	17.0	26.0	–	7.7	230.0	137.0	175.0	93.0	−55.0	38.0	72.0
	CF	100% N mineral fertilizer	–	0.33	0.28	33.0	–	17.0	30.0	–	9.4	211.0	130.0	167.0	81.0	−44.0	37.0	72.0
	OF	Chicken litter	–	0.36	0.31	31.0	–	17.0	23.0	–	8.2	225.0	139.0	178.0	86.0	−47.0	39.0	72.0

[a] Values summarized from Saha (2007 and 2010) and Champagne (2007).
[b] Rice yields from Saha (2007) were summarized based on data from two different nutrient source levels: 5 and 10 Mg ha^{-1}, and rice yields from Saha (2010) were summarized based on data from four different nutrient sources level: 5, 10, 15 and 20 Mg ha^{-1}.
[c] Mineral contents in brown rice from Saha (2007) were summarized based on data from two different nutrient sources level: 5 and 10 Mg ha^{-1}, and mineral contents from Saha (2010) were summarized based on data from four different nutrient sources level: 5, 10, 15 and 20 Mg ha^{-1}.
[d] PV, peak viscosity; HPV, hot paste viscosity; CPV, cool paste viscosity; BD, breakdown; SB, setback; CS, consistency; PT, pasting temperature.

found similar levels of vitamins such as vitamin E but slightly lower flavonoid and phenolic content in organic rice compared to conventional rice.

Some studies have found no apparent difference in protein and lipid concentration between organic and conventional rice grain (Table 1) (Champagne et al., 2007; Na et al., 2007). However, others have observed lower mean protein content in rice grown organically (Champagne et al., 2007; Kaur et al., 2015; Saha et al., 2010), whereas higher protein content was observed in organic rice in the work of Kaur et al. (2015). Some observed variations in proteins content are influenced by nitrogen supply through organic fertilization. As an example, the use of rice–soya bean rotation has been shown to result in higher levels of protein content (Bryant et al., 2009), because of the high levels of nitrogen fixated in soil by these legumes. Different organic composts have also been shown to affect rice nutritional qualities (Saha et al., 2007).

5.2 Eating and cooking quality

Champagne et al. (2007) indicated that no difference in amylose content between organic and conventional rice grain (Table 1), whereas Kaur et al. (2015) reported that organic rice had higher amylose content. Some studies have shown a higher PV, HPV, CPV and BD (Champagne et al., 2007), and lower SB (Champagne et al., 2007; Kaur et al., 2015) in organic rice compared to traditionally grown rice (Table 1). On the contrary, Kaur et al. (2015) have shown lower PV and CPV in cooked organic rice and rice starch. However, no significant difference of PT between traditional and organic rice has been found (Bryant et al., 2009; Champagne et al., 2007; Kaur et al., 2015). Better texture intensities, significant higher slickness as well as lower gruel solid loss, cohesiveness, roughness and hardness have been observed in cooked rice grown organically, though with no significant differences in flavour (Champagne et al., 2007; Kaur et al., 2015). Nonetheless, Kaur et al. (2015) have shown higher hardness, elongation and width expansion ratio, and lower swelling power and solubility of milled organic rice grains.

6 Conclusions

Global food demand and deteriorated ecosystems in the coming decades are still the huge challenges; incentive agricultural practices towards increased efficiency of fertilizer and water use and judicious use of pesticides and antibiotics are required to maintain agriculture sustainability (Tilman et al., 2002). Of the factors affected rice quality and yield potential, cultivars and growing environments were greater than management systems, but organic management may further improve rice nutritional composition (Chen and McClung, 2015). Nonetheless, only 1% of the global agricultural land was used for organic production (Crowder and Reganold, 2015). Therefore, to cope with global warming and ensure rice food security, on the one hand, climate-resilient rice varieties coupled with superior grain quality and high yield potential should be first developed (Sreenivasulu et al., 2015). On the other hand, considering the multiple sustainable benefits, it was profitable to expand organic rice production coupled with suitable cultivars high in grain quality and yield potential (Crowder and Reganold, 2015; Ghorbani et al., 2010; Mazzoncini et al., 2015).

7 Where to look for further information

There are a lot of research articles and reviews in peer-reviewed journals in this field, such as *Cereal Chemistry*, *Journal of Cereal Science*, *Food Chemistry* and *Field Crops Research*.

Furthermore, there is a great deal of useful information about rice production and cultivation available on the websites of the professional organizations. For example,

International Rice Research Institute, http://www.irri.org.

American Association of Cereal Chemists (AACC) International, http://www.aaccnet.org/.

Food and Agriculture Organization of the United Stations, http://faostat3.fao.org/home/E.

Intergovernmental Panel on Climate Change (IPCC), http://www.ipcc.ch, a good place to learn and view the scientific knowledge and reports in global climate change and its potential environmental impacts on worldwide crops growth and agricultural production.

8 References

Aboubacar, A., Moldenhauer, K. A., McClung, A. M., Beighley, D. H. and Hamaker, B. R. (2006). Effect of growth location in the United States on amylose content, amylopectin fine structure, and thermal properties of starches of long grain rice cultivars. *Cereal Chem.*, *83*(1), 93–8.

Ainsworth, E. A. (2008). Rice production in a changing climate: a meta-analysis of responses to elevated carbon dioxide and elevated ozone concentration. *Global Change Biol.*, *14*(7), 1642–50.

Ambardekar, A. A., Siebenmorgen, T. J., Counce, P. A., Lanning, S. B. and Mauromoustakos, A. (2011). Impact of field-scale nighttime air temperatures during kernel development on rice milling quality. *Field Crops Res.*, *122*(3), 179–85.

Ashida, K., Araki, E., Maruyama-Funatsuki, W., Fujimoto, H. and Ikegami, M. (2013). Temperature during grain ripening affects the ratio of type-II/type-I protein body and starch pasting properties of rice (Oryza sativa L.). *J. Cereal Sci.*, *57*(1), 153–9.

Baxter, G., Blanchard, C. and Zhao, J. (2004). Effects of prolamin on the textural and pasting properties of rice flour and starch. *J. Cereal Sci.*, *40*(3), 205–11.

Bouman, B. A. M., Peng, S., Castañeda, A. R. and Visperas, R. M. (2005). Yield and water use of irrigated tropical aerobic rice systems. *Agr. Water Manage.*, *74*(2), 87–105.

Brar, S., Mahal, S., Brar, A., Vashist, K., Sharma, N. and Buttar, G. (2012). Transplanting time and seedling age affect water productivity, rice yield and quality in north-west India. *Agr. Water Manage.*, *115*, 217–22.

Bryant, R. J., Anders, M. and McClung, A. M. (2009). Effect of cultural management practices on grain quality of two rice cultivars. *Cereal Chem.*, *86*(4), 405–9.

Bryant, R. J., Yeater, K. M. and McClung, A. M. (2015). Effect of nitrogen rate and the environment on physicochemical properties of selected high amylose rice cultivars. *Cereal Chem.*, *92*(6), 604–10.

Champagne, E. T., Bett-Garber, K. L., Grimm, C. C. and McClung, A. M. (2007). Effects of organic fertility management on physicochemical properties and sensory quality of diverse rice cultivars. *Cereal Chem.*, *84*(4), 320–7.

Champagne, E. T., Bett-Garber, K. L., Thompson, J., Mutters, R., Grimm, C. C. and McClung, A. M. (2005). Effects of drain and harvest dates on rice sensory and physicochemical properties. *Cereal Chem.*, *82*(4), 369–74.

Champagne, E. T., Bett-Garber, K. L., Thomson, J. L. and Fitzgerald, M. A. (2009). Unraveling the impact of nitrogen nutrition on cooked rice flavor and texture. *Cereal Chem.*, *86*(3), 274–80.

Chen, C., Huang, J. L., Zhu, L. Y., Shah, F., Nie, L. X., Cui, K. H. and Peng, S. B. (2013). Varietal difference in the response of rice chalkiness to temperature during ripening phase across different sowing dates. *Field Crops Res., 151*, 85–91.

Chen, M. H. and McClung, A. M. (2015). Effects of cultivars, organic cropping management and environment on antioxidants in whole grain rice. *Cereal Chem., 92*(4), 364–9.

Coast, O., Ellis, R. H., Murdoch, A. J., Quinones, C. and Jagadish, K. S. V. (2015). High night temperature induces contrasting responses for spikelet fertility, spikelet tissue temperature, flowering characteristics and grain quality in rice. *Funct. Plant Biol., 42*(2), 149–61.

Cooper, N. T. W., Siebenmorgen, T. J. and Counce, P. A. (2008). Effects of nighttime temperature during kernel development on rice physicochemical properties. *Cereal Chem., 85*(3), 276–82.

Cooper, N. T. W., Siebenmorgen, T. J., Counce, P. A. and Meullenet, J. F. (2006). Explaining rice milling quality variation using historical weather data analysis. *Cereal Chem., 83*(4), 447–50.

Crowder, D. W. and Reganold, J. P. (2015). Financial competitiveness of organic agriculture on a global scale. *PNAS, 112*(24), 7611–16.

Dang, J. M. C. and Copeland, L. (2004). Genotype and environmental influences on pasting properties of rice flour. *Cereal Chem., 81*(4), 486–9.

Degenkolbe, T., Do, P. T., Zuther, E., Repsilber, D., Walther, D., Hincha, D. K. and Köhl, K. I. (2009). Expression profiling of rice cultivars differing in their tolerance to long-term drought stress. *Plant Mol. Biol., 69*(1), 133–53.

Delmotte, S., Tittonell, P., Mouret, J. C., Hammond, R. and Lopez-Ridaura, S. (2011). On farm assessment of rice yield variability and productivity gaps between organic and conventional cropping systems under Mediterranean climate. *Eur. J. Agron., 35*(4), 223–36.

Denyer, K., Hylton, C. and Smith, A. (1994). The effect of high temperature on starch synthesis and the activity of starch synthase. *Funct. Plant Biol., 21*(6), 783–9.

Dhaliwal, Y. S., Nagi, H. P., Sidhu, G. S. and Sekhon, K. S. (1986). Physicochemical, milling and cooking quality of rice as affected by sowing and transplanting dates. *J. Sci. Food Agr., 37*(9), 881–7.

Dilday, R. H. (1989). Milling quality of rice - cylinder speed vs grain-moisture content at harvest. *Crop Sci., 29*(6), 1532–5.

Finamore, A., Britti, M. S., Roselli, M., Bellovino, D., Gaetani, S. and Mengheri, E. (2004). Novel approach for food safety evaluation. Results of a pilot experiment to evaluate organic and conventional foods. *J. Agr. Food Chem., 52*(24), 7425–31.

Frei, M., Kohno, Y., Tietze, S., Jekle, M., Hussein, M. A., Becker, T. and Becker, K. (2012). The response of rice grain quality to ozone exposure during growth depends on ozone level and genotype. *Environ. Pollut., 163*, 199–206.

Gelders, G. G., Goesaert, H. and Delcour, J. A. (2006). Amylose – lipid complexes as controlled lipid release agents during starch gelatinization and pasting. *J. Agr. Food Chem., 54*(4), 1493–9.

Ghorbani, R., Koocheki, A., Brandt, K., Wilcockson, S. and Leifert, C. (2010). Organic agriculture and food production: ecological, environmental, food safety and nutritional quality issues. In *Sociology, Organic Farming, Climate Change and Soil Science*, Springer.

Ghosh, M., Mandal, B. K., Mandal, B. B., Lodh, S. B. and Dash, A. K. (2004). The effect of planting date and nitrogen management on yield and quality of aromatic rice (Oryza sativa). *J. Agr. Sci., 142*, 183–91.

Gomez, K. (1979). Effect of environment on protein and amylose content of rice. In *Proceedings of the Workshop on Chemical Aspects of Rice Grain Quality*, pp. 59–68. International Rice Research Institute, Los Baños, Laguna, Philippines.

Goufo, P., Falco, V., Brites, C., Wessel, D. F., Kratz, S., Rosa, E. A., Carranca, C. and Trindade, H. (2014). Effect of elevated carbon dioxide concentration on rice quality: nutritive value, color, milling, cooking, and eating qualities. *Cereal Chem., 91*(5), 513–21.

Goufo, P., Ferreira, L. M., Carranca, C., Rosa, E. A. and Trindade, H. (2014). Effect of elevated carbon dioxide concentration on rice quality: Proximate composition, dietary fibers, and free sugars. *Cereal Chem., 91*(3), 293–9.

Gregory, P. J., Ingram, J. S. I. and Brklacich, M. (2005). Climate change and food security. *Philosophical Transactions of the Royal Society B-Biological Sciences, 360*(1463), 2139–48.

Grigg, B. C., Beyrouty, C. A., Norman, R. J., Gbur, E. E., Hanson, M. G. and Wells, B. R. (2000). Rice responses to changes in floodwater and N timing in southern USA. *Field Crops Res., 66*(1), 73–9.

Horai, K., Ishii, A. and Shimono, H. (2014). Genotypic variation in the response to high water temperature during vegetative growth and the effects on rice productivity under a cool climate. *Field Crops Res., 162*, 12–19.

Huebner, F., Bietz, J., Webb, B. and Juliano, B. (1990). Rice cultivar identification by high-performance liquid chromatography of endosperm proteins. *Cereal Chem., 67*(2), 129–35.

Inouchi, N., Ando, H., Asaoka, M., Okuno, K. and Fuwa, H. (2000). The effect of environmental temperature on distribution of unit chains of rice amylopectin. *Starch/Stärke, 52*(1), 8–12.

IPCC (2014). Climate Change 2014: Synthesis report. contribution of working groups I, II and III to the fifth assessment report of the intergovernmental panel on climate change, Core Writing Team, In R. K. Pachauri and L.A. Meyer (eds), IPCC, Geneva, Switzerland.

Ishimaru, T., Horigane, A. K., Ida, M., Iwasawa, N., San-oh, Y. A., Nakazono, M., Nishizawa, N. K., Masumura, T., Kondo, M. and Yoshida, M. (2009). Formation of grain chalkiness and changes in water distribution in developing rice caryopses grown under high-temperature stress. *J. Cereal Sci., 50*(2), 166–74.

Jiang, H. W., Dian, W. M. and Wu, P. (2003). Effect of high temperature on fine structure of amylopectin in rice endosperm by reducing the activity of the starch branching enzyme. *Phytochemistry, 63*(1), 53–9.

Jongkaewwattana, S. and Geng, S. (1991). Effect of nitrogen and water management on panicle development and milling quality of California rice (Oryza sativa L.). *J. Agron. Crop Sci., 167*(1), 43–52.

Kato, Y., Okami, M. and Katsura, K. (2009). Yield potential and water use efficiency of aerobic rice (Oryza sativa L.) in Japan. *Field Crops Res., 113*(3), 328–34.

Kaur, M., Kaur, N., Kaur, M. and Sandhu, K. S. (2015). Some properties of rice grains, flour and starches: A comparison of organic and conventional modes of farming. *LWT-Food Sci. Technol., 61*(1), 152–7.

Krishnan, P. and Rao, A. V. S. (2005). Effects of genotype and environment on seed yield and quality of rice. *J. Agr. Sci., 143*, 283–92.

Lanning, S. B. and Siebenmorgen, T. J. (2013). Effects of preharvest nighttime air temperatures on whiteness of head rice. *Cereal Chem., 90*(3), 218–22.

Lanning, S. B., Siebenmorgen, T. J., Ambardekar, A. A., Counce, P. A. and Bryant, R. J. (2012). Effects of nighttime air temperature during kernel development of field-grown rice on physicochemical and functional properties. *Cereal Chem., 89*(3), 168–75.

Lanning, S. B., Siebenmorgen, T. J., Counce, P. A., Ambardekara, A. A. and Mauromoustakos, A. (2011). Extreme nighttime air temperatures in 2010 impact rice chalkiness and milling quality. *Field Crops Res., 124*(1), 132–6.

Lansing, J. S. and Kremer, J. N. (2011). Rice, fish, and the planet. *PNAS, 108*(50), 19841–2.

Lee, M. S., Kang, B. M., Lee, J. E., Choi, W. J., Ko, J., Choi, J. E., An, K. N., Kwon, O. D., Park, H. G. and Shin, H. R. (2013). How do extreme wet events affect rice quality in a changing climate? *Agr. Ecosys. Environ., 171*, 47–54.

Leesawatwong, M., Jamjod, S., Kuo, J., Dell, B. and Rerkasem, B. (2005). Nitrogen fertilizer increases seed protein and milling quality of rice. *Cereal Chem., 82*(5), 588–93.

Liao, J. L., Zhou, H. W., Peng, Q., Zhong, P. A., Zhang, H. Y., He, C. and Huang, Y. J. (2015). Transcriptome changes in rice (Oryza sativa L.) in response to high night temperature stress at the early milky stage. *BMC genomics, 16*(1), 1222.

Lieffering, M., Kim, H. Y., Kobayashi, K. and Okada, M. (2004). The impact of elevated CO_2 on the elemental concentrations of field-grown rice grains. *Field Crops Res., 88*(2–3), 279–86.

Lin, C. J., Li, C. Y., Lin, S. K., Yang, F. H., Huang, J. J., Liu, Y. H. and Lur, H. S. (2010). Influence of high temperature during grain filling on the accumulation of storage proteins and grain quality in rice (Oryza sativa L.). *J. Agr. Food Chem., 58*(19), 10545–52.

Lisle, A., Martin, M. and Fitzgerald, M. (2000). Chalky and translucent rice grains differ in starch composition and structure and cooking properties. *Cereal Chem., 77*(5), 627–32.

Liu, D. H., Zhang, J. L., Cao, J. H., Wang, Z. H., Yu, C. and Jin, D. M. (2010). The reduction of amylose content in rice grain and decrease of Wx gene expression during endosperm development in response to drought stress. *J. Food Agr. Environ., 8*, 873–8.

Lyman, N. B., Jagadish, K. S. V., Nalley, L. L., Dixon, B. L. and Siebenmorgen, T. (2013). Neglecting rice milling yield and quality underestimates economic losses from high-temperature stress. *Plos ONE, 8*(8), e72157. doi:10.1371/journal.pone.0072157.

Madan, P., Jagadish, S., Craufurd, P., Fitzgerald, M., Lafarge, T. and Wheeler, T. (2012). Effect of elevated CO_2 and high temperature on seed-set and grain quality of rice. *J. Exp. Bot., 63*(10), 3843–52.

Magkos, F., Arvaniti, F. and Zampelas, A. (2003). Organic food: nutritious food or food for thought? A review of the evidence. *Int. J. Food Sci. Nutr., 54*(5), 357–71.

Martin, M. and Fitzgerald, M. A. (2002). Proteins in rice grains influence cooking properties. *J. Cereal Sci., 36*(3), 285–94.

Mazzoncini, M., Antichi, D., Silvestri, N., Ciantelli, G. and Sgherri, C. (2015). Organically vs conventionally grown winter wheat: Effects on grain yield, technological quality, and on phenolic composition and antioxidant properties of bran and refined flour. *Food Chem., 175*, 445–51.

McCauley, G. N. and Way, M. O. (2002). Drain and harvest timing affects on rice grain drying and whole-milled grain. *Field Crops Res., 74*(2–3), 163–72.

Morrison, W. R. (1995). Starch lipids and how they relate to starch granule structure and functionality. *Cereal Foods World, 40*, 437–46.

Mushtaq, R., Katiyar, S. and Bennett, J. (2008). Proteomic analysis of drought stress-responsive proteins in rice endosperm affecting grain quality. *J. Crop Sci. Biotechnol., 11*, 227–32.

Na, G. S., Lee, S. K. and Kim, S. Y. (2007). Antioxidative effects and quality characteristics of the rice cultivated by organic farming and ordinary farming. *J. Korean Soc. Applied Biological Chem., 50*(1), 36–41.

Nair, C. M., Salin, K. R., Joseph, J., Aneesh, B., Geethalakshmi, V. and New, M. B. (2014). Organic rice-prawn farming yields 20% higher revenues. *Agron. Sustain. Dev., 34*(3), 569–81.

Ottis, B. V. and Talbert, R. E. (2005). Rice yield components as affected by cultivar and seeding rate. *Agron. J., 97*(6), 1622–5.

Pan, S. G., Cao, C. G., Cai, M. L., Wang, J. P., Wang, R. H., Zhai, J. and Huang, S. Q. (2009). Effects of irrigation regime and nitrogen management on grain yield, quality and water productivity in rice. *J. Food Agr. Environ., 7*(2), 559–64.

Patindol, J. A., Siebenmorgen, T. J., Wang, Y.-J., Lanning, S. B. and Counce, P. A. (2014). Impact of elevated nighttime air temperatures during kernel development on starch properties of field-grown rice. *Cereal Chem., 91*(4), 350–7.

Reddy, M. D. and Mittra, B. (1985). Effects of complete plant submergence on vegetative growth, grain yield and some biochemical changes in rice plants. *Plant Soil, 87*(3), 365–74.

Perez, C. M., Juliano, B. O., Liboon, S. P., Alcantara, J. M. and Cassman, K. G. (1996). Effects of late nitrogen fertilizer application on head rice yield, protein content, and grain quality of rice. *Cereal Chem., 73*(5), 556–60.

Resurreccion, A. P., Hara, T., Juliano, B. O. and Yoshida, S. (1977). Effect of temperature during ripening on grain quality of rice. *Soil Sci. Plant Nutr., 23*(1), 109–12.

Saha, S., Gopinath, K., Mina, B., Kundu, S., Bhattacharyya, R. and Gupta, H. (2010). Expression of soil chemical and biological behavior on nutritional quality of aromatic rice as influenced by organic and mineral fertilization. *Commun. Soil Sci. Plant Anal., 41*(15), 1816–31.

Saha, S., Pandey, A., Gopinath, K., Bhattacharaya, R., Kundu, S. and Gupta, H. (2007). Nutritional quality of organic rice grown on organic composts. *Agron. Sustain. Dev., 27*(3), 223–9.

Sawada, H., Tsukahara, K., Kohno, Y., Suzuki, K., Nagasawa, N. and Tamaoki, M. (2016). Elevated ozone deteriorates grain quality of japonica rice cv. koshihikari, even if it does not cause yield reduction. *Rice, 9*(1). doi: 10.1186/s12284-016-0079-4.

Seetanun, W. and Dedatta, S. K. (1973). Grain yield, milling quality, and seed viability of rice as influenced by time of nitrogen application and time of harvest. *Agron. J., 65*(3), 390–4.

Seneweera, S., Blakeney, A. and Milham, P. (1996). Influence of rising atmospheric CO_2 and phosphorus nutrition on the grain yield and quality of rice (Oryza sativa cv. Jarrah). *Cereal Chem., 73*(2), 239–43.

Shirasawa, K., Sekii, T., Ogihara, Y., Yamada, T., Shirasawa, S., Kishitani, S., Sasaki, K., Nishimura, M., Nagano, K. and Nishio, T. (2013). Identification of the chromosomal region responsible for high-temperature stress tolerance during the grain-filling period in rice. *Mol. Breeding, 32*(1), 223–32.

Siebenmorgen, T. J., Grigg, B. C. and Lanning, S. B. (2013). Impacts of preharvest factors during kernel development on rice quality and functionality. *Annu. Rev. Food Sci. Technol., 4*, 101–15.

Singh, N., Paul, P., Virdi, A. S., Kaur, P. and Mahajan, G. (2014). Influence of early and delayed transplantation of paddy on physicochemical, pasting, cooking, textural, and protein characteristics of milled rice. *Cereal Chem., 91*(4), 389–97.

Singh, S., Mackill, D. J. and Ismail, A. M. (2009). Responses of SUB1 rice introgression lines to submergence in the field: yield and grain quality. *Field Crops Res., 113*(1), 12–23.

Slaton, N. A., Linscombe, S. D., Norman, R. J. and Gbur, E. E. (2003). Seeding date effect on rice grain yields in Arkansas and Louisiana. *Agron. J., 95*(1), 218–23.

Song, Y. L., Wang, C. Y., Ren, G. Y., Zhao, Y. X. and Linderholm, H. W. (2015). The relative contribution of climate and cultivar renewal to shaping rice yields in China since 1981. *Theor. Appl. Climatol., 120*(1–2), 1–9.

Sreenivasulu, N., Butardo, V. M., Misra, G., Cuevas, R. P., Anacleto, R. and Kishor, P. B. K. (2015). Designing climate-resilient rice with ideal grain quality suited for high-temperature stress. *J. Exp. Bot., 66*(7), 1737–48.

Su, J., Hu, C., Yan, X., Jin, Y., Chen, Z., Guan, Q., Wang, Y., Zhong, D., Jansson, C. and Wang, F. (2015). Expression of barley SUSIBA2 transcription factor yields high-starch low-methane rice. *Nature. 523*(7562), 602–6.

Taub, D. R., Miller, B. and Allen, H. (2008). Effects of elevated CO_2 on the protein concentration of food crops: a meta–analysis. *Global Change Biol., 14*(3), 565–75.

Tilman, D., Cassman, K. G., Matson, P. A., Naylor, R. and Polasky, S. (2002). Agricultural sustainability and intensive production practices. *Nature, 418*(6898), 671–7.

Tong, C., Liu, L., Waters, D. L. E., Rose, T. J., Bao, J. S. and King, G. J. (2014). Genotypic variation in lysophospholipids of milled rice. *J. Agr. Food Chem., 62*(38), 9353–61.

Umemoto, T., Nakamura, Y. and Ishikura, N. (1995). Activity of starch synthase and the amylose content in rice endosperm. *Phytochemistry, 40*(6), 1613–16.

Wang, L., Siebenmorgen, T. J., Matsler, A. D. and Bautista, R. C. (2004). Effects of rough rice moisture content at harvest on peak viscosity. *Cereal Chem., 81*(3), 389–91.

Wang, S. J., Liu, L. F., Chen, C. K. and Chen, L. W. (2006). Regulations of granule-bound starch synthase I gene expression in rice leaves by temperature and drought stress. *Biol. Plantarum, 50*(4), 537–41.

Wang, Y. X., Song, Q. L., Frei, M., Shao, Z. S. and Yang, L. X. (2014). Effects of elevated ozone, carbon dioxide, and the combination of both on the grain quality of Chinese hybrid rice. *Environ. Pollut., 189*, 9–17.

Xie, J., Hu, L. L., Tang, J. J., Wu, X., Li, N. N., Yuan, Y. G., Yang, H. S., Zhang, J. E., Luo, S. M. and Chen, X. (2011). Ecological mechanisms underlying the sustainability of the agricultural heritage rice-fish coculture system. *PNAS, 108*(50), E1381–7.

Xu, K. N., Xu, X., Fukao, T., Canlas, P., Maghirang-Rodriguez, R., Heuer, S., Ismail, A. M., Bailey-Serres, J., Ronald, P. C. and Mackill, D. J. (2006). Sub1A is an ethylene-response-factor-like gene that confers submergence tolerance to rice. *Nature, 442*(7103), 705–8.

Xu, L. J., Xie, J. K., Kong, X. L. and Bao, J. S. (2004). Analysis of genotypic and environmental effects on rice starch. 2. Thermal and retrogradation properties. *J. Agr. Food Chem.*, *52*(19), 6017–22.
Yamakawa, H. and Hakata, M. (2010). Atlas of rice grain filling-related metabolism under high temperature: joint analysis of metabolome and transcriptome demonstrated inhibition of starch accumulation and induction of amino acid accumulation. *Plant Cell Physiol.*, *51*(5), 795–809.
Yang, J. C., Zhang, J. H., Wang, Z. Q., Liu, L. J. and Zhu, Q. S. (2003). Postanthesis water deficits enhance grain filling in two-line hybrid rice. *Crop Sci.*, *43*(6), 2099–108.
Yang, L. X., Wang, Y. L., Dong, G. C., Gu, H., Huang, J. Y., Zhu, J. G., Yang, H. J., Liu, G. and Han, Y. (2007). The impact of free-air CO_2 enrichment (FACE) and nitrogen supply on grain quality of rice. *Field Crops Res.*, *102*(2), 128–40.
Zhang, Z. C., Zhang, S. F., Yang, J. C. and Zhang, J. H. (2008). Yield, grain quality and water use efficiency of rice under non-flooded mulching cultivation. *Field Crops Res.*, *108*(1), 71–81.

Index

Lightning Source UK Ltd.
Milton Keynes UK
UKOW06n1541280517
302163UK00001B/2/P

9 781786 760241